FUNDAMENTAL ISSUES OF NONLINEAR LASER DYNAMICS

Related Titles from AIP Conference Proceedings

524 Nonlinear Acoustics at the Turn of the Millennium: ISNA 15; 15th International Symposium on Nonlinear Acoustics
Edited by Werner Lauterborn and Thomas Kurz, July 2000, 1-56396-945-9

517 Computing Anticipatory Systems: CASYS'99—Third International Conference
Edited by Daniel M. Dubois, June 2000, 1-56396-933-5

511 Unsolved Problems of Noise and Fluctuations: UPoN'99: Second International Conference
Edited by Derek Abbott and Lazlo B. Kish, March 2000, 1-56396-826-6

502 Stochastic and Chaotic Dynamics in the Lakes: STOCHAOS
Edited by David S. Broomhead, Elena A. Luchinskaya, Peter V. E. McClintock, and Tom Mullin, February 2000, 1-56396-915-7

501 Stochastic Dynamics and Pattern Formation in Biological and Complex Systems: The APCTP Conference
Edited by Seunghwan Kim, Kyoung J. Lee, and Wokyung Sung, January 2000, 1-56396-914-9

469 Slow Dynamics in Complex Systems: Eighth Tohwa University International Symposium
Edited by Michio Tokuyama and Irwin Oppenheim, April 1999, 1-56396-811-8

465 Computing Anticipatory Systems (CASYS-98): Second International Conference
Edited by Daniel M. Dubois, March 1999, 1-56396-863-0

411 Applied Nonlinear Dynamics and Stochastic Systems Near the Millennium
Edited by James B. Kadtke and Adi Bulsara, November 1997, 1-56396-736-7

To learn more about these titles, or the AIP Conference Proceedings Series, please visit the webpage **http://www.aip.org/catalog/aboutconf.html**

Cover and Title Page: A nearfield image with six-fold symmetry of a Vertical Cavity Surface Emitting Laser measured by Christian Degen (Technical University of Darmstadt) rendered as level sets of the intensity.

FUNDAMENTAL ISSUES OF NONLINEAR LASER DYNAMICS

Concepts, Mathematics, Physics, and Applications
International Spring School

Texel, The Netherlands   16–19 April 2000

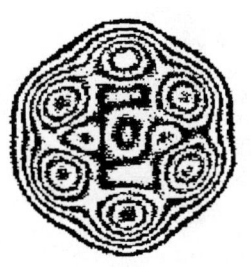

EDITORS
Bernd Krauskopf
University of Bristol
United Kingdom

Daan Lenstra
Vrije Universiteit Amsterdam
The Netherlands

Melville, New York, 2000
AIP CONFERENCE PROCEEDINGS ■ VOLUME 548

Editors:

Bernd Krauskopf
Department of Engineering Mathematics
University of Bristol
Queen's Building
Bristol BS8 1TR
UNITED KINGDOM

E-mail: b.krauskopf@bristol.ac.uk

Daan Lenstra
Vrije Universiteit
FEW, Division of Physics and Astronomy
De Boelelaan 1081
1081 HV Amsterdam
THE NETHERLANDS

E-mail: lenstra@nat.vu.nl

Authorization to photocopy items for internal or personal use, beyond the free copying permitted under the 1978 U.S. Copyright Law (see statement below), is granted by the American Institute of Physics for users registered with the Copyright Clearance Center (CCC) Transactional Reporting Service, provided that the base fee of $17.00 per copy is paid directly to CCC, 222 Rosewood Drive, Danvers, MA 01923. For those organizations that have been granted a photocopy license by CCC, a separate system of payment has been arranged. The fee code for users of the Transactional Reporting Service is: 1-56396-977-7/00/$17.00.

© 2000 American Institute of Physics

Individual readers of this volume and nonprofit libraries, acting for them, are permitted to make fair use of the material in it, such as copying an article for use in teaching or research. Permission is granted to quote from this volume in scientific work with the customary acknowledgment of the source. To reprint a figure, table, or other excerpt requires the consent of one of the original authors and notification to AIP. Republication or systematic or multiple reproduction of any material in this volume is permitted only under license from AIP. Address inquiries to Office of Rights and Permissions, Suite 1NO1, 2 Huntington Quadrangle, Melville, N.Y. 11747-4502; phone: 516-576-2268; fax: 516-576-2450; e-mail: rights@aip.org.

L.C. Catalog Card No. 00-110162
ISBN 1-56396-977-7
ISSN 0094-243X
Printed in the United States of America

Contents

Preface vii
Acknowledgment ix

Bifurcation Analysis of Laser Systems 1
    B. Krauskopf
Chaos in Periodically Driven Systems 31
    H. W. Broer and B. Krauskopf
Multiple Time Scale Analysis of Lasers 54
    T. Erneux
**The Mathematics of Delay Equations with an Application
to the Lang-Kobayashi Equations** 66
    S. M. Verduyn Lunel and B. Krauskopf
Theory of Delayed Optical Feedback in Lasers 87
    D. Lenstra and M. Yousefi
Applications of Semiconductor Lasers to Secure Communications 112
    C. R. Mirasso
Theory and Simulation of Spatially Extended Semiconductor Lasers 128
    O. Hess
Semiconductor Laser Device Modeling 149
    J. V. Moloney
High Brightness Semiconductor Lasers with Reduced Filamentation 173
    J. G. McInerney, P. O'Brien, P. Skovgaard, M. Mullane,
    J. Houlihan, E. O'Neill, J. V. Moloney, and R. A. Indik
**Nonlinear Dynamics of Semiconductor Lasers:
Theory and Experiments** 191
    A. Gavrielides
**Emission Dynamics of Semiconductor Lasers Subject
to Delayed Optical Feedback: An Experimentalist's Perspective** 218
    I. Fischer, T. Heil, and W. Elsäßer
Excitability in Laser Systems: the Experimental Side 238
    J. R. Tredicce
**Coherence, Chaos and Communication: Exploring
and Applying Nonlinear Laser Dynamics** 260
    R. Roy
Vertical Cavity Surface Emitting Lasers for Communications 279
    J. M. Rorison

Author Index 303

PREFACE

The International Spring School on Fundamental Issues of Nonlinear Laser Dynamics was held from 16 to 19 April 2000 on the Dutch Island of Texel. It was aimed at young researchers who are interested in working at the forefront of Applied Nonlinear Mathematics and Nonlinear Laser Dynamics. In a highly interdisciplinary spirit, there were tutorial presentations from 14 internationally recognized top experts from applied mathematics, theoretical and experimental physics, and engineering disciplines. The meeting was of tutorial scope in the first place, but also had very much the character of a scientific workshop, since many of the young participants took the opportunity to present and discuss their own research as well.

The editors realized already at an early stage, long before the Spring School was actually held, that the concept of this meeting would offer the unique possibility of putting together all tutorial lectures in a book. This book would be special in its kind, being the first collection of tutorials on Nonlinear Dynamics of Lasers. While it would be an excellent form of Lecture Notes to be send to all participants, the ambition was to serve an audience much wider than the participants of the Spring School alone.

The idea for this book of tutorials was well received by the prospective authors and the editors are now very proud to present the book in its final form. It presents a unique combination of 14 tutorial contributions which all have nonlinear dynamics as the central theme, and together cover the full range from modern mathematical theory and applied mathematics, through physics and engineering up to present day applications. We are extremely grateful to our collegues for putting a lot of effort and enthusiasm into this project.

We sincerely hope that this book will find its way to an interdisciplinary readership of researchers, lecturers and students in the fields of applied mathematics, physics and electrical engineering. Much in the spirit of the Texel meeting, we hope that the reader will be caught by the excitement and fascination of nonlinear dynamics as it manifests itself in lasers.

Bernd Krauskopf   Daan Lenstra
Bristol   Amsterdam

September 2000

ACKNOWLEDGMENT

The Spring School on Fundamental Issues of Nonlinear Laser Dynamics

was made possible through support from

Vrije Universiteit Amsterdam,

Netherlands Organisation for Scientific Research (NWO),

Foundation for Fundamental Research on Matter (FOM),

Samenwerkingsverband Mathematische Fysica FOM/SWON,

Stichting Physica,

the Dutch Research Schools, COMPAS, MRI and Stieltjes,

and the University of Bristol.

We thank all participants for making the Spring School a success with

their enthusiasm and energy!

Bifurcation Analysis of Laser Systems

Bernd Krauskopf

Department of Engineering Mathematics, University of Bristol, Bristol BS8 1TR, UK

Abstract. This exposition explains the basic ideas and methods of bifurcation analysis, where the rate equation model of an optically injected semiconductor laser is used as an example throughout. Bifurcations of equilibria and periodic orbits that occur when a single parameter is changed are introduced one by one. For each of them we show what the bifurcation actually looks like for the injected laser. The transitons to chaos via period-doublings and the break-up of a torus are explained as well. Finally, a brief discussion points to topics not covered here.

INTRODUCTION

Bifurcation analysis is a very powerful tool for the study of dynamical systems. The aim of this exposition is to introduce the basic bifurcations of equilibria and periodic orbits in dynamical systems, where we restrict our attention to the important class of systems defined by ordinary differential equations (ODEs). We concentrate on the ideas and methods and do not attempt to be mathematically very rigorous. The emphasis is on geometrical concepts, but in order to apply such concepts we introduce so called bifurcation conditions and the related concept of normal forms. Perhaps most importantly from the point of view of applications, we illustrate how these bifurcation 'live' in a real system. To this end, we use the rate equations of a semiconductor laser with optical injection as the illustrating example throughout. The concepts discussed here can be found in any introductory book on bifurcation theory and dynamical systems. For further reading we refer to, for example, [1–4].

Our main object of study is a dynamical system in the form of a system of autonomous ODEs. Here autonomous means that the right hand side does not depend explicitely on time. This defines a *vector field*, that is, a force field prescribing a direction and strength of the force at any point in phase space. Because the system is autonomous, the direction and strength of the force are independent of time. Solutions (or orbits or trajectories) for a given initial condition then exist and are unique. (This is true under the mathematical condition that the vector field is smooth enough, say, twice differentiable, which we assume from now on.) An orbit can be thought of as the curve that a particle, placed at an initial position in the force field, traces out over time. In general, there are no explicit expressions for solution curves, but approximations can be computed numerically by integration.

To set the stage and for future reference let us introduce a vector field in general form as

$$\dot{x} = f(x, \lambda). \tag{1}$$

Here x is a state in the phase space $\mathbb{R}^n$ and $\lambda \in \mathbb{R}^m$ is a (multi-dimensional) parameter. (We will typically use only a single real parameter in later sections, that is, $\lambda \in \mathbb{R}$.) The function f is a vector valued function

$$\begin{aligned} f : \mathbb{R}^n \times \mathbb{R}^m &\to \mathbb{R}^n \\ (x, \lambda) &\mapsto f(x, \lambda) \end{aligned}$$

that defines an n-dimensional vector, the force, at any point x in the phase space $\mathbb{R}^n$. This vector, and the force field as a whole, depends on the value of λ. We will sometimes simply write $f(x)$ when λ is fixed and we are not interested in the dependence on the parameter λ.

We now introduce the example of a vector field that we will use throughout this exposition. Note that there are many other examples, such as other low-dimensional laser models, the Lorenz system, and the driven pendulum; see also the chapter by *Broer & Krauskopf*.

Semiconductor Laser with Optical Injection

When a semiconductor laser (also called the slave laser) is subjected to optically injected light from a second laser (the master laser) then an enormous wealth of phenomena can be observed, depending on the strength and detuning (with respect to the solitary laser frequency of the free-running master laser) of the input light. Originally, optical injection was used for injection locking and frequency stabilization. It is now known to produce more complicated dynamics, such as periodicity, quasiperiodic motions and chaos. For more details, an overview over the literature and further references we refer to [5]; see also the chapters by *Erneux* and *Gavrielides*.

The system is described well by rate equations for the master laser, that is, by ODEs for its complex electric field and its population inversion. The injected light enters in the form of a periodic forcing term in the equation for the electric field. In dimensionless form and after changing to the reference frequency of the input light [5] the system can be written as

$$\dot{E} = K + \left(\frac{1}{2}(1 + i\alpha)n - i\omega\right) E \tag{2}$$

$$\dot{n} = -2\Gamma n - (1 + 2Bn)(|E|^2 - 1) .$$

Here E is the complex electric field, n is the population inversion, K is the injected field strength, and ω the detuning of the injected field from the solitary laser frequency. Furthermore, α is the linewidth enhancement factor, B is the cavity life time, and Γ is the damping rate. The parameters K and ω are the main parameters that may be controlled in experiments. The parameters α, B and Γ, on the other hand, describe material properties of the laser, and they are set to the fixed, realistic values of $\alpha = 2$, $B = 0.015$ and $\Gamma = 0.035$ from now on.

From the mathematical point of view, Eqs. (2) describe a dynamical system on the three-dimensional (E, n)-space, the phase space of the system. It is an exercise to identify f, x and λ to see that Eqs. (2) are indeed of the general form (1). The injected laser is not only an exciting system to study in its own right, but also provides the ideal example to illustrate the concepts and techniques of bifurcation analysis.

The Task

In order to understand the behavior of the system one needs to answer the following question:

What are all possible dynamics of system (2), and how do they depend on the parameters K and ω?

Note that the term 'all possible behavior' implicitly requires a definition of when we regard two types of behavior to be the same, and when we regard them to be different. In bifurcation theory there is a clear notion of 'qualitatively the same', which is called *topological equivalence*. The basic idea is that two phase portraits are called topoligically equivalent if they can be mapped onto each other by a continuous map that respects the direction of time; for details see [2,4].

This notion of topological equivalence allows us to replace the above question by the following task:

Find all regions in the (K, ω)-plane corresponding to qualitatively different dynamics of Eqs. (2) in the (E, n)-space!

This is the typical task one is confronted with when one wants to analyze any given model from applications.

Where are Scotland and Wales?

A straightforward method for studying the dynamics of a system like Eqs. (2) is to do an extensive search by simulation. An admittedly less straightforward method is that of bifurcation analysis advocated here. To contrast the two different approaches let us consider a much simpler situation.

Suppose that you want to find Scotland and Wales on the map of Great Britain in Fig. 1(a). An extensive search by simulation corresponds to laying a grid over

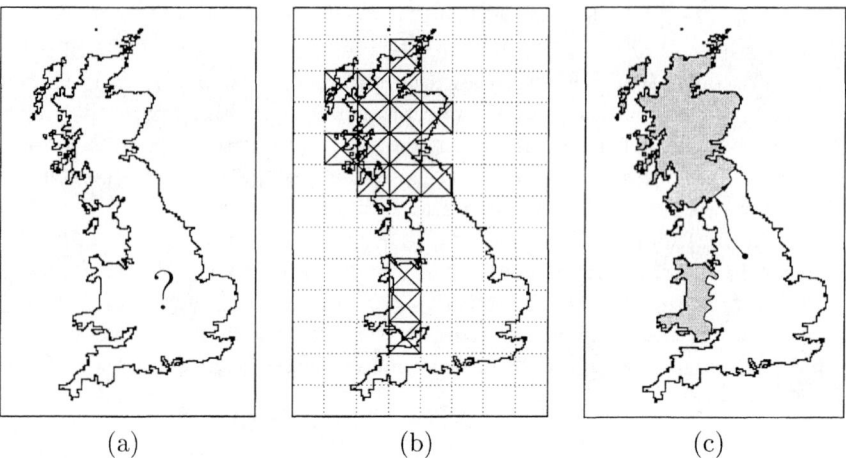

(a) (b) (c)

FIGURE 1. Where are Scotland and Wales on a map of Great Britain (a)? An extensive search on a grid gives a result as in (b), while bifurcation analysis corresponds to moving along a path, until a border is found, which is then followed directly (c).

Great Britain and marking all squares whose middle points satisfy the condition of belonging to Scotland or Wales as in Fig. 1(b). Bifurcation analysis, on the other hand, corresponds to starting somewhere, say in Sheffield in England, and then moving along a path, until a change in 'Englishness' is detected, which means that a border was found. In the next step the border is then followed directly as sketched in Fig. 1(c).

Let us get back to the problem of analyzing Eqs. (2). An extensive search by simulation requires that one numerically integrates the equations for many parameter values on some grid, and then records what dynamics was found. Clearly, in order to get good resolution this method is generally very time-consuming. Furthermore, it is not simple to automatically distinguish between different dynamics. To make matters even more complicated, there may be more than one possible observed behavior present for a given choice of K and ω, so that one will need to use many different initial conditions in phase space to be reasonably sure that all attractors are found. It should be pointed out that the only method to find the regions in the (K, ω)-plane corresponding to different behaviour of the laser *experimentally* is an extensive experimental search. By carefully recording the dynamics for a large number of (K, ω)-values, such a bifurcation diagram was produced for the injected laser by Simpson et al. [7].

Bifurcation analysis is very different in its approach and uses the idea of detecting and following boundary curves in the (K, ω)-plane between regions with qualitatively different phase portraits. While this is inherently more efficient, one needs to answer the following question. What kind of 'qualitative changes' should one look out for, and how can they be detected and followed?

Bifurcation theory answers these questions in very general terms. This theory can be seen as a collection of puzzle pieces of possible qualitative changes, called bifurcations, that must be expected *in any system* of the form (1) with the right number of parameters. Furthermore, normal form theory, a part of bifurcation theory, allows one to detect and indeed compute and, more generally, follow numerically these different types of bifurcations. This allows one to piece together the puzzle of the dynamics and find what is called the *bifurcation diagram*, the web of boundary curves between regions of different behavior of the system.

It is the purpose of this exposition to introduce the basic puzzle pieces. They are bifurcations of equilibira and periodic orbits as one parameter is changed. After introducing them in their 'pure form' via their normal forms, we illustrate what these bifurcations look like in the 'real system' Eqs. (2). We will then briefly explain how these bifurcations can be detected and followed numerically to obtain the bifurcation diagram of Eqs. (2).

BIFURCATIONS OF EQUILIBRIA

An *equilibrium* (also called a *steady state*) of a vector field in the form (1) is a point x_0 in phase space where the right-hand-side is zero, that is, where $f(x_0) = 0$. This means that there is no force at x_0, so that a particle at initial position x_0 will not move. An important question concerns the stability of the equilibrium x_0. It is determined by the Jacobian $Df(x_0)$, which is simply an $(n \times n)$ matrix.

The stability boundary for an equilibrium x_0 of an ODE is the imaginary axis in the complex plane of eigenvalues of $Df(x_0)$. It is common to call an equilibrium x_0 *hyperbolic* if there are no eigenvalues of $Df(x_0)$ on the imaginary axis. Hyperbolic equilibria come in three sorts. A hyperbolic equilibrium x_0 is called an *attractor* if all eigenvalues of $Df(x_0)$ have negative real part, that is, are in the (open) left half plane, and a *repellor* if all eigenvalues of $Df(x_0)$ have positive real part, that is, are in the right half plane. A hyperbolic equilibrium x_0 is called a *saddle point* if $Df(x_0)$ has some eigenvalues in the left and some in the right half plane. The three different types of hyperbolic equilibria are sketched in Fig. 2. Notice that we make no distinction between equilibria with spiralling directions (a) and those without (b).

According to the so-called Hartman-Grobman Theorem, the dynamics of (1) near a hyperbolic equilibrium x_0 is topologically equivalent to that of the linearized vector field

$$\dot{x} = Df(x_0)(x - x_0).$$

In other words, to understand the dynamics near a hyperbolic equilibrium x_0 one only neads to look at the linear part $Df(x_0)$ of f.

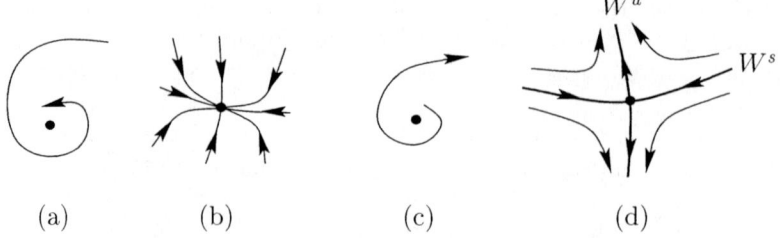

FIGURE 2. The basic types of hyperbolic equilibria are an attractor (a,b), a repellor (c) and a saddle point (c). The latter comes with stable and unstable manifolds W^s and W^u.

Saddle points are unstable and cannot be found by simulation in forward or backward time. However, they are very important for organising the overall dynamics of the system. To understand this, we need to introduce *stable and unstable manifolds*. The stable manifold $W^s(x_0)$ is the set of all points in phase space that end up at the saddle point x_0 in forward time; mathematically

$$W^s(x_0) = \{x \in \mathbb{R}^n \mid \lim_{t \to \infty} \phi^t(x) = x_0\},$$

where ϕ^t is the flow of (1). Similarily one defines the unstable manifold $W^u(x_0)$ as the set of all points in phase space that end up at x_0 in backward time, mathematically,

$$W^u(x_0) = \{x \in \mathbb{R}^n \mid \lim_{t \to -\infty} \phi^t(x) = x_0\};$$

see Fig. 2(d). According to the Stable Manifold Theorem, $W^s(x_0)$ is tangent at x_0 to the linear eigenspace $E^s(x_0)$ spanned by all (generalized) eigenvectors of the eigenvalues in the left half plane. Similarily, $W^u(x_0)$ is tangent at x_0 to $E^u(x_0)$.

The phase portrait near an equilibrium changes qualitatively, when one or more eigenvalues move through the imaginay axis. When this happens, hyperbolicity is lost and the dynamics is no longer determined by the linear part alone. Upon variation of a single parameter there are two possibilities. First, a real eigenvalue can pass through 0, which leads to a saddle-node bifurcation. Second, a pair of complex conjugate eigenvalues can cross the imaginary axis, which leads to a Hopf bifurcation.

We will now introduce these bifurcations. To this end, we present what are called *bifurcation conditions*, which are necessary conditions under which the bifurcation 'typically' occurs. For the bifurcations discussed here, the bifurcation conditions are essentially conditions on the linearization, which are quite straightforward to check. To make sure that one is in a 'typical' or generic situation, one needs to verify the so-called *genericity conditions*, which involve information on the nonlinear part of the vector field. They are harder to formulate and to check, but are satisfied in a 'typical' situation.

We will present normal forms for the basic bifurcations. A normal form is the simplest vector field with the lowest possible phase space that contains the bifurcation at hand. In a larger dimensional system, the bifurcation takes place on a family of lower dimensional subspaces, generally called the *center manifold*. Each of these subspaces has the dimension of the phase space of the respective normal form. The dynamics on the center manifold is in fact qualitatively the same as that of the normal form! So the normal form is truly a universal model of the bifurcation. The concept of a center manifold is admittedly quite abstract, but in applications one can 'see' the bifurcations 'living' on these lower dimensional subspaces. This will be illustrated with the example of the injection laser.

 Saddle-Node Bifurcation

When a single, real eigenvalue of an equilibrium passes through the imaginary axis, one typically finds a *saddle-node bifurcation* (also called *fold bifurcation* or *limit point bifurcation*). In other words, the general system Eq. (1) must have an equilibrium with an eigenvalue 0. It is convenient to formulate this as two bifurcation conditions that need to be satisfied:

(B1) $f(x_0, \lambda_0) = 0$, that is, there is an equilibrium at (x_0, λ_0).

(B2) The Jacobian $Df(x_0, \lambda_0)$ has an eigenvalue zero.

The bifurcation conditions (B1) and (B2) are necessary conditions for a saddle-node bifurcation. However, in order to make sure that there is really a saddle-node bifurcation present one needs to check certain genericity conditions. They are discussed below, because it requires a bit of notation to formulate them. However, in a 'typical' (or generic) situation the genericity conditions are satisfied, which means that one typically finds a saddle-node bifurcation if (B1) and (B2) are satisfied. Notice that it easy to check (B1) and (B2) because this involves only eigenvalue computations on the Jacobian $Df(x_0, \lambda_0)$. This means that saddle-node bifurcations can be detected and followed efficiently.

The simplest model that contains a saddle-node bifurcation is the one-dimensional saddle-node normal form

$$\dot{x} = \lambda - x^2, \tag{3}$$

which is a vector field on the real line $\mathbb{R}$, depending on the single parameter $\lambda \in \mathbb{R}$. Its phase portaits before, at and after the bifurcation are sketched in Fig. 3. In the bifurcation two new equilibria are born, which is the hallmark of this bifurcation. Notice that the two equilibria are at $\pm\sqrt{\lambda}$ for $\lambda > 0$.

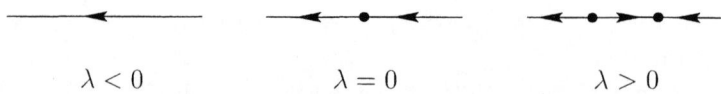

$\lambda < 0 \qquad\qquad \lambda = 0 \qquad\qquad \lambda > 0$

FIGURE 3. Saddle-node bifurcation as phase portraits of (3).

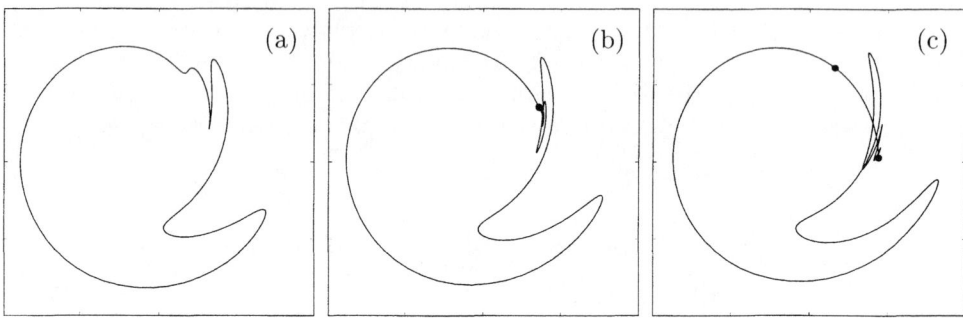

FIGURE 4. A saddle-node bifurcation of Eqs. (2). The bifurcation actually takes place on an attracting periodic orbit, seen in projection onto the complex E-plane. (The self-intersections are due to the projection.) On the periodic orbit (a), a new equilibrium, a saddle-node, appears (b), which then splits up into an attractor and a saddle point (c). Both branches of the unstable manifold of the saddle point (left point) go to the attractor (right point). The parameter values are $\omega = -0.5$, and from (a) to (c) $K = 0.2$, $K = 0.22675$, and $K = 0.25$.

The question is now: what does a saddle-node bifurcation look like in a phase space with more than one dimension? It turns out that any saddle-node bifurcation 'lives' on a one-dimensional curve, where it looks just like that of the normal form (3). Each of the other $(n-1)$ phase space directions will be either attracting or repelling directions with respect to this one-dimensional curve. Before we get more technical let us consider the example of a saddle-node bifurcation in the injected laser system Eqs. (2) shown in Fig. 4. Notice that the saddle-node bifurcation actually takes place on an attracting periodic orbit. Nevertheless, when one restricts to a small neighborhood of where the new equilibria appear then one recognises exactly the situation sketched in Fig. 3. The two additional phase space directions are attracting directions, which is why one of the new equilibria is an attractor, and the other is a saddle point. In a system with a phase space of dimension two one always sees the creation of a saddle point and a node, that is, an attractor or repellor, which is the reason for the name of this bifurcation. Notice that for any saddle-node bifurcation the equilibria are at distance $\approx \sqrt{\varepsilon}$ from each other, where

ε is the distance in parameter space from the bifurcation point. This is a direct consequence of what happens in the normal form (3).

We now need to be a bit more technical, which will allow us to formulate the genericity conditions mentioned earlier. It can be proved that there is a family $\mathcal{C}_\lambda$, parametrized by the bifurcation parameter λ, of one-dimensional curves in phase space on which the saddle-node bifurcation looks qualitatively as sketched in Fig. 3 for the normal form (3). The family $\mathcal{C}_\lambda$ forms a smooth surface in (x, λ)-space and it is called the *center manifold*. It is tangent to the eigenspace E^c of the eigenvalue zero of $Df(x_0, \lambda_0)$. The additional directions in phase space (off the center manifold) are divided into attracting and repelling directions, depending on the nonzero eigenvalues of $Df(x_0, \lambda_0)$. This is true under the following genericity condition, which is quite easy to check.

(G0) There are no other eigenvalues of $Df(x_0, \lambda_0)$ on the imaginary axis.

Furthermore, one can find the reduced vector field $\dot{r} = g(r, \lambda)$ on the center manifold $\mathcal{C}_\lambda$, which is the force field on $\mathcal{C}_\lambda$ induced by the vector field $f(x, \lambda)$. For this one-dimensional vector field the saddle-node bifurcation takes place at, say, (r_0, λ_0). The remaining genericity conditions can be summarized as saying that $g(r, \lambda)$ can be brought to the normal form (3) by suitable coordinate changes. More technically, this requires that the following genericity conditions are satisfied.

(G1) $\dfrac{\partial^2 g}{\partial r^2}(r_0, \lambda_0) \neq 0$.

(G2) $\dfrac{\partial g}{\partial \lambda}(r_0, \lambda_0) \neq 0$, which means that the bifurcation value λ_0 is crossed with positive speed.

The conditions (G0)–(G2) are called genericity conditions, because they are generically or typically satisfied. Another way of looking at this is the following. If (B1) and (B2) and (G0)–(G2) are satisfied for some (x_0, λ_0) then the entire scenario of the saddle-node bifurcation is *structurally stable*. This means that a small perturbation of (3) will also have a generic or typical saddle-node bifurcation at some perturbed point $(\tilde{x}_o, \tilde{\lambda}_0)$ close to (x_0, λ_0).

We finally remark that it is possible but not straightforward to verify the genericity conditions (G1) and (G2). This involves computing first a sufficient approximation to the center manifold, and then a sufficient approximation to $g(r, \lambda)$ near (r_0, λ_0). From a practical point of view, (G1) and (G2), and also (G0), are less important than (B1) and (B2) because they are typically satisfied anyway. We will come back to this point later on when we discuss numerical methods for detecting and following bifurcations.

Hopf Bifurcation

When a pair of complex conjugate eigenvalues passes through the imaginary axis, one typically finds a *Hopf bifurcation*. In order for a Hopf bifurcation to occur in the general system Eq. (1) again two bifurcation conditions need to be satisfied.

(B1) $f(x_0, \lambda_0) = 0$, that is, there is an equilibrium at (x_0, λ_0).

(B2) The Jacobian $Df(x_0, \lambda_0)$ has a pair of complex conjugate eigenvalues on the imaginary axis.

In a typical situation, again under the genericity conditions discussed below, there is a Hopf bifurcation in the system.

The simplest model that contains a Hopf bifurcation is the two-dimensional Hopf normal form

$$\begin{bmatrix} \dot{x}_1 \\ \dot{x}_2 \end{bmatrix} = \begin{bmatrix} \lambda & -\omega \\ \omega & \lambda \end{bmatrix} \begin{bmatrix} x_1 \\ x_2 \end{bmatrix} + a\left(x_1^2 + x_2^2\right) \begin{bmatrix} x_1 \\ x_2 \end{bmatrix} \qquad (4)$$

which is a vector field on $\mathbb{R}^2$, depending on the single parameter $\lambda \in \mathbb{R}$. Here $\omega > 0$ is a speed of rotation and $a \neq 0$ is called the Lyapunov quantity. In fact, the only important feature of a is its sign, and there are two cases of Hopf bifurcation. The *supercritical* case $a < 0$, in which an attracting periodic orbit bifurcates, and the *subcritical* case $a > 0$, in which a repelling periodic orbit bifurcates. The phase portaits of (4) before, at and after the bifurcation are sketched in Fig. 5 for the supercritical case. In the bifurcation, a stable equilibrium loses its stability and a small attracting periodic orbit is born.

Figure 6 shows what a Hopf bifurcation looks like in the injected laser system Eqs. (2). Before the bifurcation, a spiral attractor, the right point in Fig. 6(a), attracts both branches of the unstable manifold of a saddle point. Notice that panel (a) shows the same phase portrait as Fig. 4(c), but now in projection onto the $(\text{Re}(E), n)$-plane. The attractor loses its stability via a slowing down of the attraction. At the bifurcation point in Fig. 6(b) the attraction is extremely slow, and after the bifurcation an attracting periodic orbit appears as is shown in Fig. 6(c). The periodic orbit attracts both branches of the unstable manifold of the saddle point, because it is now the only attractor.

The Hopf bifurcation again 'lives' on a center manifold $\mathcal{C}_\lambda$, which is now a family of two-dimensional surfaces, tangent at x_0 to the eigenspace E^c spanned by the generalized eigenvectors of the complex conjugate eigenvalues that move through the imaginary axis. The two-dimensional subspaces (for each fixed λ) comprising the center manifold $\mathcal{C}_\lambda$ are visible in Fig. 6, where they are 'visualized' by spiralling trajectories. As before, the additional directions in phase space are divided into attracting and repelling directions, depending on the other eigenvalues of $Df(x_0, \lambda_0)$.

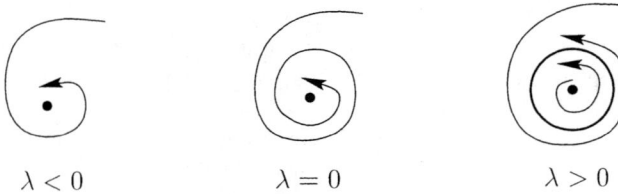

$\lambda < 0$ $\lambda = 0$ $\lambda > 0$

FIGURE 5. Supercritical Hopf bifurcation ($a < 0$) as phase portraits of (4).

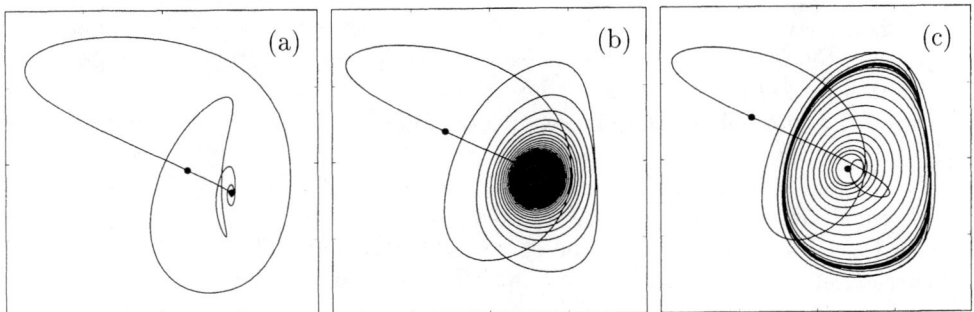

FIGURE 6. A supercritical Hopf bifurcation of Eqs. (2). A spiral attractor (right point) attracts both branches of the unstable manifold of a saddle point (left point) (a) and then loses its stability via a slowing down of the attraction. At the bifurcation point the attraction is extremely slow (b), and after the bifurcation an attracting periodic orbit appears (c) which attracts both branches of the unstable manifold of the saddle point. The phase portraits are seen in projection onto the $(\text{Re}(E), n)$-plane. The parameter values are $K = 0.25$, and from (a) to (c) $\omega = -0.5$, $\omega = -0.24$, and $\omega = -0.15$.

In Fig. 6 the additional directions are attracting, which is why we find an attracting periodic orbit in Fig. 6(c). For $\mathcal{C}_\lambda$ to exist one again needs

(G0) There are no other eigenvalues of $Df(x_0, \lambda_0)$ on the imaginary axis.

There is now a two-dimensional reduced vector field $\dot{r} = g(r, \lambda)$ on the center manifold $\mathcal{C}_\lambda$, for which the Hopf bifurcation takes place at, say, (r_0, λ_0). The other genericity conditions are difficult to state explicitly, but they can be phrased as follows.

(G1) By suitable coordinate changes $g(r, \lambda)$ can be brought to the Hopf normal form (4). The resulting Lyapunov quantity a must be nonzero.

(G2) The bifurcation value λ_0 must be crossed with positive speed.

The fact that the Hopf bifurcation lives on a two-dimensional centre manifold, where it is topologically equivalent to the normal form (4), has an important, and maybe somewhat astonishing physical consequence. In any system, the onset of small amplitude oscillations through a Hopf bifurcation is via a periodic orbit, which *lies in some plane* to first approximation close to the Hopf bifurcation. The radius of this periodic orbit is $\sqrt{\varepsilon}$, where ε is the distance from the bifurcation. Furthermore, the motion along this periodic orbit is constant to first approximation close to the Hopf bifurcation, which means that *the oscillations must be sinosoidal near their onset*. A good example is the disappearance of self-pulsations in a laser with absorber; see [6] and the chapter by *Erneux*. The pulses are very spiky, but just before they disappear in a Hopf bifurcation they become not only smaller in amplitude, but also sinosoidal. It is easy to overlook this effect, but without any calculation we know that this must be the case!

Again, it is possible to compute a suitable approximation of the reduced vector field $\dot{r} = g(r, \lambda)$ on the center manifold. The Lyapunov quantity a can then be computed, but this is truly quite involved; we refer to [2,4] for details. The punchline is that (G1) and (G2) can be checked, but are typically satisfied anyway. As for the saddle-node bifurcation, the conditions (B1) and (B2) (and also (G0)) can be checked efficiently, so that Hopf bifurcations can be detected and followed.

BIFURCATIONS OF PERIODIC ORBITS

A *periodic orbit* (or *limit cycle*) is an orbit that closes up after some time. More formally, $\Gamma(t)$ is a periodic orbit of Eqs. (1) if $\Gamma(t + T) = \Gamma(t)$ for some (smallest) time $T > 0$, which is called the period of Γ. Similar to equilibria, a typical or hyperbolic periodic orbit is either attracting, repelling or of saddle-type; see Fig. 7. In order to define these notions properly we need to introduce an important tool: *the Poincaré map*.

Suppose that we choose a (finite) cross section Σ, called a Poincaré section, transverse to the periodic orbit Γ. The section Σ intersects Γ in a point x_0, and we assume here that Σ is so small that it does not intersect Γ in a different point. This is sketched in Fig. 8 for the case that the phase space is three-dimensional, in which case the section Σ is two-dimensional. We remark that the construction of the Poincaré map works in general when the phase space is n-dimensional, in which case Σ is an $(n-1)$-dimensional hyperplane. Then the *Poincaré map* (also called *first-return map*)

$$P : \Sigma \to \Sigma \quad (5)$$
$$x \mapsto P(x) \quad (6)$$

is defined as follows. For a point $x \in \Sigma$ the point $P(x)$ is the next intersection point of the orbit through x with Σ; see Fig. 8. Note that the intersection point x_0 of Γ with Σ is a *fixed point* of P, which simply means that $P(x_0) = x_0$.

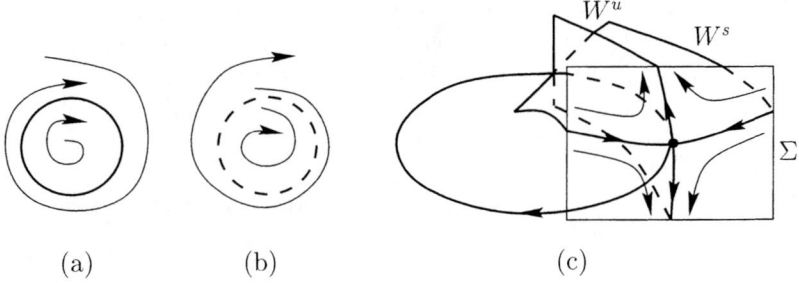

(a)   (b)   (c)

FIGURE 7. The basic types of hyperbolic periodic orbits are an attracting periodic orbit (a), a repelling periodic orbit (b), and a periodic orbit of saddle-type (c). The latter comes together with stable and unstable manifolds W^s and W^u. In the Poincaré section Σ, the saddle-type periodic orbit corresponds to a saddle fixed point of the Poincaré map; see also Fig. 8.

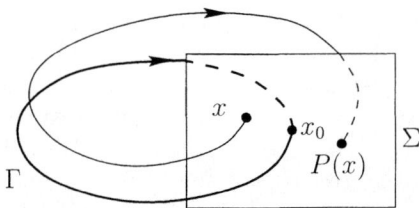

FIGURE 8. The Poincaré map P is defined on a suitable Poincaré section Σ locally near the intersection point x_0 of the periodic orbit Γ with Σ. To compute $P(x)$ for some point $x \in \Sigma$ one simply follows the flow to the next intersection with Σ.

The map P is a well-defined differentiable map with a differentiable inverse in some neighborhood of x_0. (Such a map is also called a *diffeomorphism*.) Notice that the Poincaré map P is defined only locally near x_0. Generally, it may not be defined for a point too far away from x_0, because the orbit through this point may not hit the section Σ again. We remark that in the special case of a system with an explicit periodic forcing, such as the driven pendulum in the chapter by *Broer & Krauskopf*, the Poincaré map can be defined as the stroboscopic map of the forcing. In this case it is defined globally, and not just in some small neighborhood.

Integrating from an initial condition x and recording intersection points with the section Σ corresponds to iterating the Poincaré map P. In other words, we need to find the discrete orbit $\{x, P(x), P^2(x), \ldots\}$ of points hopping around Σ. Here $P^2(x)$ stands for the second iterate $P^2(x) = P(P(x))$. Studying the dynamics of the Poincaré map P on Σ is equivalent to studying the dynamics of the vector field (1) near the periodic orbit. It feels a bit like we traded one hard problem for another by setting up a lot of new notation. In fact, considering the Poincaré map is actually one of the key tools of bifurcation theory.

The stability of the periodic orbit Γ can now be studied by determining the stability of the fixed point x_0 of P, and this allows us to be a more precise. Consider the linearization $DP(x_0)$ of P at x_0, which is an $(n-1) \times (n-1)$ matrix. The stability boundary of a fixed point x_0 for any map P is the unit circle in the complex plane of eigenvalues of $DP(x_0)$, because P can be written as

$$P(x) = DP(x_0)(x - x_0) + \text{higher-order-terms}.$$

In other words, the lowest-order effect of P is multiplication by $DP(x_0)$. Therefore, for stability it is required that $|DP(x_0)| < 1$, which means that all eigenvalues of $DP(x_0)$ must lie inside the unit circle of the complex plane of eigenvalues.

The fixed point x_0 and the periodic orbit Γ are said to be *hyperbolic* if there are no eigenvalues of $DP(x_0)$ on the unit circle in the complex plane. (The eigenvalues of $DP(x_0)$ are also called *Floquet multipliers* of the periodic orbit Γ. If one considers the full linearization around Γ including the direction transverse to Σ then there is always the additional Floquet multiplier 1 corresponding to the dynamics tangent to the periodic orbit.) Hyperbolic periodic orbits also come in three types as is sketched in Fig. 7. If x_0 is hyperbolic then we call it (and Γ) an *attractor* if all eigenvalues of $DP(x_0)$ are inside the unit circle, a *repellor* if all eigenvalues of $DP(x_0)$ are outside the unit circle, and a *saddle point* if some eigenvalues of $DP(x_0)$ are inside the unit circle and some are outside. As was the case for equilibria of vector fields, the dynamics near a hyperbolic fixed point x_0 is determined completely by the linear part $DP(x_0)$.

Again, saddle-type periodic orbits are very important for organising the overall dynamics of the system. We can define the stable and unstable manifolds $W^s(\Gamma)$ and $W^u(\Gamma)$ of a saddle-type periodic orbit Γ, just like for equilibria, as the set of points in phase space that end up at Γ in forward and backward time, respectively. Locally near x_0, the stable and unstable manifolds intersect the section Σ and give rise to stable and unstable manifolds $W^u(x_0)$ and $W^u(x_0)$ of the saddle fixed point x_0 as is sketched in Fig. 7(c). According to the Stable Manifold Theorem, $W^s(x_0)$ is tangent at x_0 to the linear stable eigenspace $E^s(x_0)$ spanned by all (generalized) eigenvectors of the eigenvalues inside the unit circle, and $W^s(x_0)$ is tangent at x_0 to the unstable eigenspace $E^u(x_0)$ spanned by all (generalized) eigenvectors of the eigenvalues outside the unit circle.

It must be pointed out that the construction of the Poincaré map is not explicit. It is not possible in general to come up with a nice formula for P even if there is a nice formula for f in Eq. (1)! However, it is possible to compute P numerically by integrating the vector field and detecting the next intersection with Σ exactly as sketched in Fig. 8. Furthermore, $DP(x_0)$ can be computed by solving what is called the first variational equation, that is, by integrating the derivative Df along Γ. In other words, from a practical point of view, x_0, P and $DP(x_0)$ are known.

The phase portrait near a periodic orbit changes qualitatively, when one or more eigenvalues of $DP(x_0)$ move through the unit circle. Then hyperbolicity of the fixed point is lost and it is not sufficient to consider only the linear part DP of P. Upon variation of a single parameter there are three possibilities. First, a real eigenvalue can pass through 1, which leads to a saddle-node of limit cycle bifurcation. Second, a real eigenvalue can pass through -1, which leads to a period-doubling bifurcation. Third, a pair of complex conjugate eigenvalues can cross the unit circle, which leads to a torus bifurcation.

As we have done for bifurcations of equilibria, we will now introduce these bifurcations of periodic orbits by presenting the respective bifurcation and genericity conditions. We will also give their normal forms, which must be formulated as normal forms of the Poincaré map P. The associated center manifolds are now families of lower dimensional subspaces in the Poincaré section Σ. This will again be illustrated with the example of the injection laser.

 Saddle-Node Bifurcation of Limit Cycles

When a single eigenvalue of $DP(x_0)$ passes through $+1$ then one typically has a *saddle-node bifurcation of limit cycles* (also called *fold bifurcation of periodic orbits*). In order for a saddle-node bifurcation of limit cycles to occur in the general system Eq. (1) two bifurcation conditions need to be satisfied. They can only be formulated in terms of the Poincaré map P.

(B1) $P(x_0, \lambda_0) = x_0$, that is, there is a fixed point of P at (x_0, λ_0).

(B2) The Jacobian $DP(x_0, \lambda_0)$ has an eigenvalue $+1$.

Again, under genericity conditions discussed below, there is a saddle-node bifurcation of limit cycles in the system.

The normal form for this bifurcation can also only be given for the Poincaré map P. It is the one-dimensional map

$$P : \mathbb{R} \times \mathbb{R} \to \mathbb{R}$$
$$(x, \lambda) \mapsto \lambda + x - x^2, \qquad (7)$$

which is a map on the real line $\mathbb{R}$, depending on the single parameter $\lambda \in \mathbb{R}$. In the bifurcation, an attracting and a repelling periodic orbit are born as sketched in Fig. 9(a). This corresponds to the phase portraits of the normal form (7) as shown in Fig. 9(b). Note the similarity with the phase portraits of the saddle-node bifurcation in Fig. 3, but recall that points hop under the map (7), rather than flow under a vector field.

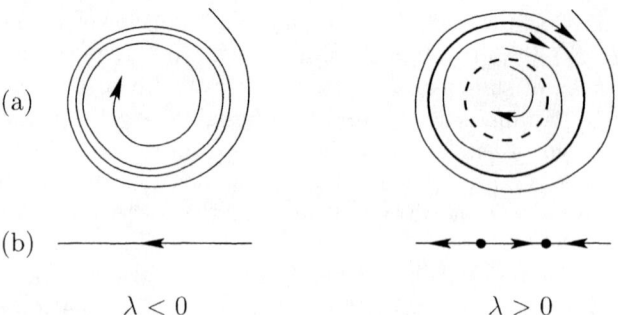

FIGURE 9. Saddle-node bifurcation of limit cycles in the phase space of the vector field (a), and on the level of the normal form (7) of the Poincaré map (b).

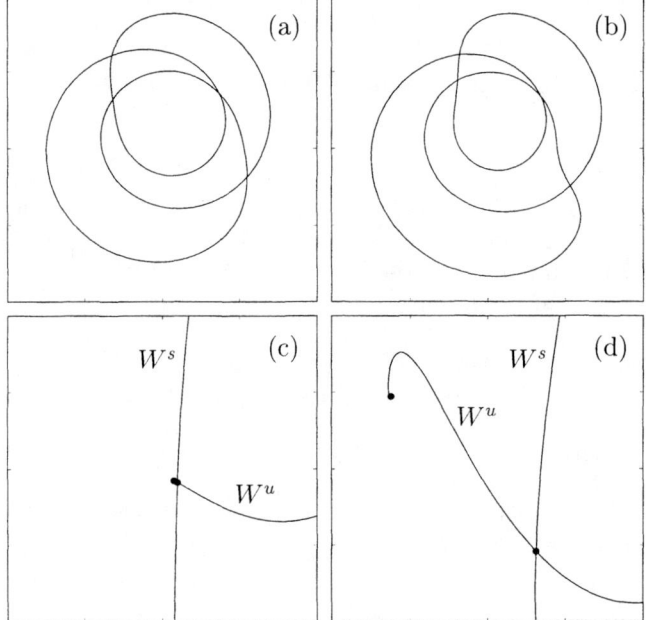

FIGURE 10. A saddle-node bifurcation of limit cycles of Eqs. (2) seen in projection onto the complex E-plane (a,b), and in the Poincaré section $\{\text{Re}(E) = 0\}$ (c,d). In panels (a) and (c) the system is very close to, but just past the bifurcation, while panels (b) and (d) show a situation well away from the bifuircation point. Note that before the bifurcation one finds no limit cycles or fixed points of the Poincaré map. The parameters are $\omega = -1.36$, and $K = 0.20233$ (a,c) and $K = 0.22$ (b,d).

A saddle-node bifurcation of limit cycles in the injected laser system Eqs. (2) is shown in Fig. 10 in projection onto the complex E-plane in panels (a) and (b) and in the Poincaré section $\{\text{Re}(E) = 0\}$ in panels (c) and (d). At the bifurcation value (a) an attracting and a saddle-type periodic orbit are born. (The panels (a) and (c) are for parameter values very close to, but just after the saddle-node bifurcation of limit cycles.) The two periodic orbits then move away from each other, and in (b) only the stable periodic orbit is shown. In the Poincaré section a saddle-node appears at the bifurcation point (c), which then splits into a saddle point and an attracting fixed point (d). One branch of the unstable manifold W^u of the saddle point goes to the attractor. Like for a saddle-node bifurcation of equilibria, the two fixed points are at distance $\approx \sqrt{\varepsilon}$, where ε is the distance in parameter space from the bifurcation point.

As was the case for the bifurcations of equilibria, one can find a center manifold, a family $\mathcal{C}_\lambda$ of one-dimensional curves, but now in the Poincaré section Σ, on which the saddle-node bifurcation looks qualitatively as sketched in Fig. 9(b) for the normal form (7). This is visible in Fig. 10(c) and (d). The family $\mathcal{C}_\lambda$ is a smooth surface in $(\Sigma \times \lambda)$-space and it is tangent to the eigenspace E^c of the eigenvalue $+1$ for λ_0. The additional directions in Σ are divided into attracting and repelling directions, depending on the other eigenvalues of $DP(x_0, \lambda_0)$. The center manifold exists under the condition

(G0) There are no other eigenvalues of $DP(x_0, \lambda_0)$ on the unit circle.

One can now find the reduced Poincaré map $R(r, \lambda)$ on the center manifold $\mathcal{C}_\lambda$. For this one-dimensional map the bifurcation takes place at, say, (r_0, λ_0). We can now formulate the genericity conditions, which again guarantee that $R(r, \lambda)$ can be brought to the normal form (7).

(G1) $\dfrac{\partial^2 R}{\partial R^2}(r_0, \lambda_0) \neq 0$.

(G2) $\dfrac{\partial R}{\partial \lambda}(r_0, \lambda_0) \neq 0$, which means that the bifurcation value λ_0 is crossed with positive speed.

Notice the similarity of the bifurcation and genericity conditions with those of the saddle-node bifurcation of equilibria. As mentioned earlier, it is possible to compute x_0, P and DP numerically. This allows one to compute a suitable approximation to the center manifold $\mathcal{C}_\lambda$ and to the reduced map $R(r, \lambda)$ so that (G1) and (G2) can be checked. However, this is very involved. On the other hand, it is relatively easy to check (B1), (B2) and (G0) because this involves only eigenvalue computations on the Jacobian $DP(x_0, \lambda_0)$. Nevertheless, the Jacobian needs to be computed numerically. In any case, a saddle-node bifurcation of limit cycles can be detected and followed efficiently numerically.

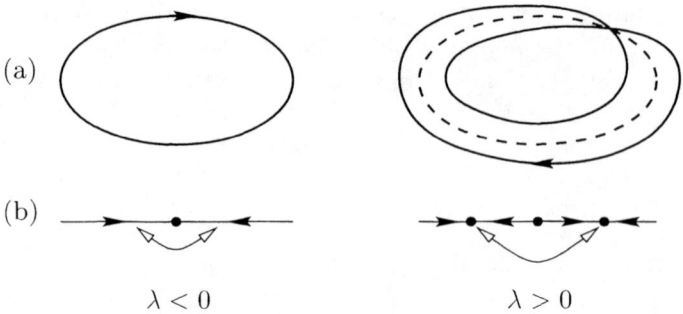

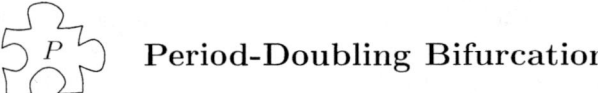

λ < 0   λ > 0

FIGURE 11. Period-doubling bifurcation in the phase space of the vector field (a), and on the level of the normal form (8) of the Poincaré map (b).

Period-Doubling Bifurcation

When a single eigenvalue of $DP(x_0)$ passes through -1, one typically has a *period-doubling bifurcation* (also called *flip bifurcation*). In order for a period-doubling bifurcation to occur in the general system Eq. (1) two bifurcation conditions need to be satisfied, which again must be formulated in terms of the Poincaré map P.

(B1) $P(x_0, \lambda_0) = x_0$, that is, there is a fixed point of P.

(B2) $DP(x_0, \lambda_0)$ has an eigenvalue -1.

Under the genericity conditions discussed below, there is a period-doubling bifurcation in the system.

There is no normal form of a vector field with such a bifurcation, but it is possible to write down a normal form of the Poincar'e map P. It is given by the one-dimensional map

$$\begin{aligned} P : \mathbb{R} \times \mathbb{R} &\to \mathbb{R} \\ (x, \lambda) &\mapsto -(1+\lambda)x - x^3, \end{aligned} \qquad (8)$$

which is a map on the real line $\mathbb{R}$, depending on the single parameter $\lambda \in \mathbb{R}$. In the bifurcation, an attracting periodic orbit loses its stability, which gives rise to a new periodic orbit with twice the period as is sketched in Fig. 11(a). This corresponds to the phase portraits of the normal form (8) as shown in Fig. 11(b). Notice that points hop from one side of the origin to the other under iteration, so that the two new points lie on a period-two orbit.

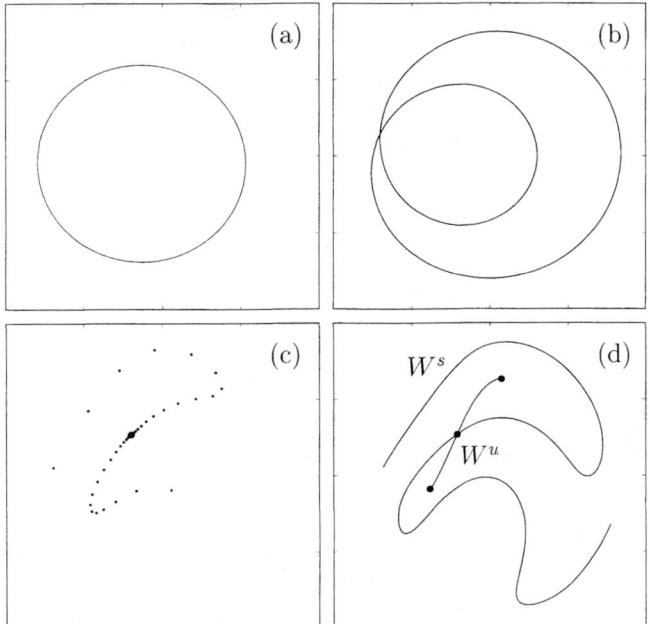

FIGURE 12. A period-doubling bifurcation of Eqs. (2) seen in projection onto the complex E-plane (a,b), and in the Poincaré section $\{\text{Re}(E) = 0\}$ (c,d). An attracting one-periodic orbit (a) loses its stability and gives rise to a two-periodic orbit (b). In the Poincaré section a stable fixed point (c) loses its stability. After the bifurcation the one-periodic orbit is of saddle type, and its unstable manifold converges to the two-periodic orbit (d). The parameters are $\omega = 1.8$, and $K = 0.15$ (a,c) and $K = 0.17$ (b,d).

A period-doubling bifurcation in the injected laser system Eqs. (2) is shown in Fig. 12 in projection onto the complex E-plane in panels (a) and (b) and in the Poincaré section $\{\text{Re}(E) = 0\}$ in panels (c) and (d). An attracting periodic orbit (a) gives rise to a period-doubled orbit (b). In the Poincaré section this corresponds to an attracting fixed point (c), which then gives rise to a period-two point (d). Notice that the points on the unstable manifold W^u hop between its two branches. We remark that the points on W^s must also hop between its two branches, because the product of the eigenvalues of $DP(x_0)$ must be positive. This is so because any Poincaré map must be orientation preserving.

As was the case for the saddle-node bifurcation of limit cycles, there is a center manifold $\mathcal{C}_\lambda$ consisting of one-dimensional curves in the Poincaré section Σ parametrized by λ, which exists under the genericity condition

(G0) There are no other eigenvalues of $DP(x_0, \lambda_0)$ on the unit circle.

On $\mathcal{C}_\lambda$ the period-doubling bifurcation looks qualitatively as sketched in Fig. 11(b) for the normal form (8). Let $R(r, \lambda)$ be again the reduced Poincaré map on $\mathcal{C}_\lambda$, with the bifurcation taking place at, say, (r_0, λ_0). The additional directions in Σ are divided into attracting and repelling directions, depending on the other eigenvalues of $DP(x_0, \lambda_0)$. We can now formulate the other genericity conditions

(G1) $\dfrac{\partial^3 R}{\partial R^3}(r_0, \lambda_0) \neq 0$.

(G2) $\dfrac{\partial^2 R}{\partial R \partial \lambda}(r_0, \lambda_0) \neq 0$, which means that the bifurcation value λ_0 is crossed with positive speed.

Again it is very tedious but possible to check (G1) and (G2), while verifying (B1), (B2) and (G0) is relatively easy. This means that a period-doubling bifurcation can be detected and followed efficiently numerically.

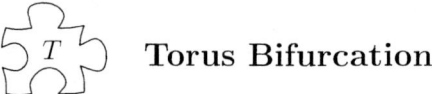 Torus Bifurcation

When a pair of complex conjugate eigenvalues of $DP(x_0)$ passes through the unit circle then one typically has a *torus bifurcation*. (This bifurcation is also called *Hopf bifurcation for maps* or *Neimark-Sacker bifurcation*.) In order for a torus bifurcation to occur in the general system Eq. (1) two bifurcation conditions, which again must be formulated in terms of the Poincaré map P, need to be satisfied.

(B1) $P(x_0, \lambda_0) = x_0$, that is, there is a fixed point of P.

(B2) $DP(x_0, \lambda_0)$ has the pair of complex conjugate eigenvalues $e^{\pm 2\pi i \omega}$.

There is no normal form of a vector field with such a bifurcation, but it is possible to write down a normal form of the Poincar'e map P. It is given by the two-dimensional map [4]

$$P\left(\begin{bmatrix} x_1 \\ x_2 \end{bmatrix}, \lambda\right) = \begin{bmatrix} \cos(\omega) & -\sin(\omega) \\ \sin(\omega) & \cos(\omega) \end{bmatrix}$$
$$\times \left((1+\lambda)\begin{bmatrix} x_1 \\ x_2 \end{bmatrix} + (x_1^2 + x_2^2)\begin{bmatrix} a & -b \\ b & a \end{bmatrix}\begin{bmatrix} x_1 \\ x_2 \end{bmatrix}\right) \quad (9)$$

which is a map on $\mathbb{R}^2$, depending on the single parameter $\lambda \in \mathbb{R}$. Here $\omega > 0$ is the speed of rotation under iteration of the map and $a \neq 0$ is a constant, which is very similar to the Lyapunov quantity in the normal form (4) of the Hopf bifurcation. (The constant $b \in \mathbb{R}$ is in fact not important.) Again, the only important feature of a is its sign, and there are two cases of torus bifurcations. The *supercritical* case

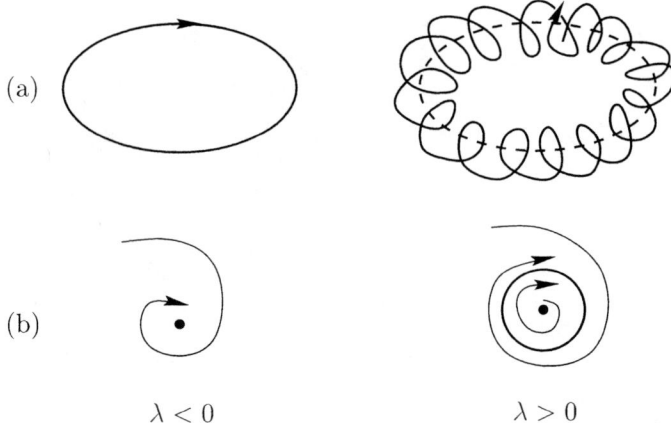

FIGURE 13. Torus bifurcation in the phase space of the vector field (a), and on the level of the normal form (9) of the Poincaré map (b).

$a < 0$, in which an attracting invariant circle bifurcates, and the *subcritical* case $a > 0$, in which a repelling invariant circle bifurcates.

The torus bifurcation is the most complicated of the puzzle pieces we consider. The dynamics on this invariant circle can be *locked*, in which case there is an attracting periodic orbit on the invariant circle, or *quasiperiodic*, when any point on the invariant circle fills the invariant circle densely under iteration of P. For the dynamics of the corresponding torus this means that there is an attracting periodic orbit on the torus in the locked case, while in the quasiperiodic case trajectories fill the torus densely. The boundaries between locking and quasiperiodicity are saddle-node bifurcations of limit cycles, which form *resonance tongues* (also called *Arnold tongues*) in the parameter plane [2,4].

The torus bifurcation is sketched in Fig. 13(a) for the supercritical case. This corresponds to the phase portraits of the normal form (9) as shown in Fig. 13(b). Notice that points hop around the invariant circle under iteration.

A supercritical torus bifurcation in the injected laser system Eqs. (2) is shown in Fig. 14 in projection onto the complex E-plane in panels (a) and (b) and in the Poincaré section $\{\text{Re}(E) = 0\}$ in panels (c) and (d). An attracting periodic orbit (a) gives rise to an attracting invariant torus (b). The motion on this torus appears to be quasiperiodic for the chosen parameter values. In the Poincaré section we see an attracting fixed point (c), which then gives rise to an invariant circle (d).

It is possible to find a center manifold $\mathcal{C}_\lambda$ consisting of two-dimensional surfaces in the Poincaré section Σ parametrized by λ. On it the torus bifurcation looks qualitatively as sketched in Fig. 13(b) for the normal form (9). The additional

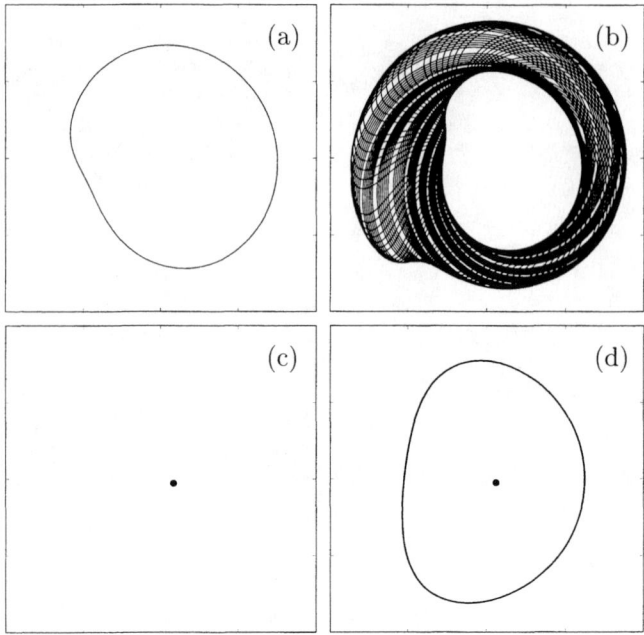

FIGURE 14. A torus bifurcation of Eqs. (2) seen in projection onto the complex E-plane (a,b), and in the Poincaré section $\{\text{Re}(E) = 0\}$ (c,d). An attracting periodic orbit (a) loses its stability, which gives rise to an attracting torus (b). In the Poincaré section a stable fixed point (c) loses its stability, which leads to the birth of an attracting invariant circle surrounding the now unstable fixed point (d). The parameters are $\omega = 0.8$, and $K = 0.1$ (a,c) and $K = 0.113$ (b,d).

directions in Σ are divided into attracting and repelling directions, depending on the other eigenvalues of $DP(x_0, \lambda_0)$. This is possible under the condition

(G0) There are no other eigenvalues of $DP(x_0, \lambda_0)$ on the unit circle.

It turns out that there is an additional genericity condition for the Jacobian $DP(x_0, \lambda_0)$.

(GR) $\omega \notin \{1, \frac{1}{2}, \frac{1}{3}, \frac{2}{3}, \frac{1}{4}, \frac{3}{4}\}$, where $e^{\pm 2\pi i \omega}$ are the eigenvalues of $DP(x_0, \lambda_0)$ on the unit circle.

The condition (GR) is important in order to exclude what are called the *strong resonances*. At strong resonances it is not true that an invariant torus bifurcates, but the dynamics is much more complicated; see [2,4].

Let $R(r, \lambda)$ be again the reduced Poincaré map on $\mathcal{C}_\lambda$, with the bifurcation taking place at, say, (r_0, λ_0). We can now formulate the last two genericity conditions, which are very similar to those of the Hopf bifurcation.

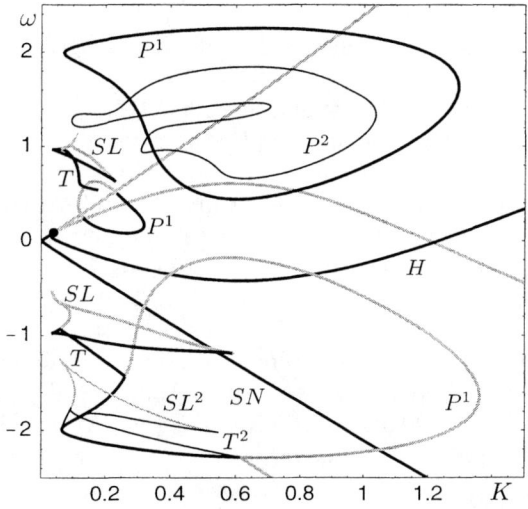

FIGURE 15. The bifurcation diagram in the (K, ω)-plane of Eqs. (2).

(G1) After suitable coordinate changes, the coefficient a in the normal form (9) is nonzero.

(G2) The bifurcation value λ_0 is crossed with positive speed.

Again it is very tedious but possible to check (G1) and (G2), while verifying (B1), (B2), (G0) and (GR) is relatively easy. This means that a torus bifurcation can be detected and followed efficiently.

ASSEMBLING THE PUZZLE

We have now introduced *all* basic bifurcations of equilibria and periodic orbits that can occur as a single parameter is varied. (We assumed that there is no extra structure, such as symmetry of the equations.) As was mentioned earlier, the task is now to find the respective bifurcation curves in the (K, ω)-plane of Eqs. (2). The idea is to find and follow an equilibrium or periodic orbit, that is, points where the bifurcation condition (B1) is satisfied, in one parameter, for example, in ω. While doing this, one monitors (B2) to find the respective bifurcation, say, at ω_0. This is like finding the border between England and Schotland. In the next step one can now follow in the two parameters K and ω the curve in parameter space where both (B1) and (B2) are satisfied. In this way, one traces out a curve of the respective bifurcation. Such a curve can generally be followed up to a point where one of the genericity conditions is not satisfied, which constitutes what is called a codimension-two bifurcation near which other bifurcation curves are known to exists. We do not discuss codimension-two bifurcations here; for more details see [2,4].

It is possible to solve the bifurcation equations (B1) and (B2) for the equilibrium bifurcations, the saddle-node and Hopf bifurcations, as a system of algebraic equations. However, for systems from applications this often leads to impractical expressions that need to be evaluated numerically. The bifurcation conditions (B1) and (B2) for the bifurcations of periodic orbits, the saddle-node bifurcation of limit cycles, period-doubling and torus bifurcations, are stated in terms of the Poincaré map P, which is not known explicitely. As a result, in this case the bifurcation conditions can be checked only numerically to begin with.

There are several packages that allow one to detect and follow the basic bifurcations discussed here [8–10]. We have used the package AUTO [8] to find and follow the bifurcations described here. We remark that AUTO does not check the respective genericity conditions. However, is not a big problem here because the loss of genericity generally leads to a bifurcation curve ending at some special point, which is picked up during AUTO computations. (If one wants to detect and follow curves of codimension-two bifurcations then one needs to use a program such as Content [10] that does check the respective genericity conditions.)

As an example of the type of bifurcation diagrams one can compute we show in Fig. 15 the bifurcation diagram of Eqs. (2). This shows how the different bifurcation curves fit together in a consistent picture, describing the dynamics of the injected laser. The different bifurcations are labeled with the letters shown in the puzzle pieces of the headings of the respective sections. Supercritical bifurcations leading to attractors are in black, and subcritical bifurcations leading to repellors are in grey. The superscript in the labeling distinguish between bifurcation curves of periodic orbits of basic period and those that have already undergone a period-doubling. Figure 15 shows an intruiging array of bifurcation curves in the (K,ω)-plane. We remark that not all bifurcation curves are drawn in these figures in order to keep them managable. In fact, there are infinitely many bifurcation curves, and we only show here the 'backbone' of the bifurcation diagram.

TRANSITIONS TO CHAOS

The bifurcation diagram in Fig. 15 allows one to explain many complicated effects. In order to 'read it' one also needs to know what the phase portraits look like in the different regions between curves. They can be computed numerically by simulation. We remark that the bifurcation diagram already tells us where to expect what kind of phase portrait, so that the simulation is not done 'blindly'. For illustration we now present two transitions to chaos by a sequence of attractors in phase space and in suitable Poincaré sections, which we computed with the package DsTool [11].

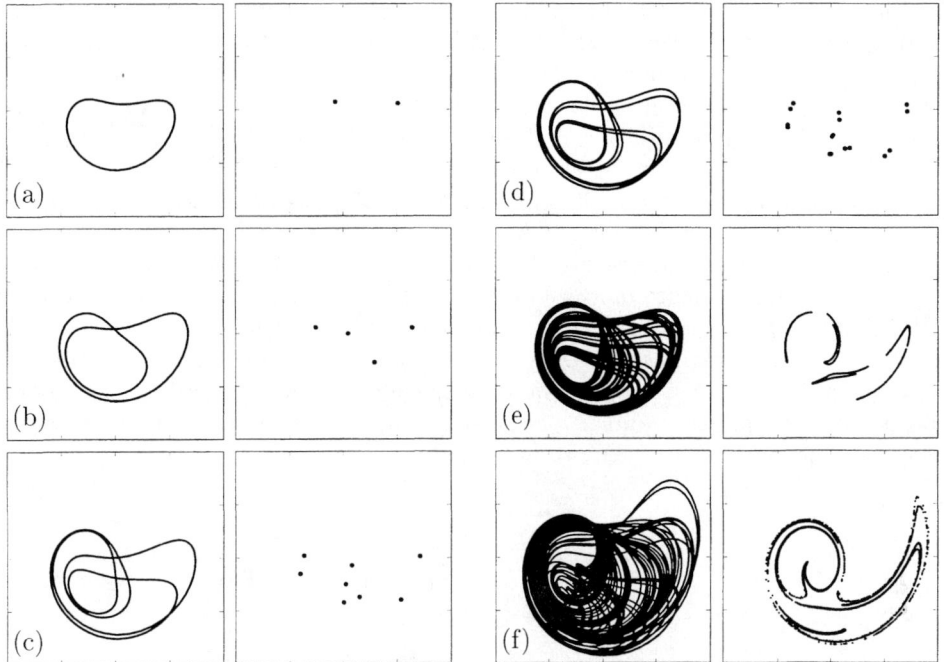

FIGURE 16. Transition to chaos via a sequence of period-doublings, shown in projection onto the complex E-plane (left panels) and in the Poincaré section $\{n = 0.1\}$ (right panels). A one-periodic orbit (a) undergoes successive period-doublings (b)-(d) until a chaotic attractor appears (e), which then grows in size (f). The parameters are $K = 0.62$, and from (a) to (f) ω takes the values 0.3, 0.5, 0.7, 0.71, 0.78, and 1.1.

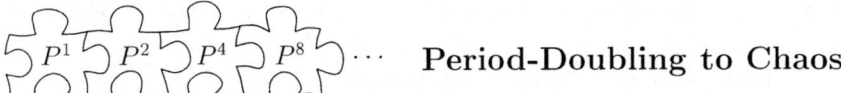 **Period-Doubling to Chaos**

It is well-known that, as one parameter is changed, a periodic orbit can undergo a sequence of period-doublings, the parameter values of which accumulate as described by the universal Feigenbaum constant. In the process one sees the emergence of a chaotic attractor. The period-doubling route to chaos is arguably the best known transition to chaos and it has been found in numerous systems, ranging from animal populations to a dripping tap. Therefore, it is not too surprising that it has also been found in the injection laser [12].

The nested islands of period-doubling curves, P^1 and P^2 in Fig. 15, already hint at the presence of a period-doubling route to chaos. Such a transition is illustrated in Fig. 16. An attracting periodic orbit undergoes consecutive period-doubling until it apparently becomes chaotic. The three-dimensional phase portraits are projected onto the complex E-plane. It should be noted, however, that it is very difficult to decide by looking at the three-dimensional phase portrait in panel (e) whether it shows a periodic orbit of high period or a chaotic attractor. The matter is settled by the respective panel of the Poincaré map: the attractor in panel (e) shows the typical almost one-dimensional shape of a chaotic attractor that appears after successive period-doublings; see the chapter by *Broer & Krauskopf*. Note that the section we chose intersects all periodic orbits in Fig. 12 twice, while in the definition of the Poincaré map we only considered only a locally defined section.

A final word of warning is that the chaos inside the islands of period-doubling curves is not necessarily reached by period doubling. The transition to chaos may be quite different, for example, via what is called a boundary crisis. The story is much more complicated and involves homoclinic bifurcations. This is beyond the scope of this exposition; for more information see [13].

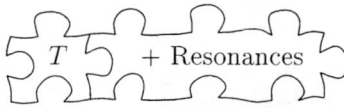 ## Break-up of a Torus

In a torus bifurcation an attracting or repelling torus is born from an attracting or repelling periodic orbit. As was mentioned earlier, the dynamics on this torus can be locked or quasiperiodic. Which situation occurs is organised by resonance tongues. In fact, every torus curve in the parameter plane comes with infinitely many resonance tongues attached to it at every point where there is a rational ratio between the two frequencies on the torus [2,4]. These resonance tongues can be quite narrow, and they are not shown in Fig. 15. When resonance tongues start to overlap each other at some distance from the torus bifurcation curve, there are additional bifurcations, such as homoclinic bifurcations not discussed here, which lead to the break-up of the torus resulting in a transition to chaos. This transition is rather spectacular, and it results in a 'much larger' chaotic attractor than the one immediately after the accumulation of period-doublings in Fig. 16(e).

To find this transition one can start near any of the black torus curves in Fig. 15 and change a parameter. An example is shown in Fig. 17. Again, we show attractors in phase space projected onto the complex E-plane and also the attractors of the Poincaré map on a suitable cross section. An attracting periodic orbit (a) loses its stability and a smooth attracting torus with quasiperiodic motion appears (b). The smooth torus in (b) starts to lose its smoothness by forming self-similar protrusions (c) forming a chaotic attractor as is evidenced by the fractal structure of the attractor of the Poincaré map, which nevertheless is still close to the original smooth torus. As the parameter is changed one can find locking on this chaotic attractor: in a certain parameter window we find locking onto a period-three orbit (d), so that there is no chaotic dynamics observable in the system. When the

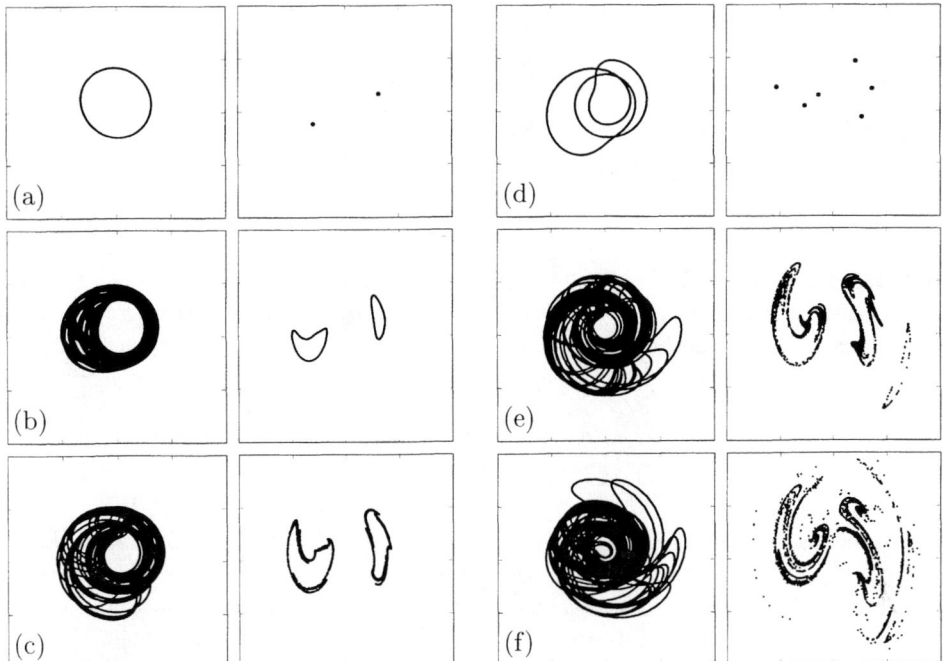

FIGURE 17. Transition to chaos via the break-up of a torus, shown in projection onto the complex E-plane (right panels) and in the Poincaré section $\{n = 0.1\}$ (right panels). A one-periodic orbit (a) undergoes a torus bifurcation (b), and then the torus begins to break up (c). Locking on the breaking torus can occur (d) while the chaotic attractor becomes more fractal (e) and (f). The parameters are $\omega = -1.2$, and from (a) to (f) K takes the values 0.15, 0.19, 0.23, 0.25, 0.29 and 0.44.

parameter is changed further the chaotic attractor reappears (e) and it becomes bigger (f) and does not resemble the original torus any longer.

We finally remark that detailed knowledge of properties of chaotic attractors is important for applications, such as communication schemes using chaotic carrier signals; see the chapter by *Mirasso*. Notice the differences between the chaotic attractors in Figs. 16 and 17 which manifest themselves in different output characteristics in terms of the time series of the power and the optical spectrum [13].

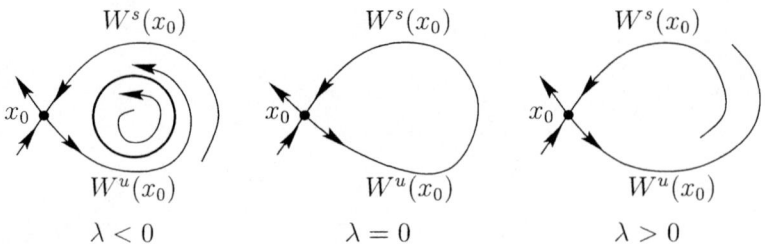

FIGURE 18. Example of a homoclinic bifurcation. An attracting periodic orbit existing for $\lambda < 0$ collides with a saddle point x_0 for $\lambda = 0$, leading to a homoclinic orbit where $W^s(x_0)$ and $W^u(x_0)$ coincide. After the bifurcation for $\lambda > 0$ the periodic orbit disappeared and $W^s(x_0)$ and $W^u(x_0)$ have changed their relative positions.

THERE IS MORE ...

In the bifurcation diagram in Fig. 15 bifurcation curves meet each other at special points. These are points where one of the genericity conditions is not satisfied, and they are (typically) codimension-two bifurcations. This means that two parameters need to be varied in order to find all behaviour in a neighborhood of such a point. As was the case for codimension-one bifurcations, there is a whole theory of codimension-two bifurcations, together with normal forms and new bifurcation and genericity conditions. These codimension-two bifurcations are important if one wants to check whether the bifurcation diagram is consistent. The theory of codimension-two bifurcations is beyond the scope of this exposition and we refer to the literature [1,2,4]. Still, this is not the end of it: one gets codimension-three bifurcations when genericity conditions of codimension-two bifurcations are not satisfied, and so on. The study of bifurcations of higher codimension is ongoing research, which is often motivated by applications such as laser systems.

An important class of bifurcations not treated here are homoclinic and heteroclinic bifurcations. They are not 'basic' bifurcations, because they do not allow for normal forms in the vicinity of an equilibrium of fixed point. Rather, they are *global bifurcations* that have to do with changes of the relative positions of stable and unstable manifolds. The simplest example is the homoclinic bifurcation sketched in Fig. 18, where, as one parameter is changed, a large periodic orbit collides with a saddle equilibrium and then disappears. Global bifurcations are important for understanding the overall dynamics, but they can only be understood through knowledge of the vector field in a 'large region' of phase space. It should be noted, however, that the package AUTO contains the part HomCont with routines for finding and following homoclinic and heteroclinic bifurcations [8]. Furthermore, there are numerical methods for computing global stable and unstable manifolds [14], so that their relative positions can be studied.

We have studied the basic bifurcations of ODEs in a generic setting. However, in many situations one is dealing with models from a class with special properties, such as systems with a particular symmetry. Roughly speaking, one also gets a list of basic bifurcations, whose normal forms have the required property; examples are the transcritical bifurcation and the pitchfork bifurcation. There is extensive literature on such systems, and we refer to, for example, [15] for an introduction.

This exposition introduced the basic bifurcations in the setting of low dimensional ODE models, illustrated with the injection laser Eqs. (2). However, there are other classes of dynamical systems, which are also used to describe different types of laser systems. First, there are dynamical systems with delay, given by delay differential equations (DDEs), the most well-known example are the Lang-Kobayashi equations describing a semiconductor laser with optical feedback; see the chapters by *Fisher et al.*, *Gavrielides* and *Verduyn Lunel & Krauskopf*. Second, there are partial differential equation (PDE) models used to describe spatially extended lasers, such as broad area lasers and VCSELs; see the chapters by *Hess*, *McInerney et al.* and *Moloney*. Both DDEs and PDEs have infinite dimensional phase spaces. Nevertheless, the basic bifurcations introduced here are also the basic bifurcations in these more complicated systems. In fact, one can also construct suitable center manifolds on which these bifurcations live and where they look just like described in the sections above! In other words, bifurcation theory is not restricted to low-dimensional ODE models, but is valid in greater generality. However, finding and following these bifurcations is more difficult for PDEs and DDEs. On the other hand, very recently there have been important developments in numerical analysis and packages for finding and following bifurcations in infinite-dimensional systems are becoming available [16,17].

In summary, even though there is a lot more out there, what is presented here forms the basis for unraveling the behavior of dynamical systems in general!

ACKNOWLEGDEMENTS

The author thanks Alan Champneys, Hinke Osinga and Sebastian Wieczorek for helpful comments.

REFERENCES

1. Chow, S.N., and Hale, J.K., *Methods of Bifurcation Theory*, Springer 1982.
2. Guckenheimer, J., and Holmes, P., *Nonlinear Oscillations, Dynamical Systems and Bifurcations of Vector Fields*, Springer 1983.
3. Strogatz, S. *Nonlinear Dynamics and Chaos*, Addison-Wesley 1994.
4. Kuznetsov, Yu.A., *Elements of Applied Bifurcation Theory*, Springer 1995.
5. Wieczorek, S.M., Krauskopf, B., and Lenstra, D., "A unifying view of bifurcations in a semiconductor laser subject to optical injection", *Opt. Comm.* **172**(1-6), 279 (1999).

6. Dubbeldam, J.L.A., and Krauskopf, B., "Self-pulsations of lasers with saturable absorber: dynamics and bifurcations", *Opt. Comm.* **159**, 325 (1999).
7. Simpson, T.B., Liu, J.M., Huang, K.F., and Tai, K., "Nonlinear dynamics induced by external optical injection in semiconductor laser" *Quant. Semiclass. Opt.* **9**(5), 765 (1997).
8. Doedel, E.J., Champneys, A.R., Fairgrieve, T.F., Kuznetsov, Yu.A., Sandstede, B., and Wang, X., "AUTO97: continuation and bifurcations software for ordinary differential equations (with HomCont)" (1997). (code of AUTO available at `http://indy.cs.concordia.ca/auto/`)
9. Khibnik, A.I., Kuznetsov, Yu.A., Levitin, V.V., and Nikolaev, E.V., "Continuation techniques and interactive software for bifurcation analysis of ODEs and iterated maps", *Physica D* **62**(1-4), 360 (1993).
10. Kuznetsov, Yu.A., Levitin, V.V., and Skovoroda, A.R. "Continuation of stationary solutions to evolution problems in CONTENT", Report AM-R9611, Centrum voor Wiskunde en Informatica, Amsterdam 1996. (code of CONTENT available at `ftp://ftp.cwi.nl/pub/CONTENT`)
11. Back, A., Guckenheimer, J., Myers, M.R., Wicklin, F.J., Worfolk, P.A., "DsTool: Computer assisted exploration of dynamical systems", *Notices Amer. Math. Soc.* **39**, 303 (1992). (code of DsTool available at `http://www.geom.umn.edu/software/dstool/`)
12. Simpson, T.B., Liu, J.M., Gavrielides, A., Kovanis, V. and Alsing, P.M., "Period-doubling route to chaos in a semiconductor laser subject to optical injection" *Appl. Phys. Lett.* **64**(26), 3539 (1994).
13. Krauskopf, B., Wieczorek, S.M., and Lenstra, D., "Different types of chaos in an optically injected semiconductor laser", to appear in *Appl. Phys. Lett.*; Wieczorek, S.M., Krauskopf, B., and Lenstra, D., "Mechanisms for multistability in a semiconductor laser with optical injection", to appear in *Opt. Comm.*; "Sudden chaotic transitions in an optically injected semiconductor laser", submitted.
14. Krauskopf, B., and Osinga, H.M., "Growing 1D and quasi 2D unstable manifolds of maps" *J. Comp. Phys.* **146**(1), 404 (1998); "Globalizing two-dimensional unstable manifolds of maps", *Int. J. Bif. Chaos* **8**(3), 483 (1998); "Two-dimensional global manifolds of vector fields" *CHAOS* **9**(3), 768 (1999).
15. Golubitsky, M., and Schaeffer, D.G., *Singularities and Groups in Bifurcation Theory I*, Springer 1985; Golubitsky, M., Stewart, I., and Schaeffer, D.G., *Singularities and Groups in Bifurcation Theory II*, Springer 1988.
16. Lust, K., Roose, D., Spence, A., and Champneys, A.R., "An adaptive Newton-Picard algorithm with subspace iteration for computing periodic solutions", *SIAM J. Scient. Computing*, **19**(4), 1188 (1998). (code of PDECONT available at `http://www.cs.kuleuven.ac.be/~kurt/r_bifcode.html`)
17. Engelborghs, K., Luzyanina, T., Roose D., "Numerical bifurcation analysis of delay differential equations using DDE-BIFTOOL", submitted. (code of DDE-BIFTOOL available at `http://www.cs.kuleuven.ac.be/~koen/delay/ddebiftool.shtml`)

Chaos in Periodically Driven Systems

Henk W. Broer* and Bernd Krauskopf[†]

*Department of Mathematics and Computing Science, Groningen University, PO Box 800, NL-9700 AV Groningen

[†]Department of Engineering Mathematics, University of Bristol, Bristol BS8 1TR, UK

Abstract. We give a review of the most well–known examples of dynamical systems with chaotic dynamics. After a phenomenological introduction, a definition of chaos is 'deduced'. One of the examples concerns the Hénon–attractor, that has only recently been characterized as the closure of a coiling curve. We will explain why it is plausible that this characterization also holds for more realistic cases and, in particular, for periodically forced system. As examples we use the driven damped pendulum and the laser with optical injection. Finally, we indicate some links between chaos and probability.

DETERMINISM, CHAOS AND CHANCE

Chaos is one of the intriguing features of nonlinear deterministic dynamics. Dynamical systems in principle model anything that moves in time. Examples are as various as simple contraptions of springs and pendula, or the whole solar system, or the world economy, or even the atmospheric circulation which produces the weather. In this context, the only important thing is that the system is deterministic: there is some *state space*, or *phase space*, containing all possible states of the system and an evolution law that prescribes the whole future when the present state is given. Of course it is not completely clear that all these examples should be deterministic.

Anyway, it turns out that even in extremely simple deterministic dynamical systems, that is, those with a low dimensional state space, the phenomenon of chaos can occur. Apart from the fact that this gives rise to interesting geometries in the state space, it also leads to basic unpredictability of the future dynamics.

Setting of the Problem

The first entry to the subject matter is phenomenological, we just study a few examples in low–dimensional phase spaces, or with 'few degrees of freedom'. The interest is with the long-term, or asymptotic, dynamical behaviour, that is, the regime where the initial and transient phenomena have disappeared.

As we shall see, there is a wide variety of possibilities. In the simplest cases the system is asymptotically at rest. We shall also encounter asymptotic periodic or multiperiodic behaviour. In those cases, a point or (multi–) periodic attractor exists in the phase space, to which nearby evolutions converge. All of this is still reckoned to be orderly dynamics. A first, rough definition classifies all remaining cases to be 'chaotic', where the attractors in phase space usually are called 'strange'. After a while we shall arrive at a more sophisticated definition of chaos.

Elsewhere, unpredictable systems are often described by probabilistic models. A good example is the dice, which by nature forms a highly chaotic, deterministic system. The description in these terms is so complicated though, that we usually disguise our incompetence and ignorance by using statistics.

Outline

We start with a few classical examples, including the Logistic (or Quadratic) family, the Hénon family, the Baker transformation and the Tent map family. In these examples time is discrete and the dynamics mathematically amounts to the iteration of a map.

Next we present the famous Lorenz system that has continuous time and is generated by an autonomous system of ordinary differential equations (ODE's). After this we arrive at a definition of chaos and at the characterization of the strange attractor of a 2-dimensional map like the Hénon map.

The second part of the paper focuses on periodically driven systems. We shall describe some of the background physics and consider the Poincaré map of such a system. It seems that the strange attractors of these maps have the same kind of characterization as the Hénon map.

We continue by explicitly introducing some probabilistic aspects by defining an invariant distribution, hinting in the direction of Ergodic Theory. We also briefly mention the case of fractal basin boundaries, where we return to the dice.

THE LOGISTIC FAMILY

The first example is the well-known Logistic or Quadratic family; e.g., see [1]. It is a discrete dynamical system, given by the iterations

$$x_{n+1} = \mu x_n(1 - x_n), \qquad (1)$$

$x_n \in [0, 1]$, $n = 0, 1, 2, \ldots$. Moreover $\mu \in [0, 4]$ is a parameter. The simplest interpretation of (1) is in terms of population dynamics. In that case an evolution $x_0, x_1, x_2, \ldots$ is interpreted as the (scaled) size of a certain population as a function of the time n. Let us discuss this in some detail.

The Linear Growth Model

We begin by considering the linear growth model

$$x_{n+1} = \mu x_n, \quad x_n \in \mathbb{R},$$

$n = 0, 1, 2, \ldots$. The parameter μ in this model is the constant ratio between the sizes of successive generations, and can be interpreted as a net growth rate of the population at hand. It is easily seen that in terms of the initial 'population' x_0 we have

$$x_1 = \mu x_0,$$
$$x_n = \mu^n x_0,$$

$n = 0, 1, 2, \ldots$, which clearly is an exponential behaviour. Regarding the parameter μ we distinguish between the case where $0 < \mu < 1$, corresponding to exponential decay, and the case where $1 < \mu$, which corresponds to exponential growth. There are not too many circumstances in which this linear model is realistic. The Logistic family (1) can be seen as its simplest nonlinear modification.

Graphical Analysis

As an intermezzo we study the dynamics of more general iterations

$$x_{n+1} = \Phi(x_n),$$

$n = 0, 1, 2, \ldots$, where $\Phi : \mathbb{R} \to \mathbb{R}$ is a smooth function. In Fig. 1(a) we sketched the graph of such a function as well as the graph of the identity map (that is, of $y = \Phi(x)$ and $y = x$). It is now easy to describe the iteration–step from x_n to x_{n+1}: indeed given x_n on the x–axis we read off the next iterate x_{n+1} on the y–axis. With the help of the diagonal $y = x$ we next reflect the value x_{n+1} to the x–axis and we can repeat the process to obtain x_{n+2}.

In Fig. 1(b) this idea has been applied to the linear growth model for $\mu > 1$. In this case we simply have $\Phi(x) = \mu x$.

Overpopulation

Let us consider the dynamics of the Logistic family itself, which can now be described as

$$x_{n+1} = \Phi_\mu(x_n) \text{ with } \Phi_\mu(x) = \mu x(1 - x),$$

the Logistic map. First observe that, for small values of x_n, equation (1) is well–approximated by the linear growth model.

Next however, we observe that in the nonlinear model there is a maximal size of the population, namely $x_n = 1$. Indeed, if $x_n = 1$ it follows that $x_{n+1} = 0$, and so

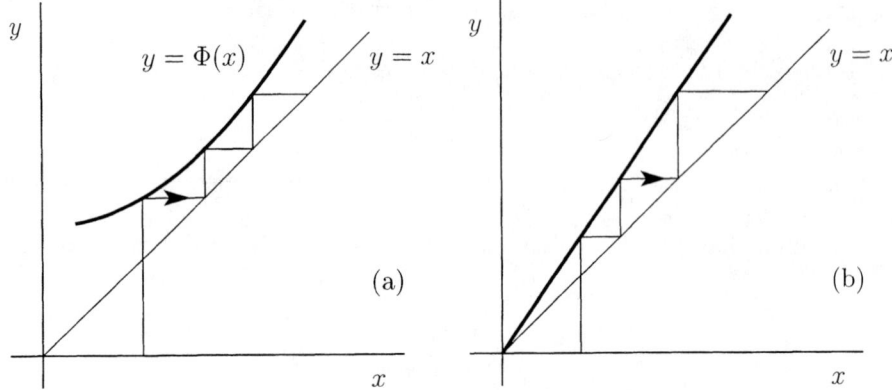

FIGURE 1. Graphical explanation of the general iteration $x_{n+1} = \Phi(x_n)$ (a), and applied to the linear growth model (b). Reprinted from [27] with permission from *Het Wiskundig Genootschap*.

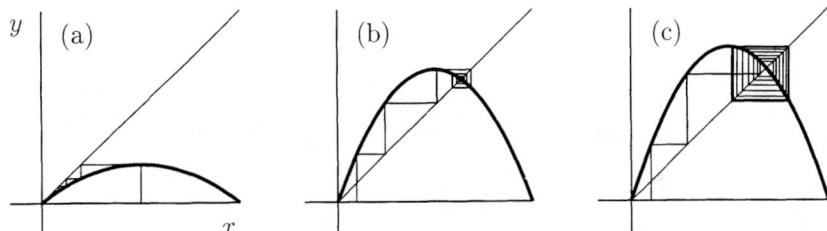

FIGURE 2. Dynamics of the Logistic family for different values of μ, namely for $0 < \mu < 1$ (a), $1 < \mu < 3$ (b), and $0 < \mu - 3 \ll 1$ (c). Reprinted from [27] with permission from *Het Wiskundig Genootschap*.

the next generation is extinct. This is one of the simplest ways to model the effect of overpopulation. Of course the maximal size 1 is artificial, but any other bound may be brought to 1 by an appropriate scaling.

As before, we vary the parameter μ and consider the dynamics. To this end we perform the graphical analysis described in the previous subsection; see Fig. 2. As in the linear case, for $0 < \mu < 1$ the population becomes extinct, where for larger values of n we almost are in the case of exponential decay. In a more geometrical language we say that $x = 0$ is a point, or equilibrium, attractor. This kind of dynamics is still quite uninteresting from the view point of population dynamics.

For $1 < \mu < 3$ we find a point attractor different from $x = 0$, indeed it is $x = p_\mu$, with

$$p_\mu := (\mu - 1)/\mu.$$

Observe that for $2 < \mu < 3$, the equilibrium p_μ is approached in a spiraling mode.

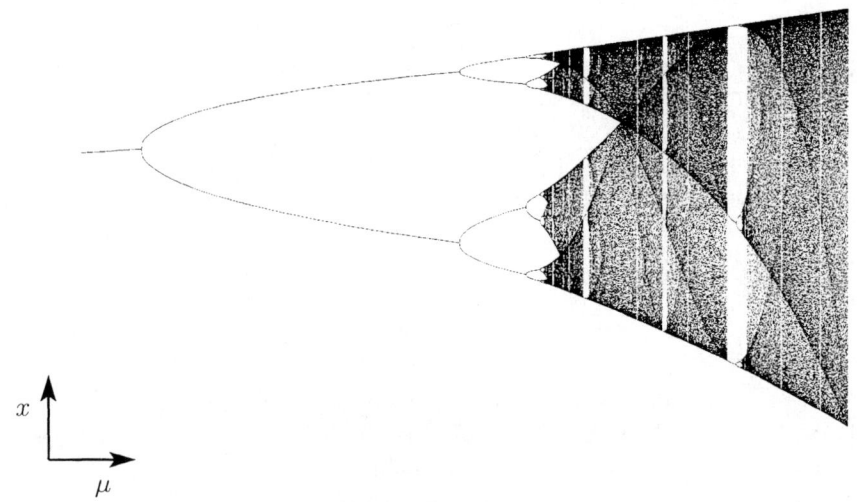

FIGURE 3. Bifurcation diagram of the Logistic family for μ between 3 and 4. Reprinted from [27] with permission from *Het Wiskundig Genootschap*.

Then, for $3 < \mu$ none of the two equilibria 0 and p_μ is attracting, so where does the population size x_n go as $n \to \infty$? The answer for $0 < \mu - 3 \ll 1$ is given in Fig. 2(c): the system now has a period–2 attractor. Intuitively such a situation can be described as follows. Suppose the generations with even n are smaller than the ones with odd n. Then each odd generation exhausts the natural resources so much that the next generation is smaller, which allows the next generation to be larger again, etc.

For increasing values of μ an interesting scenario arises, which is described often in the literature; see Fig. 3 and, e.g., [1,2]. To obtain it, for each value of μ, about 200 iterations are considered of which only the last 100 are plotted. In this way the transient phenomena are ignored and only the asymptotic dynamics is observed. The first thing to notice is that for increasing μ the system doubles its period. So the period–2 attractor is replaced by a period–4 attractor, then period–8, etc. It has been shown that this process goes with geometric progression: to be precise, the corresponding sequence $\mu_1, \mu_2, \mu_3, \ldots$ of parameter-values converges (almost) geometrically to a limit μ_∞. Names like Feigenbaum and Tresser are related to these phenomena, compare [2,3]. For the parameter-value $\mu = \mu_\infty$ that limits this progression there is no longer any periodicity. For such aperiodic dynamics we shall use the adjective *chaotic*, but below we shall arrive at a more precise definition.

In the range $3 < \mu < 4$ many chaotic and periodic regimes can be distinguished along with many additional period-doubling sequences.

Remarks.

1. One of the significant features of the present chaos is the stretching of the x–axis by the Logistic map $\Phi(x) = \mu x(1-x)$, except very near $x = 1/2$, for larger values of μ. By stretching alone the points all would leave the interval $[0,1]$. However, the map also folds the interval over itself. It is this combination of stretching and folding that we shall see again when meeting other chaotic systems.

2. The above (almost) geometric progression of the sequence μ_n can be expressed by saying that
$$\lim_{n\to\infty} \frac{\mu_{n-1} - \mu_{n-2}}{\mu_n - \mu_{n-1}} = \delta,$$
for a positive constant $\delta = 4.6692016091029\ldots$. This constant δ turns out to be *universal* in the sense that it also occurs for similar unimodal maps Φ, that is, maps with one 'hump', again see [2,3].

3. The Logistic family, innocent though it looks, is hard to digest mathematically, as will be illustrated now. Again consider the bifurcation diagram of the Logistic family in Fig. 3. One may easily guess that the union of 'windows' in the μ–axis, corresponding to periodic attractors, densely fills the interval $[0,4]$. This fact, however, has only been proven quite recently [4–6].

4. The arguments of this section have the same character as in the derivation of the Volterra–Lotka systems for population dynamics. Their aim is not so much to precisely model the evolution of a given population, but rather to give systems with certain qualitative properties that are as transparant as possible; compare [7,8]. In Mathematical Biology, however, often more realistic models are used, with similar qualitative properties as the Logistic family.

THE HÉNON FAMILY

This family of maps was introduced by Hénon in 1976, intended as a simple (polynomial) example of a planar, invertible and differentiable map (a diffeomorphism) with a strange attractor. Only quite recently the structure of this family has been understood better, mathematically speaking; compare [1–3,9].

Just Iterating

Let us first describe what is the case. Given is the 2–dimensional map
$$\Phi_{a,b}(x,y) = (1 - ax^2 + y, bx),$$
where a and b are real parameters. Iterating
$$(x_{n+1}, y_{n+1}) = \Phi_{a,b}(x_n, y_n),$$

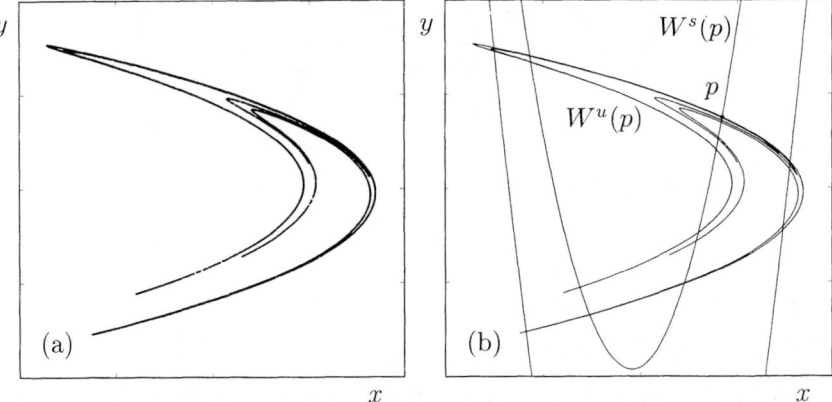

FIGURE 4. The Hénon attractor in the (x,y)-plane for $a = 1.4$, $b = 0.3$ (a), and the unstable manifold $W^u(p)$ of the fixed point p accumulating on it (b). Notice how the stable manifold intersects the unstable manifolds in what is called a homoclinic tangle.

$n = 0, 1, 2, \ldots$, then produces 2–dimensional evolutions of the form

$$(x_0, y_0), (x_1, y_1), (x_2, y_2), \ldots.$$

In Fig. 4(a) the result of this iteration has been plotted (after ignoring transients) for parameter values $a = 1.4$ and $b = 0.3$. This is the celebrated Hénon attractor $\mathcal{H}$, which attracts an open subset of initial points in $\mathbb{R}^2$. The set $\mathcal{H}$ appears to be a fractal, the local geometry of which is the product of a curve and a Cantor set. Numerical estimates of a suitable fractal dimension (limit capacity, box counting dimension) of $\mathcal{H}$ give approximately 1.2. For background information see [3], especially the chapters II and III, and also [2] for general reference. It is interesting to study what happens upon variation of the parameters (a, b). In fact, the strange attractor $\mathcal{H}$ is not persistent for small perturbations, since for nearby values of (a, b) periodic attractors turn up. As we shall argue now, the complexity of the bifurcation diagram in Fig. 3 also is to be expected here: windows with periodic attractors, period doubling sequences, strange attractors, etc.

Mathematical Intermezzo

Let us make this a bit more precise. To begin with, notice that the following two versions exist of the Logistic family:

$$x_{n+1} = \mu x_n (1 - x_n) \text{ and}$$
$$x_{n+1} = 1 - a x_n^2.$$

The former we already know, and the latter can be derived from this by the affine scaling of x from $[0, 1]$ to $[-1, +1]$ and by reparametrizing

$$\mu \in [0, 4] \longleftrightarrow a \in [0, 2] : a = \frac{1}{4}\mu(\mu - 2).$$

This gives rise to the following observation. Consider the Hénon family for $b = 0$:

$$\Phi_{a,0}(x, y) = (1 - ax^2 + y, 0).$$

Notice that this map is no longer an invertible map. Clearly the line $y = 0$, or the x-axis, is invariant[1] and the restriction $\Phi_{a,0}(x, 0) = (1 - ax^2, 0)$ is nothing but the newly begotten version of the Logistic family. A natural question now is which aspects of the dynamics of the Logistic family are inherited by the Hénon family for small (nonzero) values of b.

At the level of periodic attractors a lot is known about this; see [10] and also [11]. Indeed, it turns out that the periodic windows of the Logistic family as shown in Fig. 3 can be continued for small values of b.

The chaotic dynamics is a lot harder to continue. To understand what is going on here, we note that the Hénon map $\Phi_{a,b}$ has a saddle fixed point $p_{a,b}$. Such a point is unstable. Let us consider the set

$$W^u(p_{a,b}) = \{(x, y) \in \mathbb{R}^2 \mid \lim_{n \to \infty} \Phi_{a,b}^{-1}(x, y) = p_{a,b}\},$$

that is, all points that by backward iteration of $\Phi_{a,b}$ go to $p_{a,b}$. Clearly one has $p_{a,b} \in W^u(p_{a,b})$. It can be shown that $W^u(p_{a,b})$ (locally) is a smooth curve, called the unstable manifold of $p_{a,b}$.

Remarks.

1. Similarly the stable manifold $W^s(p_{a,b})$ exists, consisting of all points converging to $p_{a,b}$ by forward iteration of $\Phi_{a,b}$. Both curves in $p_{a,b}$ are tangent to the corresponding eigenspaces of the derivative map.

2. To compute stable and unstable manifolds it is not sufficient to simply iterate points. Rather, one needs to employ specially designed methods. The figures of stable and unstable manifolds in this paper were computed with the algorithm described in [12,13].

3. For the Hénon map the inverse $\Phi_{a,b}^{-1}$ can be directly computed as

$$\Phi_{a,b}^{-1}(x, y) = (\frac{1}{b}y, x + \frac{a}{b^2}y^2 - 1).$$

---

[1] In fact, $\Phi_{a,0}$ maps $\mathbb{R}^2$ onto the x-axis, and in the x-axis we have 1-dimensional dynamics as before.

One of the features of the Hénon family is its constant Jacobian determinant: $|\det D\Phi_{a,b}| \equiv b$, meaning that for $|b| < 1$ the map $\Phi_{a,b}$ contracts area. This property is also called *dissipative*. From this fact one easily derives that the attractor $\mathcal{H}_{a,b}$ of $\Phi_{a,b}$ is contained in the closure of the unstable manifold $\overline{W^u(p_{a,b})}$. All this was already known in the beginning of the eighties, proven by Tangerman (unpublished).

The difficult part is the converse, namely to show that the closure of the unstable manifold is contained in the attractor, which mathematically amounts to

$$\mathcal{H}_{a,b} = \overline{W^u(p_{a,b})}.$$

In 1991 Benedicks & Carleson [9], also see [2], showed that this is indeed true for a subset of the (a, b)–plane of positive area. The proof again is based on perturbation arguments. One remaining question is whether the 'original' Hénon attractor $\mathcal{H} = \mathcal{H}_{1.4,0.3}$ has this property or not. Comparing Fig. 4(a) with Fig. 4(b) makes this very believable. A definite answer, however, may never be found.

THE BAKER TRANSFORMATION

One of the few ways to get a handle onto chaos is symbolic dynamics. This can be best explained with the help of a toy model called the Baker transformation. This transformation is given by

$$x_{n+1} = 2x_n (\text{mod } 1), \tag{2}$$

again with $x_n \in [0, 1)$, $n = 0, 1, 2, \ldots$. Clearly the map Φ in this example is given by $\Phi(x) = 2x (\text{mod } 1)$; for its graph see Fig. 5. Note that here the interval $[0, 1)$ firstly is stretched by a factor of two and secondly cut, after which $[0, 1)$ covers itself twice. This reminds one of kneading dough, which explains the name. In this form the tranformation is discontinuous, but when $2\pi x$ is interpreted as an angle we can view the Baker transformation as the complex map

$$z \mapsto z^2$$

restricted to the unit circle, in which case it is continuous.

Remark. The projection of this circle dynamics onto the real axis gives rise to the Logistic map with $\mu = 4$; e.g., see [1].

Let us consider an evolution $x_0, x_1, x_2, \ldots$, given by iteration of the Baker transformation, so with $x_{n+1} = 2x_n (\text{mod } 1)$. In binary expansion the initial point x_0 has the form

$$x_0 = \sum_{j=1}^{\infty} a_j 2^{-j},$$

where $a_j \in \{0, 1\}$ for all $j = 1, 2, 3, \ldots$. In the usual notation $x_0 = 0.a_1 a_2 a_3 \ldots$, the Baker transformation becomes simple:

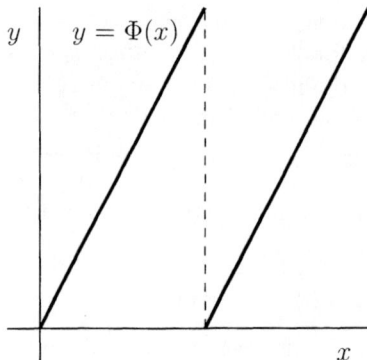

FIGURE 5. The graph $y = \Phi(x)$ of the Baker transformation. Reprinted from [27] with permission from *Het Wiskundig Genootschap*.

$$x_0 = 0.a_1a_2a_3\ldots,$$
$$x_1 = 0.a_2a_3a_4\ldots,$$
$$x_2 = 0.a_3a_4a_5\ldots,$$
$$x_n = 0.a_{n+1}a_{n+2}a_{n+3}\ldots,$$

an operation called (Bernoulli–) *shift*. Observe that the above has the following geometric meaning. If $a_{n+1} = 0$ the n–th iterate x_n is in the left half interval $I_0 = [0, \frac{1}{2})$, while for $a_{n+1} = 1$ we have that $x_n \in I_1 = [\frac{1}{2}, 1)$, the right half interval. Conversely the positioning of the $x_0, x_1, x_2, \ldots$ with respect to the intervals I_0 and I_1 determines[2] the binary expansion of the initial point x_0. We now are in the realm of *symbolic dynamics*, a powerful tool for understanding chaotic dynamics.

One of the first suprising consequences of the above is the following. If the initial point x_0 is known only with finite precision, this precision decreases when iterating. After some time all precision even totally disappears! This is the heart of the unpredictability aspect of chaos, and it is of importance since in practice one is always dealing with finite precision.

A second direct consequence is the following. An initial condition $x_0 \in [0,1]$ has a periodic evolution of period N, precisely if the binary expansion $x_0 = 0.a_1a_2a_3 \cdots a_N a_{N+1} \ldots$ repeats with period N. We now can formulate

Theorem 1 (Denseness of periodic points) *The set of periodic points of the iterated Baker transformation is dense in $[0, 1)$. This means that arbitrarily close to any point in $[0, 1)$ one can find a periodic point.*

Proof. Let $p \in [0, 1)$ be an arbitrary number, with a binary expansion $p = 0.a_1a_2a_3 \cdots a_N a_{N+1} \ldots$. Then consider the repeating binary expansion $q = $

---

[2] For the moment ignoring ambiguities like $0.01111\ldots = 0.10000\ldots = \frac{1}{2}$.

$0.a_1a_2a_3\cdots a_Na_1a_2a_3\ldots a_Na_1\ldots$. Clearly the point q is periodic with period N, where $|p-q| \leq 2^{-N}$. In other words, we can contruct a periodic point q as close to the original $p \in [0,1)$ as we want, and this is the statement of the theorem. □

All these periodic points are unstable: the system in practice will behave in an aperiodic fashion. This can be seen by observing the graph of the Baker transformation and its iterates; compare Fig. 5.

Speaking of dense properties, in general a point $x_0 \in [0,1]$ is called eventually periodic for the map Φ, if some finite iterate $\Phi^N(x_0)$ is a periodic point under Φ. In the present case, where Φ is the Baker transformation, the set of points that are eventually periodic with period 1, that is, the eventual fixed points, also is dense in $[0,1]$. Indeed, the eventual fixed points are exactly the finite binary expansions, which evidently fill the interval densely. The reader is invited to find other dense properties for him– or herself ....

A less intuitive consequence of the binary expansions is the following

Theorem 2 (Existence of a dense orbit) *There exists a point $x_0 \in [0,1)$ such that its orbit $\{x_0, x_1, x_2, \ldots\}$ under iteration of the Baker transformation is dense in $[0,1)$, that is, its orbit comes arbitrarily close to any other point in $[0,1)$.*

Proof. Consider the binary expansion

$$x_0 = 0.\ 0\ 1$$

$$00\ 01\ 10\ 11$$

$$000\ 001\ 010\ 011\ 100\ 101\ 110\ 111$$

$$\ldots,$$

where all possible truncated expansions are listed. By a similar argument as in the previous proof, this point comes arbitrarily close to any given point $p \in [0,1)$. □

The last theorem says that the 'attractor' $[0,1)$ of the Baker transformation is connected: it cannot be decomposed into smaller subattractors.

Remarks.

1. Consider the family of Tent maps

$$\Phi_a(x) = \begin{cases} 2ax, & 0 \leq x \leq \frac{1}{2} \\ 2a(1-x), & \frac{1}{2} \leq x \leq 1, \end{cases}$$

where a is a real parameter. The Tent map Φ_1, that is, with $a = 1$ behaves strikingly similarly to the Baker transformation. In particular, it has strongly related symbolic dynamics. Moreover, Φ_1 has the same dynamics as the Logistic map $\Phi_\mu(x) = 4x(1-x)$ for $\mu = 4$; see [1].

2. In general the Tent maps Φ_a, as well as the Logistic maps Φ_μ, can be attacked succesfully with symbolic dynamics. Special care has to be taken with the orbit of the central point $x_0 = \frac{1}{2}$. Moreover, general symbolic dynamics is defined on Cantor sets, which is nicely illustrated by considering the Tent map family for $a > 1$. In this case the invariant set

$$\bigcap_{n=0}^{\infty} \Phi^{-n}([0,1]),$$

consisting of the initial points, the orbits of which stay inside $[0,1]$ for all time, is a Cantor set; compare [1–3,14].

MATHEMATICAL INTERMEZZO: TOWARDS A DEFINITION OF CHAOS

Summarizing the above, we met dynamical systems with discrete time, generated by a map Φ by iteration $\vec{x}_{n+1} = \Phi(\vec{x}_n)$, $n = 0, 1, 2, \ldots$. Here $\vec{x}$ varies over some finite dimensional state space. Let $\vec{x}_0, \vec{x}_1, \vec{x}_2, \ldots$ be an evolution of this system, then

$$\vec{x}_1 = \Phi(\vec{x}_0),$$
$$\vec{x}_2 = \Phi(\vec{x}_1) =$$
$$= (\Phi \circ \Phi)(\vec{x}_0) = \Phi^2(\vec{x}_0)$$
$$\ldots$$

The examples we saw are the Logistic family $\Phi_\mu(x) = \mu x(1-x)$, the Hénon family $\Phi_{a,b}(x,y) = (1 - ax^2 + y, bx)$ and the Baker transformation $\Phi(x) = 2x \pmod 1$.

Until now we discussed chaos in a heuristic manner, saying that it is aperiodic, or just 'weird'. Now we shall give a somewhat more rigid definition of a chaotic attractor, compare [1], although we shall not be exeedingly mathematical. First we have to say what will be understood by an attractor.

Definition 3 (Attractor, [3]) *A subset $\mathcal{H}$ of the phase space of the system generated by Φ is called an attractor if*

1. *$\mathcal{H}$ is invariant, that is, if $\vec{x}_0 \in \mathcal{H}$, then also $\Phi(\vec{x}_0) \in \mathcal{H}$;*

2. *There exists a neighhourhood $\mathcal{U}$ of $\mathcal{H}$, such that for all initial states $\vec{x}_0 \in \mathcal{U}$, for the corresponding evolutions $\vec{x}_0, \vec{x}_1, \vec{x}_2, \ldots$ (with $x_n = \Phi^n(\vec{x}_0)$, $n = 0, 1, 2, \ldots$), one has that $\vec{x}_n \to \mathcal{H}$;*

3. *In $\mathcal{H}$ the iterated map Φ has a dense orbit.*

We repeat that the existence of a dense orbit means that $\mathcal{H}$ is indecomposable or that it is minimal with respect to the former two properties. We met this property before in the Baker transformation.

Next consider an evolution $\vec{x}_0, \vec{x}_1, \vec{x}_2, \ldots$, with $\vec{x}_{n+1} = \Phi(\vec{x}_n)$, $n = 0, 1, 2, \ldots$, inside the attractor $\mathcal{H}$. For nearby initial points $\vec{y}_0 \in \mathcal{H}$, we also consider the evolution $\vec{y}_0, \vec{y}_1, \vec{y}_2, \ldots$, with $\vec{y}_{n+1} = \Phi(\vec{y}_n)$, $n = 0, 1, 2, \ldots$.

Definition 4 (Sensitivity) *The restriction $\Phi|_{\mathcal{H}}$ has sensitive dependence on initial values iff there exists a positive constant δ, such that the following holds. For any initial value $\vec{x}_0 \in \mathcal{H}$ and any (small) positive number ε, there exists an initial value $\vec{y}_0 \in \mathcal{H}$, with $|\vec{y}_0 - \vec{x}_0| < \varepsilon$, and there exists a number $N \in \mathbb{N}$, such that $|\vec{y}_N - \vec{x}_N| > \delta$.*

The property of sensitive dependence also holds for the Baker transformation, it has to do with the stretching that leads to loss of information. In the practical case when one is working with finite precision, this leads to unpredictability of the evolutions on a longer time scale. For further discussion of this notion and possible refinements, see [3], chapters I and III, and also [15], chapter VI. For more background information see [14]. Finally we have:

Definition 5 (Chaos) *The attractor $\mathcal{H}$ of the dynamical system generated by the map Φ is chaotic iff the restriction $\Phi|_{\mathcal{H}}$ has sensitive dependence on initial values.*

It is mathematically more or less reassuring to have the above conditions available. However, we then carry the burden of proving that in the above examples these definitions play a significant role. In the case of the Baker transformation this can be done. (Hint: try $\delta = \frac{1}{4}$.) In the other cases this is not easy at all, and sometimes only vague statements are proven like 'for a set of parameter values of positive measure the system has a chaotic attractor'.

Remark. In the literature many variations exist on the above definitions. For example, an attractor does not necessarily have to attract a full neighbourhood, but a set of full (or just positive) measure will do. Sometimes it is additionally required that the periodic points should be dense in the attractor; again see the example of the Baker transformation.

One reason for this is provided by the unwieldy examples, that look chaotic, but can only be forced to satisfy alternate definitions. Another reason is that there is at least one other entrance to chaotic dynamics, which uses probabilistic or measure theoretical methods. At the end we shall come back to this.

EXAMPLES WITH CONTINOUS TIME

The scope of the above seems to be somewhat limited, although we shall see that the case of the Hénon family represents a large class of systems that occur in applications.

One of the icons of the theory is the Lorenz attractor shown in Fig. 6. It is generated by the following autonomous system of ODE's in $\mathbb{R}^3$ in the same way as before: let the system run and omit all transient information. An evolution now simply is a solution or integral curve of

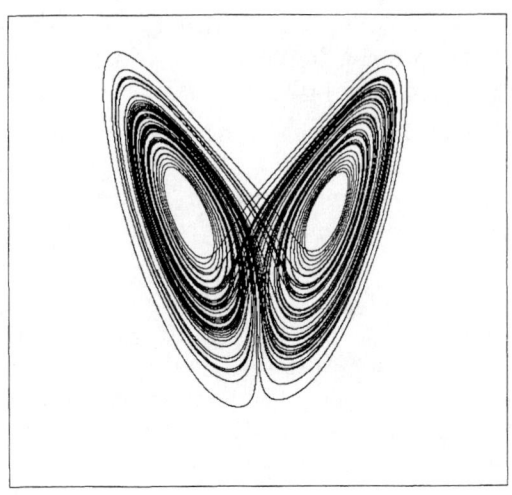

FIGURE 6. The Lorenz attractor projected onto the (x,z)-plane. Reprinted from [27] with permission from *Het Wiskundig Genootschap*.

$$\dot{x} = -\sigma x + \sigma y$$
$$\dot{y} = Rx - y - xz$$
$$\dot{z} = -Bz + xy,$$

where $\sigma = 10$, $B = \frac{8}{3}$ and $R = 28$. Here and elsewhere we use the notation $\dot{x}(t) = \frac{d}{dt}x(t)$. The example was published in 1963 in a meteorological journal, for details and background, e.g., see [2,16]. We just mention here that the system has to do with long term behaviour of the atmospheric circulation and indicates that even at the level of simplified climate models chaos may occur. Again chaos is related to unpredictability. The attractor locally seems to be the product of a surface and a Cantor set. Numerically estimated fractal dimensions are between 2 and 3; for background information see [2,3,16,17]. This attractor has been a challenge for mathematicians for a long time. Only quite recently it has been shown to be chaotic according to the definitions; see [18,19].

There are many more examples with continious time in the literature. They include periodically forced systems on which we will concentrate next.

HÉNON-TYPE ATTRACTORS IN PERIODICALLY FORCED SYSTEMS

The examples of dynamical systems we met until now all had a somewhat artificial character. In this section we introduce two examples in which we shall produce

Hénon–like attractors of a 2–dimensional invertible and differentiable map (a diffeomorphisms). The first example is the driven damped pendulum, and the second the laser with optical injection; see also the chapter by *Krauskopf*. We aim to illustrate that the theorem of Benedicks and Carleson regarding strange attractors for such maps seems to be quite a universal characterization; also see [20,21] for extensions. In other words, we demonstrate that apparently the respective strange attractor is the closure of the unstable manifold $W^u(p)$ of an appropriate saddle point p. It has to be noted here that the mathematics of the matter is not so far yet that these illustrations can be made precise.

Mechanical Background

As a starting point we take the planar pendulum without friction, as it occurs in almost all Mechanics textbooks; see also [15,23]. Consider a mathematical pendulum of length ℓ with mass m that moves in a vertical plane, under the influence of graviation with acceleration g. Let x be the deviation from the downward vertical equilibrium position, measured in radians. As before $\dot{x}$ and $\ddot{x}$ denote the derivatives with respect to the time t, in this case called velocity and acceleration, respectively. The equation of motion then reads

$$\ddot{x} = -\omega^2 \sin x, \tag{3}$$

where $\omega^2 = g/\ell$. This is a direct consequence of Newton's second law $F = m \times \ell\ddot{x}$, where $F = F(x)$ is the component of the gravitational force in the direction of motion. We note the established fact that this equation of motion is independent of the mass m, which was known experimentally already to Galileo.

The above considerations hold in the *conservative* case without friction. However, it is easy to introduce some friction or damping in the equation of motion as follows

$$\ddot{x} = -\omega^2 \sin x - c\dot{x}, \tag{4}$$

where $c > 0$ is a positive damping coefficient.

Phase Plane and Energy

One tool when dealing with nonlinear systems like the pendulum is the phase plane. Indeed, introducing $y = \dot{x}$ as an independent variable we consider the Cartesian (x, y)–plane. The equations of motion (4) then turn into the system of ODE's

$$\begin{aligned}\dot{x} &= y \\ \dot{y} &= -\omega^2 \sin x - cy.\end{aligned} \tag{5}$$

Compare the Lorenz systems from the previous section. This vector field gives rise to integral curves, the projection of which to the x–axis corresponds to the physical

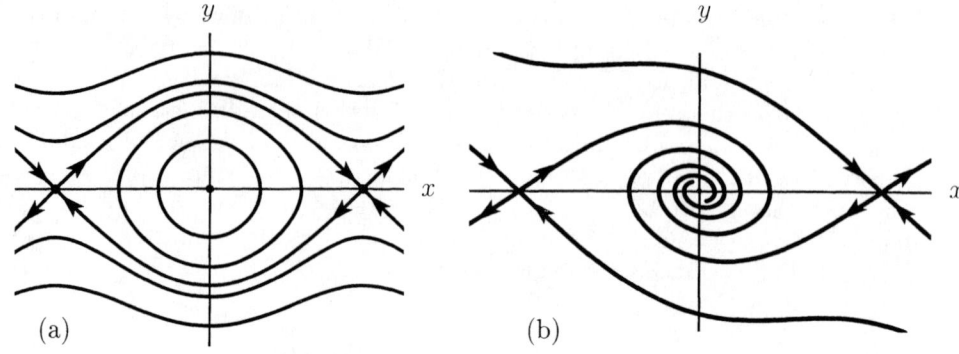

FIGURE 7. Phase portrait of the pendulum equation (5) for the undamped case $c = 0$ (a), and the damped case $c > 0$ (b). Reprinted from [27] with permission from *Het Wiskundig Genootschap*.

dynamics of this model. In Fig. 7 we depicted the corresponding phase portraits for the undamped case $c = 0$ and the damped case $c > 0$, respectively. In the damped case almost all evolutions end in the lower equilibrium point $(x, y) = (0, 0)$, which is the unique attractor.

For the understanding of these phase portraits it is helpful to consider the *Hamiltonian* or *energy* [3] function

$$H(x, y) = \frac{1}{2}y^2 + \omega^2(1 - \cos x). \qquad (6)$$

The infinitesimal change of H along the evolution of (5) is given by the directional derivative

$$\begin{aligned}\dot{H}(x, y) &= \frac{\partial H}{\partial x}\dot{x} + \frac{\partial H}{\partial y}\dot{y} \\ &= \omega^2 \sin x \times y - y \times (\omega^2 \sin x + cy) \\ &= -cy^2.\end{aligned}$$

From this we can see the established fact that for $c = 0$ the energy H is conserved, while for $c > 0$ the energy H is decreasing. Therefore, the integral curves of system (5) for $c = 0$ exactly are the level sets of the function H, as can be observed from Fig. 7(a). In the case of Fig. 7(b) the integral curves spiral inward to the attractor $(0, 0)$.

---

[3] Ignoring the matter of physical dimensions ...

Remarks.

1. The Hamiltonian (6) is of the form $H(x,y) = \frac{1}{2}y^2 + V(x)$, where the first term represents the kinetic energy and the second the potential energy. In that case the system (5) for $c = 0$ reads

$$\begin{aligned} \dot{x} &= y \\ \dot{y} &= -\frac{dV}{dx}(x). \end{aligned}$$

2. For a general Hamiltonian $H = H(x,y)$ the equations of motion read

$$\begin{aligned} \dot{x} &= \frac{\partial H}{\partial y}(x,y) \\ \dot{y} &= -\frac{\partial H}{\partial x}(x,y), \end{aligned}$$

In this general setting conservation of energy is just as easily proven as in the special case above. As before, see (5), one can introduce a damping term $-cy$ in the right hand side of the second equation.

Periodic Forcing and the Poincaré Map

In the damped case generally all energy dissipates away and the pendulum comes to complete rest. Now what happens if we introduce energy into the system, e.g., by also varying the pendulum–length periodically in time? Just consider the equation of motion

$$\ddot{x} + c\dot{x} + (\omega^2 + \varepsilon\sin(\Omega t))\sin x = 0 \tag{7}$$

which leads to the 3–dimensional vector field

$$\begin{aligned} \dot{x} &= y \\ \dot{y} &= -(\omega^2 + \varepsilon\sin(\Omega t))\sin x - cy \\ \dot{t} &= 1, \end{aligned} \tag{8}$$

which is $2\pi/\Omega$–periodic in t. Let $(x(t), y(t), t)$ be an integral curve of the system (8) and consider instants $t_n = 2\pi n/\Omega$, $n = 0, 1, 2, \ldots$. We call $(x(t_n), y(t_n))_{n=0}^{\infty}$ a *stroboscopic* sequence. From the theory of ordinary differential equations, e.g., see [7,8], it follows that there exists a map $(x,y) \mapsto \Phi(x,y)$ such that the stroboscopic sequence is formed by iterations under Φ. We call Φ the stroboscopic map or Poincaré map of the time periodic vector field (8).

The Poincaré map Φ also has a nice geometric meaning when one considers the vector field (8) in the 3–dimensional (x,y,t)–space. By periodicity in t, the whole

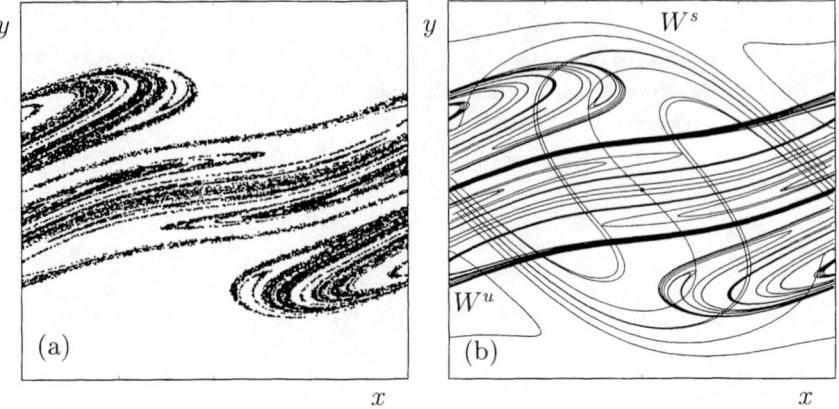

FIGURE 8. A Hénon–like strange attractor (obtained by iteration) of the Poincaré map of the parametrically forced damped pendulum (a). The attractor is conjecturally the closure of the unstable manifold $W^u((0,0))$ of the origin (b). Again, the stable manifold intersects the unstable manifolds in a homoclinic tangle, illustrating the complicated dynamics of the system.

'block' of dynamics between $t = 0$ and $t = 2\pi/\Omega$ repeats itself periodically. The Poincaré map now follows the integral curves from the section $t = 0$ to $t = 2\pi/\Omega$, thus defining the smooth planar map Φ; compare [7,8] and see also the chapter by *Krauskopf*.

In Fig. 8(a) we depicted the attractor $\mathcal{H}$ for a certain value of c, ω, Ω and ε. In the particular case we are in, a strong resonance is taking place in the sense that $\omega : \Omega \approx 1 : 2$, in which case $(x, y) = (0, 0)$ is a saddle–point of the Poincaré map Φ. This means that the downward 'equilibrium' of the pendulum has been destabilized by the resonant excitation. It is shown in Fig. 8(b) that it is a good guess that $\mathcal{H} = \overline{W^u((0,0))}$, compare the Hénon case in Fig. 4(b), although at the moment this would not be easy to prove. Locally the geometry also looks like in the Hénon case: as before it 'is' the product of curve and Cantor set. For a more extensive discussion see [20–22].

Remarks.

1. The mathematical conjecture behind all this goes much further, namely that in families of 2–dimensional dissipative diffeomorphisms strange attractors are *characterized* by $\mathcal{H} = \overline{W^u(p)}$, for a suitable periodic saddle point p.

2. The undamped case $c = 0$ is also interesting. Then the system is *conservative* and the Poincaré map Φ can be shown to preserve area[4]; see [15,23,24]. For

---

[4] A conceptual proof of this fact uses the Gauß Divergence Theorem on a flow box.

this reason there are no attractors, and also no chaos in the form of strange attractors as discussed earlier. Instead regular and irregular, or chaotic, behaviour occur for the same parameter values. This is not yet well–understood. The general conjecture is that a chaotic evolution densely fills a set with *positive* area (the so–called chaotic sea) and that the dynamics on this sea is sufficiently 'mixing'. This problem was raised in the sixties of the 20–th century, see [25], and to this day remains unsolved. For an elementary discussion also see [22]. Recently however, a result in this direction was proven in a simpler case of the so–called area preserving standard map [26].

In [27] a few more details are given with respect to the undamped case; also some extra references to relevant papers and books in Dutch have been included there.

Chaos in the Injected Laser

We finish this section with an example of a Hénon-like attractor in a laser system, namely in a semiconductor laser with optical injection. A semiconductor laser can be described well by single-mode rate equation for the complex electric field E and the population inversion n. Suppose now that a semiconductor laser receives an injected signal through one of its facet mirrors of a given amplitude K and detuning ω from the free-running laser frequency. This incoming signal enters the rate equations for the electric field E, and it is of the form $K e^{i\omega t}$. In other words, the semiconductor laser is periodically forced by the injected signal.

It turns out to be possible to reduce this system to three autonomous ODE's, two for E and one for n. The injected laser rate equations appear as Eqs. (2) in the chapter by *Krauskopf*; further details of their derivation can be found in [28]. For certain values of the parameters one finds Hénon-like attractors in the Poincaré section $\{Re(E) = 0\}$. It should be noted, however, that unlike for the periodically driven pendulum discussed earlier, the Poincaré map is not a stroboscopic map of Eqs. (2). This is explained in more detail in the chapter by *Krauskopf*.

An example of a Hénon-like attractor in the injected laser can be found in Fig. 9(a). The attractor appears to be of the same structure as that of the Hénon map; compare Fig. 4(a). Furthermore, there is a saddle point p on the attractor whose unstable manifold $W^u(p)$ accumulates on the attractor; see Fig. 9(b). It appears again to be the case that the attractor is the closure of the unstable manifold.

Finally, we remark that the dynamics of the injected laser are not necessarily chaotic. What dynamics one finds depends indeed crucially on the parameters K and ω; see [28]. It is an important question how the dynamics of a system can change from simple to chaotic as parameters are changed. This is the topic of what is known as Bifurcation Theory. An introduction to this theory is given in the chapter by *Krauskopf*, where the injection laser serves as the illustrating example throughout.

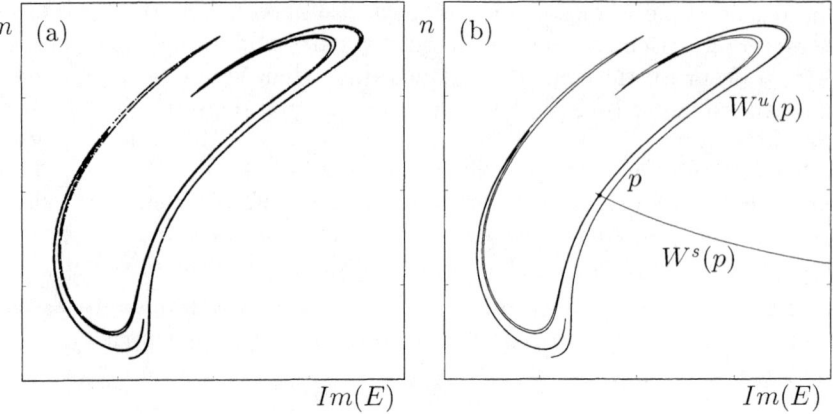

FIGURE 9. A Hénon–like strange attractor (obtained by iteration) of the Poincaré map in the plane $\{Re(E) = 0\}$ of the rate equation model of an optically injected semiconductor laser (a). The attractor is conjecturally the closure of the unstable manifold $W^u(p)$ (b). Only a piece of the stable manifold $W^s(p)$ can be computed, because the Poincaré map is only defined locally near the chaotic attractor. The parameter values are $K = 0.62$, $\omega = 0.9$, $\alpha = 2$, $B = 0.015$ and $\Gamma = 0.035$ for the Eqs (2) in the chapter by *Krauskopf*.

RANDOM ASPECTS

In this section we indicate some connections of the above with Probability Theory; compare [2,23,25].

The Dynamical Distribution

Let $\mathcal{H}$ be a (strange) attractor of the discrete dynamical system $x_{n+1} = \Phi(x_n)$, with initial state x_0. This leads to the Φ–evolution $x_0, x_1, x_2, \ldots$. For any (reasonable) subset $A \subseteq \mathcal{H}$ consider the probability

$$P(A) := \lim_{N \to \infty} \frac{\#\{n \mid 0 \leq n \leq N \,\&\, x_n \in A\}}{N+1},$$

which is nothing but the relative visiting frequency of the evolution to A. In many cases $P(A)$ is independent of the initial state x_0. Assuming this independence, we obtain a probability distribution on $\mathcal{H}$, usually called the *dynamial* (or *physical*) *distribution*. It is easy to see that the distribution is dynamically invariant, in the sense that $P(\Phi^{-1}(A)) = P(A)$, for (reasonable) subsets $A \subseteq \mathcal{H}$. This invariant distribution allows an important connection with Ergodic Theory, a subdiscipline of Probability Theory; see [29,30]. Our earlier remark about 'mixing' belongs to this theory.

For the Baker transformation the dynamical distribution is just the uniform distribution on $[0, 1]$, while for the Logistic map with $\mu = 4$ the density is given by the Bèta–function

$$\frac{1}{\pi\sqrt{x(1-x)}}.$$

For the Hénon attractor such a simple formula cannot be given, nevertheless it is worthwile to think in this way. The strange attractor $\mathcal{H}$, as depicted in Fig. 4(a), can be seen as the support of the invariant distribution. The sensitivity means that a small cloud of initial states is smeared out over the whole attractor $\mathcal{H}$. Again, more precise statements can be made with the help of Ergodic Theory.

In the general chaotic setting one can say that the sensitivity of Definition 4 means that iterates $\Phi^N(\vec{x}_0)$ and $\Phi^M(\vec{x}_0)$ are almost uncorrelated for large values of $N - M$. This seems to imply that for individual evolutions (orbits) only the short term future allows for deterministic predictions and that for the longer term future statistical methods have to be applied; see also [17].

Chaos or Chance?

Another way to deal with uncertainty is by the so-called fractal basin–boundaries. Suppose a discrete dynamical system $x_{n+1} = \Phi(x_n)$ is given, with initial state x_0. It may occur that the system has a finite number of point attractors, but that the basins of attraction have strange boundaries. Indeed, in the case of three attractors the boundaries may have the following property: as soon as two of the boundaries meet, also the third is joining in. In that case, the total boundary certainly is a fractal set. This kind of subsets of the plane was studied extensively by L.E.J. Brouwer.

An example of this phenomena can be found in [2], where the computer experiment is presented of a magnetic spherical pendulum that is attracted by three different magnets.

Such a fractal basin boundary often is a strange repellor. A class of examples can be found when iterating complex analytic maps: many so–called Julia sets belong to this class; e.g., see [2,3].

With some good will, also the good old die can be seen in this perspective. Indeed, we can view this system to have six different point attractors with complicated basin boundaries. As we said before, in such a case 'randomness' seems to be only introduced for the sake of convenience.

All this leads to a well–known philosophical question [31], namely whether the latter is not always the case: that randomness is a convenient short–cut in deterministic dynamics that are just too complicated to deal with in another way. In this respect one may think of systems like the atmospheric circulation (producing the weather). Another example comes from Statistical Physics and concerns the hard spheres approximation model for a gas, which has sensitive dependence on initial states.

On the other hand it seems that Quantum Mechanics takes care of 'true' randomness at very small precision.

ACKNOWLEDGEMENTS

H.W.B. thanks Aernout van Enter, Igor Hoveijn and Floris Takens for their help in preparing [27] from which this exposition developed. Substantial parts of the material appeared earlier in [27], and we thank the publisher *Het Wiskundig Genootschap* for the kind permission to reproduce text passages and figures 1–3 and 5–7.

REFERENCES

1. Devaney, R.L., *An Introduction to Chaotic Dynamical Systems*, (2nd edition), Addison–Wesley, 1989.
2. Peitgen, H.-O., Jürgens, H., and Saupe, D., *Chaos and Fractals, New Frontiers of Science*, Springer–Verlag, 1992.
3. Broer, H.W., and Verhulst F. (eds.), *Dynamische Systemen en Chaos, een revolutie vanuit de wiskunde*, Epsilon Uitgaven **14**, 1990.
4. Graczyk, J., and Swiatek, G., "Generic hyperbolicity in the logistic family", *Annals of Math.* **146**, 1-52 (1997).
5. Lyubich, M., "Dynamics of quadratic polynomials I-II", *Acta Math.* **178**, 185-297 (1997).
6. Levin, G., van Strien, S.J., "Local connectivity of the Julia set of real polynomials", *Annals of Math.* **147**, 471-541 (1998).
7. Hirsch, M.W., and Smale, S., *Differential Equations, Dynamical Systems, and Linear Algebra.* Academic Press, 1974.
8. Verhulst, F., *Nonlinear Differential Equations and Dynamical Systems* (2nd edition), Universitext, Springer–Verlag, 2000.
9. Benedicks, M., and Carleson, L., "The dynamics of the Hénon map", *Ann. Math.* **133**, 73–169 (1991).
10. Tatjer, J.C., *Invariant manifolds and bifurcations for one and two dimensional maps*, PhD Thesis, Universitat de Barcelona, 1991.
11. Broer, H.W., Simó, C., and Tatjer, J.C., "Towards global models near homoclinic tangencies of dissipative diffeomorphisms", *Nonlinearity* **11**, 667–770 (1998).
12. Krauskopf, B., and Osinga, H.M., "Growing 1D and quasi 2D unstable manifolds of maps", *J. Comp. Phys.* **146**(1), 404–419 (1998).
13. Krauskopf, B., and Osinga, H.M., "Investigating torus bifurcations in the forced Van der Pol oscillator", in *Numerical Methods for Bifurcation Problems and Large-Scale Dynamical Systems*, edited by E.J. Doedel and L.S. Tuckerman, IMA Volumes in Mathematics and its Applications **119**, Springer-Verlag, 2000, pp. 199–208.

14. de Melo, W.C., and van Strien, S.J., *One–Dimensional Dynamics*, Ergebnisse der Mathematik und ihrer Grenzgebiete **3**(25). Springer–Verlag, 1993.
15. Broer, H.W., Dumortier, F., van Strien, S.J., and Takens, F., *Structures in Dynamics*, Studies in Mathematical Physics **2**, North–Holland, 1991.
16. Lorenz, E.N., *The Essence of Chaos*, University of Washington/University College London Press, 1993.
17. Ruelle, D., *Chance and Chaos*, Princeton University Press, 1991.
18. Robinson, R.C., *Dynamical Systems: Stability, Symbolic Dynamics, and Chaos*, CRC–Press Inc., 1999.
19. Tucker, W., *The Lorenz attractor exists*, PhD Thesis, Uppsala University, 1999; Preprint at http://www.math.uu.se/~warwick/papers.html
20. Mora, L., Viana, M., "Abundance of strange attractors", *Acta Math.* **171**, 1–71 (1993).
21. Palis, J., and Takens, F., *Hyperbolicity & Sensitive Chaotic Dynamics at Homoclinic Bifurcations*, Cambridge Studies in Advanced Mathematics **35**, Cambridge University Press, 1993.
22. Broer, H.W., van de Craats, J., and Verhulst, F., *Het einde van de voorspelbaarheid? Chaostheorie, ideeën en toepassingen*. (Aramith Uitgevers &) Epsilon Uitgaven **35**, 1995.
23. Arnol'd, V.I., *Mathematical Methods of Classical Mechanics*. Springer–Verlag, 1980.
24. Marsden, J.E., and Ratiu, T., *Introduction to Mechanics and Symmetry*, TAM **17**, Springer–Verlag, 1994.
25. Arnol'd, V.I., Avez, A., *Ergodic Problems of Classical Mechanics*, Benjamin Inc., 1968.
26. Knill, O., "Positive Kolmogorov–Sinai entropy for the standard map", Preprint 00-203 at mp.arc, available at www.ma.utexas.edu/mp.arc/c /99/99-293.ps.gz.
27. Broer, H.W., "The how and what of Chaos", *Nieuw Arch. Wisk.* 5-th series, **1**(1), (2000), 34–43.
28. Wieczorek, S.M., Krauskopf, B., and Lenstra, D., "A unifying view of bifurcations in a semiconductor laser subject to optical injection", Optics Communications **172**(1-6), 279–295 (1999).
29. Katok, A., and Hasselblatt, B., *Introduction to the Modern Theory of Dynamical Systems*, Encyclopedia of Mathematics and its Applications **54**, Cambridge University Press, 1995.
30. Mañé, R., *Ergodic Theory and Differentiable Dynamics*, Ergebnisse der Mathematik und ihrer Grenzgebiete **8**, Springer–Verlag, 1987.
31. Wick, D., *The Infamous Boundary, Seven Decades of Controversy in Quantum Physics*, Birkhäuser, 1995.

Multiple time scale analysis of lasers

Thomas Erneux

Université Libre de Bruxelles, Optique Nonlinéaire Théorique, Campus Plaine C.P. 231, 1050 Bruxelles, Belgium

Abstract. In many areas of engineering and science the mathematical formulation leads to nonlinear problems that are difficult to solve even numerically. Many of the numerical difficulties come from the stiffness of the equations due to the presence of small or large parameters multiplying time derivatives. Very often, the solution exhibits different time scales because of these parameters. It is therefore important to identify them and possibly simplify our equations by applying asymptotic techniques appropriate for multiple scale problems. In this paper, we review a certain number of multiple time scale problems which have appeared for a variety of lasers.

MULTIPLE TIME SCALES

We begin by giving two basic examples of the consequences of different time scales in laser systems.

Laser relaxation oscillations

Practical lasers used in laboratories and in applications exhibit different time scales which have an immediate impact on their stability. For an idealized active system consisting of only two energy levels ([1], p 13), the simplest laser rate equations consist of two coupled equations for the intensity of the laser field I and the difference between the excited and ground state population densities D. They are given by

$$I' = 2I\left(-1 + AD\right), \tag{1}$$
$$D' = \gamma\left(1 - D - DI\right) \tag{2}$$

where prime means differentiation with respect to t. Furthermore, A measures the rate of stimulated emission and absorption and is called the *pump parameter*. The parameter $\gamma \equiv \gamma_{\|}/\kappa$ is a ratio of two decay rates: $\gamma_{\|}$ is defined as the decay rate of D towards its equilibrium state in the absence of laser light and κ is the decay rate of the field in the laser cavity (typical values are shown in Table 1). In Eqs. (1) and (2), time t is measured in units of the cavity constant κ^{-1} ($t \equiv \kappa t'$). As A is

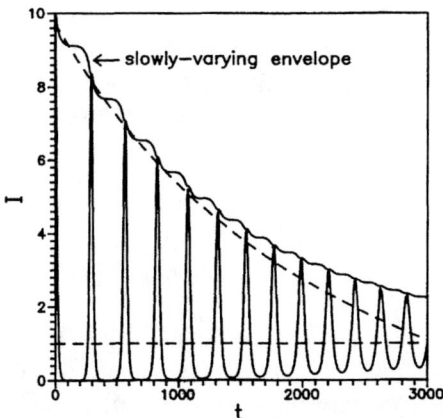

FIGURE 1. Relaxation oscillations. Eqs. (1)-(2) have been integrated numerically with $A = 2$, $\gamma = 5 \times 10^{-4}$, $I(0) = 10$ and $D(0) = 0.5$. The figure shows slowly decaying intensity oscillations towards the stable steady state $I = 1$ (horizontal dotted line). We also represent the slowly varying envelope of the maxima. It is given by $I_M \equiv (A-1)(1+y_M)$ where y_M is defined as the positive root of $y - \ln(1+y) - E = 0$, and E is the slowly varying energy defined by (5). The exponentially decaying dotted line is the large I_M approximation. It is given by $I_M = (A-1)(1+E+\log(E+1))$ where $E = (E_i + 1)\exp(-\frac{2}{3}A\gamma t) - 1$ and $E_i = E(x(0), y(0))$.

progressively increased from zero, a zero intensity steady state ($I = 0$) exchanges its stability at $A = 1$ to a stable non-zero intensity steady state ($I = A - 1$). Although there are no Hopf bifurcations, Eqs. (1) and (2) admit slowly decaying oscillations, called *relaxations oscillations*; see Fig. 1. These damped intensity oscillations have been observed as soon as lasers have been studied (see [1], p 5, for a historical review) and result from the small value of γ (see Table 1). Indeed, solving the linearized problem for the non-zero intensity steady state shows that a small perturbation oscillates on a $t_1 = \gamma^{1/2}t$ time scale and slowly decays on a $t_2 = \gamma t$ time scale. The frequency of these oscillations

$$\omega \equiv \sqrt{2\gamma(A-1)} \qquad (3)$$

is called the *relaxation oscillation frequency*.

We wish to take advantage of the fact that γ is small but setting $\gamma = 0$ in Eq. (2) will not help because we loose one equation and oscillatory solutions are clearly not possible. This is an example of singular perturbation. The proper way to resolve this problem is to introduce the new time s and the deviations from the steady state x and y defined by

$$s \equiv \omega t, \ D \equiv \frac{1}{A}(1 + \frac{\omega}{2}x) \text{ and } I \equiv (A-1)(1+y) \qquad (4)$$

where ω is given by (3). The resulting system of equations for x and y has then the form of a weakly perturbed conservative oscillator which has been discovered independently by various groups and in different formats [4–6]. But it is only recently that the asymptotic analysis of weakly perturbed strongly nonlinear oscillators has come to maturity [7,8]. Applying these multiple-scale or averaging methods to our laser equations allows us to find an equation for the slowly varying envelope of the rapid oscillations. This is possible because the function $E = E(x,y)$ defined by

$$E \equiv \frac{1}{2}x^2 + y - \ln(1+y) \tag{5}$$

is slowly varying in the time s $(dE/ds = O(\sqrt{\gamma}))$. The quantity E is called the *energy of the rapid oscillations*. For large or small E (equivalently large or small amplitude oscillations), we may determine simple analytical expressions [9]; see Fig. 1.

TABLE 1. Decay rates. Note that the values of κ and $\gamma_\parallel$ are relatively different between lasers but that $\gamma \equiv \gamma_\parallel/\kappa$ is $O(10^{-3})$ small.

laser	κ (s^{-1})	$\gamma_\parallel$ (s^{-1})
CO_2 [2]	10^8	2.5×10^5
solid state [3]	10^6	4×10^3
semiconductor laser [3]	10^{12}	10^9

Sustained laser oscillations

Sustained laser intensity oscillations are possible if the laser contains an intracavity saturable absorber [1,10]. In case of strong saturability, the saturable absorber operates like a passive switch and, under some conditions, may switch spontaneously leading to short and intense pulses; see Fig. 2.

Most experimental and theoretical studies have concentrated on gas lasers [11,12] or semiconductor lasers [13–15]. But recently, passively Q-switched solid state microchip lasers have been investigated both experimentally [16,17] and numerically [18,19]. The laser rate equations for a laser with a saturable absorber are now modified as

$$I' = I\left(-1 + AD + \overline{AD}\right), \tag{6}$$
$$D' = \gamma\left(1 - D(1+I)\right), \tag{7}$$
$$\overline{D}' = \overline{\gamma}\left(-1 - \overline{D}(1+\alpha I)\right) \tag{8}$$

where $\overline{D}$ is the dimensionless saturable absorber gain. $\overline{\gamma}$ is proportional to the decay rate of $\overline{D}$ normalized by the cavity decay time κ^{-1}. The parameter α is defined as the ratio of the gain saturation and saturable absorber saturation fluxes. $\overline{A}$ is the

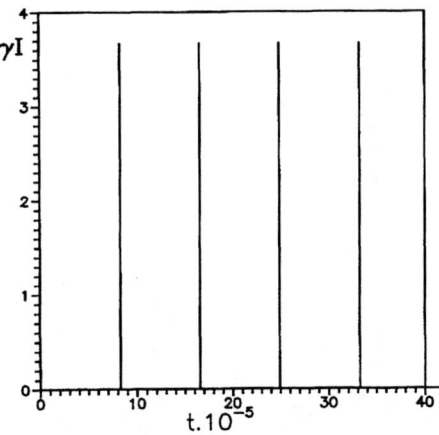

FIGURE 2. Passive Q-switching oscillations. Eq. (6)-(8) are integrated numerically using the values of the parameters shown in the first line of Table 2 and $A = 10.48$. The interpulse solution with I extremely small ($I = O(\exp(-1/\gamma))$) is computed using the asymptotic formula given in [18,19].

TABLE 2. Dimensionless parameters for microchip solid state lasers using two different absorbers.

saturable absorber	γ	$\bar{\gamma}$	$\bar{A}$	α	A/A_{th}
Cr^{4+} :YAG [16,18]	1.75×10^{-6}	6.35×10^{-5}	3.96	0.0852	~ 2
semiconductor [17]	3.7×10^{-7}	9.3×10^{-2}	0.36	4×10^{-3}	$\sim 2-3$

ratio of the saturable absorber small signal gain to the cavity losses. Values of the dimensionless parameters have been evaluated in [18,19] and are listed in Table 2.

It is interesting to discuss the values of these parameters. Similarly to gas or semiconductor lasers, microchip solid state lasers admit small values of γ and $\bar{\gamma}$ (sometimes even smaller). But by contrast to gas and semiconductor lasers, the value of α is small ($\alpha = 4 - 100$ for gas or semiconductor lasers). The small value of α suggests a weak saturable absorber but the fact that $\alpha\bar{\gamma}$ is $O(\gamma)$ or larger is sufficient to generate the pulsating oscillations. These oscillations appear from a Hopf bifurcation to a nearly vertical branch of periodic solutions [20].

Again, we wish to take advantage of the small values of γ and $\bar{\gamma}$. However, the pulsating behavior of the intensity shown in Fig. 2 is quite different from the slowly damped relaxation oscillations shown in Fig. 1. Assuming $\bar{\gamma} = O(\gamma)$, the oscillations now consist of large $O(\gamma^{-1})$ intensity pulses separated by long $O(\gamma^{-1})$ time intervals where the intensity is almost zero. Multiple scales or averaging methods are not helpful because the solution exhibits both different time scales and different amplitude scales. However, we may derive separate problems for the

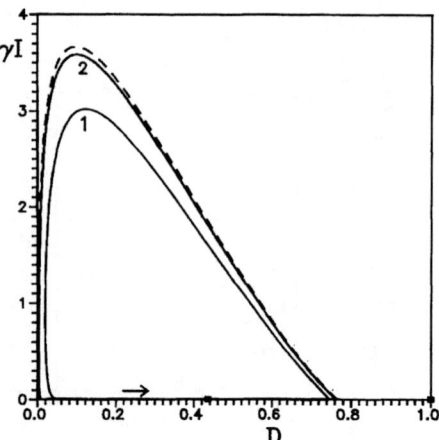

FIGURE 3. Asymptotic limit of the limit cycle as $\gamma \to 0$. The limit-cycle solution has been determined for two different values of γ. The values of the parameters other than γ and $\bar{\gamma}$ can be found in the first line of Table 2. Curve 1 and Curve 2 correspond to $\gamma = 1.75 \times 10^{-2}$, $\bar{\gamma} = 6.35 \times 10^{-1}$ and $\gamma = 1.75 \times 10^{-3}$, $\bar{\gamma} = 6.35 \times 10^{-2}$, respectively. For smaller γ and $\bar{\gamma}$, the numerical limit cycle matches its asymptotic limit shown by the broken line and given by Eq. (9) with $D_b \simeq 0.76$.

quick pulse and the interpulse period and connect their solutions by using the method of matched asymptotic expansions [22]. Figure 3 shows limit cycles for gradually decreasing values of γ and $\bar{\gamma}$. The broken line in the figure represents the leading approximation of the fast pulse given by

$$\gamma I = A(D_b - D) + \ln(\frac{D}{D_b}) + \frac{\bar{A}}{m}\left((\frac{D}{D_b})^m - 1\right) \tag{9}$$

where $m = \alpha\bar{\gamma}/\gamma$. Here $D = D_b$ is the value of the gain as the pulse starts, and it is obtained from the interpulse period equations. The function (9) is defined for the interval $D_a \leq D \leq D_b$ where $D_a < D_b$ is the second root of (9) with $I = 0$. For $\gamma = O(10^{-4})$ and $\bar{\gamma} = O(10^{-3})$ and smaller, the numerical solution is in full quantitative agreement with the asymptotic solution. The analytical theories developed by Szabo and Stein [21] and later by Degnan [23] are not based on an asymptotic limit of the laser rate equations. They correctly found the leading equations for the high intensity pulse but fail to obtain an expression of the interpulse period that depends on the laser parameters only. Accurate expressions for the repetition rate are now available from the γ small asymptotic analysis. Near the laser threshold, $A = A_{th} = 1 + \bar{A}$, the frequency of the time-periodic pulses is $O(\gamma)$ and given by

$$\omega_{pulse} \simeq 2\pi\gamma\frac{(A - A_{th})}{A_{th}\Delta D} \tag{10}$$

where $\Delta D \equiv D_b - D_a \simeq 1 - D_a$ is the gain reduction and D_a satisfies Eq. (9) with $I = 0$ and $D_b = 1$.

Although the limit γ small is now well understood, a small parameter analysis for a combination of several small parameters has not been done. This is particularly important for semiconductor lasers. Models of semiconductor lasers exhibit new parameters that are not relevant for other lasers (linewidth enhancement factor, delay of the optical feedback), and they require to solve new problems (delay differential equations).

HOPF BIFURCATION TECHNIQUES

Many of our laser problems are formulated in terms of a system of ordinary differential equations of the form

$$u' = F(u, \lambda) \tag{11}$$

where λ is a bifurcation parameter; see also the chapter by *Krauskopf*. Assume that the first bifurcation point of stable steady state, $u = u_0(\lambda)$, is a Hopf bifurcation at $\lambda = \lambda_0$ with frequency $\omega = \omega_0$. We may construct the long time solution by a multiple scale method. Specifically, our solution has the form

$$u = u_0 + \varepsilon \left[A(\nu) p \exp(i\omega_0 t) + c.c. \right] + O(\varepsilon^2) \tag{12}$$

where ε is defined by

$$\varepsilon \equiv \sqrt{\frac{\lambda - \lambda_0}{c}} \quad (c = 1 \text{ or } -1), \tag{13}$$

p is a constant vector, and ν is a slow time variable defined by

$$\nu \equiv \varepsilon^2 t. \tag{14}$$

The perturbation method is documented in many places for elementary problems ([24], p220; [7,25]). The long time behavior of the solution is determined by solving an equation for A of the form

$$\frac{dA}{d\nu} = c_1 A + c_2 A^2 A^* \tag{15}$$

where the complex coefficients c_1 and c_2 depend on all the parameters except λ. This equation is called the slow time amplitude equation or bifurcation equation. If the original equations (11) are supplemented by non autonomous terms, extra terms may appear in Eq. (15). For a periodically modulated laser, our slow time amplitude equation is given by

$$\frac{dA}{d\nu} = c_1 A + c_2 A^2 A^* + c_3 \tag{16}$$

where the complex coefficient c_3 is proportional to the modulation amplitude. For a semiconductor laser subject to an optoelectronic feedback exhibiting a large delay, we have

$$\frac{dA}{d\nu} = c_1 A + c_2 A^2 A^* + c_3 A(\nu - \nu_0) \tag{17}$$

where ν_0 is a slow time delay. Equation (15) admits only one non-zero steady state solution. By contrast to Eq. (15), Eqs. (16) and (17) admit more than one steady state solution. In particular, they may admit multiple steady state solutions as well as time-periodic solutions. Surprisingly, these amplitude equations have rarely been studied for their periodic solutions [26]. A periodically modulated laser problem is investigated in the next subsection. The multiple periodic solutions of Eq. (17) for a laser subject to a delayed opto-electronic feedback are studied in [27].

Semiconductor lasers subject to injection and the limit T large

Most of our understanding of the effects of optical injection in a single mode diode laser comes from numerical and analytical studies of simple rate equations for the complex electric field Y and the excess carrier number Z [28]; see also the chapters by *Gavrielides* and *Krauskopf*. In dimensionless form, these equations are given by

$$Y' = (1 + i\alpha)ZY + i\Omega Y + \Lambda \tag{18}$$
$$TZ' = P\left[1 + \delta \cos(\omega_m t)\right] - Z - P(1 + 2Z)\left|E\right|^2 \tag{19}$$

where prime means differentiation with respect to time t measured in units of photon lifetime τ_p. The last two terms in Eq. (18) model the injection which is controlled by changing either its amplitude Λ or the optical frequency detuning Ω. The term $\delta \cos(\omega_m t)$ models the modulation of the pump. If $\delta = 0$, these equations admit a stable limit-cycle solution that emerges and terminates at a low and a high injection Hopf bifurcation point, respectively. If $\delta \neq 0$ but small, the modulation frequency ω_m is typically chosen close to the limit-cycle frequency. The parameter α is the linewidth enhancement factor ($\alpha = 3 - 6$) which measures the degree of the amplitude-phase coupling. T is defined as the ratio of the carriers to photon lifetimes ($T = \tau_s/\tau_p = O(10^3)$) and P is the dimensionless pumping current above threshold ($|P| < 1$).

Note that the parameter T is proportional to γ^{-1} where γ appeared in our previous rate equations (1) and (2). The singular limit $\gamma \to 0$ means $T \to \infty$. However, our conclusions on this limit will be different because Eq. (18) is complex. As $T \to \infty$, Eqs. (18) and (19) reduce to

$$Y' = (1 + i\alpha)ZY + i\Omega Y + \Lambda, \tag{20}$$
$$Z' = 0. \tag{21}$$

Eq. (21) simply requires that

$$Z = Z_0 = cst \tag{22}$$

while Eq. (20) for $Y = Y_0$,

$$Y_0' = [(1 + i\alpha)Z_0 + i\Omega] Y_0 + \Lambda \tag{23}$$

is an equation for a damped linear oscillator exhibiting the new frequency

$$\sigma = \alpha Z_0 + \Omega. \tag{24}$$

Note that Z_0 is unknown which motivates a higher order analysis. Moreover, the solution may depend on different time scales, namely the original time t and the relaxation oscillation time $\nu = T^{-1/2}t$. Equation (21) then means that

$$Z = Z_0(\nu) \tag{25}$$

is an unknown function of ν rather than an unknown constant. Furthermore, Eq. (23) becomes an equation for an oscillator with a slowly varying frequency which require a more sophisticated multiple scale method [25]. Of course, this analysis makes sense only if the frequency (24) is sufficiently large so that the linear decay of the oscillations may be balanced by nonlinear terms appearing through a higher order analysis. This analysis is proposed in [29] and further developed in [30]. For large negative detunings we found the coexistence of stable steady and time-periodic states which explain the anomalous transitions from locking to unlocking observed experimentally. For moderate to small values of the detuning, stable quasiperiodic oscillations are possible which depend on the relaxation oscillation frequency, $\sigma_1 = \sqrt{2P/T}$, and the detuning frequency $\sigma_2 = |\Omega|$. We come back to the limit T large in the next section.

Periodically modulated semiconductor laser subject to injection

In this subsection, we concentrate on the periodically modulated laser. Optically injected semiconductor lasers are promising devices that can produce microwave radiation of sufficient quality and magnitude. For large injection rates and slave-master detuning the relaxation frequency of the system can reach values as high as 50-80 GHz [31]. However, even though the microwave spectra are easily induced in diode lasers under external optical injection, the bandwidth of these sidebands can be broad and of the order of a few hundreds of MHz. Phase locking to an external modulation signal at a frequency close to the intrinsic frequency of the limit cycle has been proposed as a stabilizing mechanism [32]. The experiments have produced robust phase locking even under weak external modulations and the effect of noise is considerably reduced.

In [33] we determine the conditions for periodic locking by an asymptotic analysis based on the limit α large. The main advantage of considering this limit is the fact that the amplitude of the periodic states can be arbitrary. More recently, we investigated Eqs. (18) and (19) numerically. We found that qualitative changes occur for the locking domain as α progressively increases, which question the validity of our large α analysis. This has motivated a new analysis of the laser equations. Specifically, we determine a small amplitude solution of Eqs. (18) and (19) by a two time perturbation analysis [34]. The problem is reduced to the analysis of an amplitude equation of the form (16). Steady state and time-periodic solutions of this equation correspond to periodic and quasiperiodic states of the original laser equations, respectively. We note that periodic locking is possible either through limit points (that is, through a saddle-node bifurcation) or through a Hopf bifurcation. The quasiperiodic oscillations depend on the modulation frequency ω_m and a new frequency which is $O(\delta)$ as $\delta \to 0$.

SEMICONDUCTOR LASER SUBJECT TO OPTICAL FEEDBACK

Lang and Kobayashi [35] have considered a laser diode exposed to optical feedback from a flat external mirror. For weak to moderate feedback, the laser is now modeled by the following delay differential equations [36], which are now generally called the Lang-Kobayashi (or LK) equations (see also the chapters by *Gavrielides, Lenstra & Yousefi* and *Verduyn & Krauskopf*):

$$Y' = (1 + i\alpha)ZY + \eta \exp(-i\Omega_0 \theta)Y(s - \theta), \tag{26}$$
$$TZ' = P - Z - (1 + 2Z)|Y|^2. \tag{27}$$

The delay θ is defined by $\theta \equiv \tau/\tau_p$ where τ is the external round-trip time. In general, θ has the same scale as T and is an $O(10^3)$ quantity. Furthermore, $\Omega_0 \equiv \omega_0 \tau_p$ is the angular frequency of the solitary laser ω_0 normalized by τ_p^{-1} ($\Omega_0 \theta$ (mod $2\pi) = O(1)$). $\eta \equiv \eta' \tau_p$ is the feedback rate η' normalized by τ_p^{-1} ($\eta < 1$).

A basic solution of Eqs. (26) and (27) is called an external cavity mode solution (ECM). (A different name is continuous wave solution or CW-state.) It is a single frequency solution of the form

$$Y = A \exp\left(i(\sigma - \Omega_0)t\right) \text{ and } Z = B \tag{28}$$

where A, σ and B are constants. Substituting (28) into (26) and (27) leads to conditions for A, B and σ. The ECM frequency defined by

$$\Delta \equiv \sigma \theta \tag{29}$$

satisfies the following transcendental equation

$$\Delta - \Omega_0 \theta = -\eta \theta \left(\alpha \cos(\Delta) + \sin(\Delta)\right). \tag{30}$$

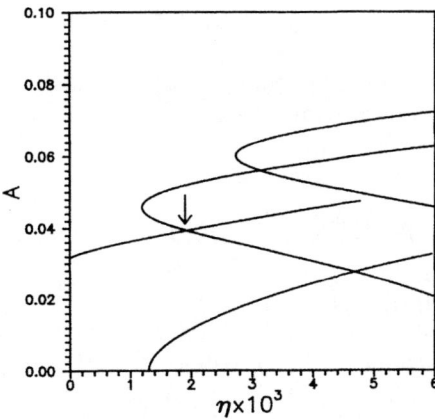

FIGURE 4. External cavity modes solutions of a laser subject to optical feedback. Branches of ECM solutions are represented as a function on the feedback rate η. The figure represents the intensity of each single ECM. The intensity is defined by $A^2 = \frac{P+\eta \cos(\Delta)}{1-2\eta \cos(\Delta)} > 0$ where Δ satisfies (30). The values of the parameters [39] are $P = 10^{-3}$, $T = \theta = 10^3$ and $\Omega_0 \theta = -1$. The arrow indicates a critical point at which two single ECMs admit the same intensity.

By analyzing Eq. (30) in the implicit form $\eta = \eta(\Delta)$ we note that the number of ECM solutions increases with η but always remains odd. Although the single ECM solution (28) satisfy the LK equations (26) and (27) a mixed modes solution of the form

$$Y_0 = A_1 \exp\left(i(\sigma_1 - \Omega_0)t\right) + A_2 \exp\left(i(\sigma_2 - \Omega_0)t\right) \tag{31}$$

is not an exact solution of the LK equations. However, it can be the leading approximation of an asymptotic solution valid for large T [37]. Without going into the details of the perturbation analysis, this can be anticipated from the large T approximation of Eqs. (26) and (27) given by

$$Y_0' = (1 + i\alpha) Z_0 Y_0 + \eta \exp(-i\Omega\theta) Y_0(s - \theta), \tag{32}$$

$$Z_0' = 0. \tag{33}$$

Eq. (33) implies that $Z_0 = cst$ while Eq. (32) admits a solution of the form (31) for particular values of the feedback rate $\eta = \eta_0$ at which two distinct single ECM solutions admit the same steady state intensity; see Fig. 4. The importance of those points has been discussed previously [38] but a higher order analysis is needed in order to determine equations for the amplitudes A_1 and A_2. From the physical point of view, the interesting result is the fact that the intensity is now given by

$$I = |Y_0|^2 = |A_1|^2 + |A_2|^2 + 2|A_1||A_2|\cos\left((\sigma_2 - \sigma_1)t + \phi\right) \tag{34}$$

where a new frequency $\sigma_2 - \sigma_1$ appears. It corresponds to the beating between two ECM frequencies and may provide an explanation to recent experimental observations of a laser subject to two optical feedbacks [40].

ACKNOWLEDGMENTS

The research was supported by the US Air Force Office of Scientific Research grant AFOSR F49620-98-1-0400, the National Science Foundation grant DMS-9973203, the Fonds National de la Recherche Scientifique (Belgium) and the Inter University Attraction Pole of the Belgian government.

REFERENCES

1. Abraham, N.B., Mandel, P., and Narducci, L.M., *Prog. in Opt.* **25**, 3 (1988).
2. Midavaine, Th., Dangoisse, D., and Glorieux, P., *Phys. Rev. Lett.* **55**, 1989 (1985).
3. Fabiny, L., Colet, P., Roy, R., and Lenstra, D., *Phys. Rev.* **A 47**, 4287 (1993).
4. Basov, N.G., Morosov, V.N., and Oraevsky, A.N., *IEEE J. Quant. Electr.* QE-**2**, 542 (1966); V.N. Morozov, *JOSA B* **5**, 909 (1988).
5. Oppo, G.L., and Politi, A., *Z. Phys. B* **59**, 111 (1985); *Phys. Rev.* **A 40**, 1422 (1989).
6. Erneux, T., Baer, S.M., and Mandel, P., *Phys. Rev.* **A 35,** 1165 (1987)
7. Kervokian, J., and Cole, J.D., *Multiple Scale and Singular Perturbation Methods*, Applied Mathematical Sciences **114**, Springer-Verlag, 1996.
8. Bourland, F.J., and Haberman, R., *SIAM J. Appl. Math.* **48**, 737 (1988); Bourland, F.J., Haberman, R., and Kath, W.L., *SIAM J. Appl. Math.* **51**, 1150 (1991).
9. Erneux, T., "Advanced Asymptotic Methods for the Applied Sciences", Lecture notes at the Université des Sciences et Technologies de Lille (1999).
10. Mandel, P., *Theoretical Problems in Cavity Nonlinear Optics*, Camb. Stud. in Modern Optics, Camb. University Press, 1997.
11. Arimondo, E., Casagrande, F., Lugiato, L.A., and Glorieux, P., *Appl. Phys.* B **30**, 5 (1983); Arimondo, E., Bootz, P., Glorieux, P., and Menchi, E., *JOSA B* **2**, 193 (1985).
12. Tachikawa, Tanii, K., and Shimizu, T., *JOSA B* **4**, 387 (1987); *JOSA B* **5**, 1077 (1988).
13. Kawaguchi, H., *Bistabilities and Nonlinearities in Laser Diodes*, Artech House, 1994.
14. Yamada, M., *IEEE J. Quant. Electron.* QE-**29**, 1330 (1993).
15. Dubbeldam, J.L.A, and Krauskopf, B., *Opt. Comm.* **159**, 325 (1999).
16. Zayhowski, J.J., *Opt. Lett.* **21**, 588 (1996); Errata *Opt. Lett.* **21**, 1618 (1996); Zayhowski, J.J., and Dill III, C., *Opt. Lett.* **19**, 1427 (1994).
17. Spühler, G.J., Paschotta, R., Fluck, R., Braun, B., Moser, M., Zhang, G., Gini, E., and Keller, U., *JOSA B* **16**, 376 (1999).

18. Peterson, P., Gavrielides, A., Sharma, M.P., and Erneux, T., *IEEE J. Quant. Electron.* **35**, 1247 (1999).
19. Erneux, T., Peterson, P., and Gavrielides, A., *Europ. J. Physics* **D10**, 423 (2000).
20. Erneux, T., and Kozyreff, G., *J. Stat. Phys.*, in press (2000).
21. Szabo, A., and Stein, R.A., *J. Appl. Phys.* **36**, 1562 (1966).
22. Erneux, T., *JOSA B* **5**, 1063 (1988).
23. Degnan, J.J., *IEEE J. of Quant. Electron.* **31**, 1890 (1995).
24. Strogatz, S.H. *Nonlinear Dynamics and Chaos*, Addison Wesley, 1995.
25. Bender C.M., and Orszag, S.A., *Advanced Mathematical Methods for Scientists and Engineers*, McGraw-Hill, 1978.
26. Kath, W.L., *Studies in Appl. Mathematics*, **65**, 95 (1981).
27. Pieroux, D., Erneux, T., Gavrielides, A., and Kovanis, V., *SIAM J. on Appl. Math.*, in press (2000).
28. Special Issue on Fundamental Nonlinear Dynamics of Semiconductor Lasers, edited by Lenstra, D., *Quantum and Semiclass. Opt.* **9**, # 5 (1997).
29. Kovanis, V., Erneux, T., and Gavrielides, A., *Opt. Comm.* **159**, 177 (1999).
30. Nizette, M., Erneux, T., Gavrielides A., and Kovanis, V., *Proc. SPIE* **3625**, 679 (1999).
31. Simpson, T.B., Liu, J.M., and Gavrielides, A., *IEEE J. Quant. Electron.* **32**, 1456 (1996); Simpson, T.B., Liu, J.M., *IEEE Photon. Technol. Lett.* **9**, 1325 (1997).
32. Simpson, T.B., and Doft, F., *Opt. Comm.*, **170**, 93 (1999).
33. Gavrielides, A., Kovanis, V., Erneux, T., and Nizette, M., *Proc. SPIE* **3944**, to appear (2000).
34. Nizette, M., Erneux, T., Gavrielides, A., and Kovanis, V., submitted for publication.
35. Lang, R., and Kobayashi, K., *IEEE J. Quant. Electron.* QE-**16**, 347 (1980).
36. Alsing, P.M., Kovanis, V., Gavrielides, A., and Erneux, T., *Phys. Rev.* **A 53**, , 4429 (1996).
37. Erneux, T., Rogister, F., Gavrielides, A., and Kovanis, V., *Opt. Comm.*, in press (2000).
38. Tager, A.A., and Elenkrig, B.B., *IEEE J. of Quant. Electron.* **29**, 2886 (1993); Tager, A.A., and Petermann, K., *IEEE J. of Quant. Electron.* **30**, 1553 (1994).
39. Hohl, A., and Gavrielides, A., *Phys. Rev. Lett.* **82**, 1148 (1999)
40. Rogister, F., Sukow, D.W., Mégret, P., Deparis, O., Gavrielides, A., and Blondel, M., *Proc. SPIE* **3944**, to appear (2000).

The Mathematics of Delay Equations with an Application to the Lang-Kobayashi Equations

Sjoerd M. Verduyn Lunel[*] and Bernd Krauskopf[†]

[*]*Mathematisch Instituut, Universiteit Leiden, P.O. Box 9512, 2300 RA Leiden, The Netherlands*
[†]*Department of Engineering Mathematics, University of Bristol, Bristol BS8 1TR, UK*

Abstract. For dynamical systems governed by feedback laws, time delays arise naturally in the feedback loop to represent effects due to communication, transmission, transportation or iternia effects. The introduction of time delays in a system of differential equations results in an infinite dimensional state space. We give an introduction to the basic theory of delay equations. After this introduction we illustrate the theory with a local analysis of the Lang-Kobayashi equations describing a laser with delayed optical feedback.

INTRODUCTION

In this paper we give an introduction to the theory of delay equations. The basic topics include the very definition, the state space approach, the solution operator and the stability properties of solutions. After this introduction we turn to the behaviour near equilibria and periodic orbits and discuss the notions of stable and unstable manifolds in the infinite dimensional state space.

Our main motivating example are the so-called Lang-Kobayashi equations describing a laser with delayed optical feedback; see also the chapters by *Erneux*, *Fischer et al.*, *Gavrielides* and *Lenstra & Yousefi*. They are a three-dimensional system of differential equations with time delays. In the physical literature one typically sees projections of solutions, obtained by numerical simulation, onto the three-dimensional space of complex electrical field and inversion. It is our aim to justify and extend this analysis in the appropriate infinite dimensional state space.

First we present a short introduction to differential delay equations. To illustrate the general theory we consider a simple nonlinear differential delay equation with one delay

$$\frac{dx}{dt}(t) = F(x(t), x(t-\tau)), \qquad t \geq 0, \tag{1}$$

where

$$F : \mathbb{R}^n \times \mathbb{R}^n \to \mathbb{R}^n$$

is a given differentiable vector field and $\tau \geq 0$ is a fixed real number. The solution of (1) is a vector-valued function $x : [0, \infty) \to \mathbb{R}^n$ which at each nonnegative point t in time assigns a vector $x(t) \in \mathbb{R}^n$.

A moment of reflection indicates that, in order for (1) to have a unique solution, the minimal amount of information of initial data one needs is a continuous function on the interval $[-\tau, 0]$. This implies that one is really dealing with an infinite dimensional phase space.

In fact, if $x(t) = \varphi(t)$ for $-\tau \leq t \leq 0$, where φ is a given continuous function, then the solution $x(t)$ for $0 \leq t \leq \tau$ satisfies the nonautonomous ordinary differential equation

$$\frac{dx}{dt}(t) = F(x(t), \varphi(t - \tau)) \qquad \text{for } 0 \leq t \leq \tau, \quad x(0) = \varphi(0). \tag{2}$$

This equation has a unique solution and the solution of (2) on $[0, \tau]$ coincides with the solution of (1) on $[0, \tau]$ with initial data φ on $[-\tau, 0]$. Once the solution x is known on $[0, \tau]$, we can repeat the same procedure: starting with the solution on $[0, \tau]$ we find the solution $x(t)$ for $\tau \leq t \leq 2\tau$, etc. This process is called *the method of steps* and yields a unique, globally defined solution $x(\,\cdot\,;\varphi)$ of (1) given an initial condition φ on $[-\tau, 0]$. Note that the solution becomes smoother in t as t increases. If we start with a continuous function on $[-\tau, 0]$, then the solution $x(\,\cdot\,;\varphi)$ is continuously differentiable on $(0, \tau)$, twice continuously differentiable on $(\tau, 2\tau)$ etc.

From the method of steps, it also follows that, in general, solutions of (1) are only defined on $[-\tau, \infty)$. Backward continuation of solutions of differential delay equations for negative time requires additional smoothness of the initial function φ on $[-\tau, 0]$. If the solution would have a backward extension to $[-2\tau, -\tau]$, then we can start with this extension to $[-2\tau, -\tau]$ as the initial data and use the method of steps to find the solution on $[-\tau, 0]$ which of course should be equal to φ. But after one step the solution becomes continuously differentiable and, therefore, φ must be continuously differentiable on $(-\tau, 0)$. Thus if φ is not differentiable, then there is no backward extension to $[-2\tau, -\tau]$. This lack of backward continuation also shows that differential delay equations are *infinite* dimensional dynamical systems.

Numerical computations of differential delay equations can also be based on the method od steps and naturally lead to complicated systems of equations. The approach is as follows. The initial interval $[-\tau, 0]$ is divided into small pieces of length h such that $N = \tau/h$ is an integer. The discrete approximation of (1) becomes

$$x_{k+1} = x_k + hF(x_k, x_{k-N}), \qquad k = 0, 1, 2, \ldots,$$

where $x_{-k} = \varphi(-kh)$ for $k = 0, 1, 2, \ldots, N$.

To illustrate the method of steps more explicitly, we consider as an example the simplest nonlinear feedback differential delay equation

$$\frac{dx}{dt}(t) = Ax(t) + G(x(t-\tau)), \qquad t \geq 0, \tag{3}$$

where A is an $n \times n$-matrix and $G : \mathbb{R}^n \to \mathbb{R}^n$ a given vector field. If $x(t) = \varphi(t)$ for $-\tau \leq t \leq 0$, then the variation-of-constants formula yields that the solution $x(t)$ for $0 \leq t \leq \tau$ is given by

$$x(t) = e^{At}\varphi(0) + \int_0^t e^{A(t-s)} G(\varphi(s-r))\, ds, \qquad 0 \leq t \leq \tau. \tag{4}$$

This is the unique solution of (3) on $[0, \tau]$ with initial data φ on $[-\tau, 0]$ and we can repeat the same procedure. Note that the solution exists without smoothness assumptions on G. The method of steps can be interpreted as follows. The differential Eq. (3) can be viewed as an iterative process on a space of continuous functions, iterating a segment of the solution. Indeed, set $x_n(\theta) = x(n\tau + \theta)$, $-\tau \leq \theta \leq 0$, $n = 0, 1, 2, \ldots$. Then

$$x_{n+1}(\theta) = e^{A\theta} x_n(0) - \int_\theta^0 e^{A(\theta-s)} G(x_n(s))\, ds, \qquad -\tau \leq \theta \leq 0. \tag{5}$$

If we linearize the nonlinear differential delay Eq. (1) around a steady state $x(t) = x_0$, we obtain an n-dimensional linear system of differential delay equations

$$\frac{dx}{dt}(t) = Ax(t) + Bx(t-\tau), \qquad t \geq 0, \tag{6}$$

where $A = D_1 F(x_0, x_0)$ is the derivative of F with respect to the first variable and $B = D_2 F(x_0, x_0)$ is the derivative of F with respect to the second variable both evaluated at (x_0, x_0). The initial data is given by $x(\theta) = \varphi(\theta)$ for $-\tau \leq \theta \leq 0$, where φ is a given continuous function on $[-\tau, 0]$ with values in $\mathbb{R}^n$. The solution with initial data $\varphi \in \mathcal{C}$ we denote by $x = x(\cdot, \varphi)$, where $\mathcal{C}$ denotes the space of $\mathbb{R}^n$-valued continuous functions on the interval $[-\tau, 0]$ provided with the sup-norm

$$\|\varphi\| = \sup_{-\tau \leq \theta \leq 0} |\varphi(\theta)|.$$

Linear differential delay equations have, in general, infinitely many independent exponential solutions

$$x_j(t) = e^{\lambda_j t}, \qquad \lambda_j \in \mathbb{C}, \quad j = 1, 2, 3, \ldots.$$

Indeed, substitution of an exponential solution $x(t) = e^{\lambda t}$ into (6) yields that for each root λ of the characteristic equation

$$\det \Delta(z) = 0 \quad \text{with} \quad \Delta(z) = \lambda I - A - Be^{-\lambda} \tag{7}$$

the exponential $x(t) = e^{\lambda t}$ is a solution of (6). Since (7) is a transcendental equation, there are, in general, infinitely many roots.

Characteristic equations of type (7) have been well-studied. We state here some basic results and refer to Chapter I and XI of [1] for further details.

Lemma 1 *The roots of the entire function (7) have the following properties*

(i) *There exists a $\eta \in \mathbb{R}$, such that the right half-plane $\{z \in \mathbb{C} \mid \operatorname{Re} z > \eta\}$ does not contain any roots of (7).*

(ii) *The number of roots of (7) in a given strip $\{z \in \mathbb{C} \mid \eta_- < \operatorname{Re} z \leq \eta_+\}$ is finite.*

(iii) *The roots of (7) in the left half-plane, necessarily satisfy*

$$|\operatorname{Im} z| \leq C e^{-\tau \operatorname{Re} z},$$

where C is a constant determined by A and B.

The asymptotic behaviour as t tends to infinity of the solutions of (6) is completely controlled by the behaviour of the roots of (7).

Theorem 1 *Let x be a solution of (6) corresponding to some initial function φ. For any $\gamma \in \mathbb{R}$ such that (7) has no roots on the line $\operatorname{Re} z = \eta$, we have the asymptotic expansion of the solution*

$$x(t) = \sum_{j=1}^{l} p_j(t) e^{\lambda_j t} + o(e^{\eta t}) \quad \text{for } t \to \infty,$$

where $\lambda_1, \ldots, \lambda_l$ are the finitely many roots of (7) with real part exceeding η and where $p_1(t), \ldots, p_l(t)$ are polynomials in t.

Corollary 1 *All solutions of (6) converge to zero exponentially as $t \to \infty$ if and only if (7) has no roots in the right half-plane $\{z \mid \operatorname{Re} z \geq 0\}$.*

Therefore, questions about exponential stability of steady states of (1) can be reduced to questions about the location of the roots of the analytic function (7). At the end of this section we give an example to illustrate this procedure. A first application is the following observation. If the delay τ is sufficiently small the roots of (7) have negative real part if and only if the roots of the polynomial

$$\det(\lambda I - (A + B))$$

all have negative real part. Thus, for Eq. (1) small delays have no effect on the stability properties of the steady states.

Large delays, however, result in a singularly perturbed (or multiple time scale) system; see also the chapter by *Erneux*. If we rescale time to normalize the time delay τ to one, we arrive at the following system of equations

$$\frac{1}{\tau} \frac{dx}{dt}(t) = F(x(t), x(t-1)), \qquad t \geq 0. \tag{8}$$

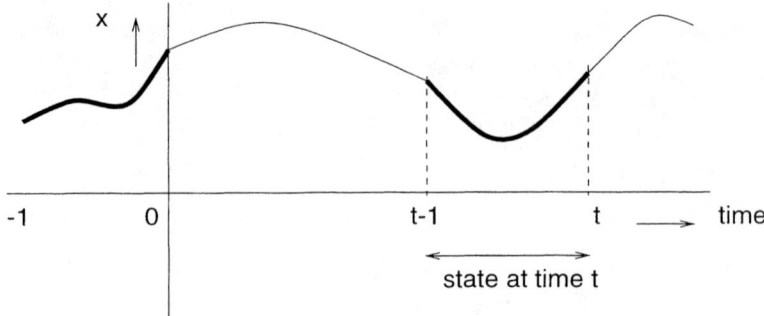

FIGURE 1. The state space approach

For the simple feedback system

$$\frac{1}{\tau}\frac{dx}{dt}(t) = -x(t) + G(x(t-1)), \qquad t \geq 0,$$

taking the singular limit as $\tau \to \infty$ results in a discrete finite dimensional dynamical system defined by iterating the map G. It is an interesting problem to determine how the dynamics of the discrete finite dimensional dynamical system mirrors the dynamics of the feedback system if the delay is sufficiently large; see [2,3] and Section 12.7 of [4] and the references given there.

The dynamical system approach to differential delay equations is to associate with (1) a semiflow defined by the time evolution of segments of solutions, as is illustrated in Fig. 1. If the solution of (1) with initial data $x(\theta) = \varphi(\theta)$, $-1 \leq \theta \leq 0$, is denoted by $x(\,\cdot\,;\varphi)$ and the state of the solution $x(\,\cdot\,;\varphi)$ is defined by

$$x_t(\theta;\varphi) = x(t+\theta;\varphi), \qquad -1 \leq \theta \leq 0,$$

then the semiflow $T(t)$ is defined by

$$T(t)\varphi = x_t(\,\cdot\,;\varphi).$$

An important observation is that the evolution of the state x_t is given by an abstract nonlinear ordinary differential equation in the infinite dimensional state space $\mathcal{C}$ of continuous functions defined on the interval $[-1,0]$. Given Eq. (1) this abstract ordinary differential equation can be computed explicitly and is given by

$$\frac{du}{dt} = Du, \qquad u(0) = \varphi, \ \varphi \in \mathcal{C}, \tag{9}$$

where $D : \text{Dom}(D) \to \mathcal{C}$ is an unbounded operator defined by

$$D\varphi = \frac{d\varphi}{d\theta}.$$

The domain of definition of D is given by

$$\text{Dom}(D) = \{\varphi \in \mathcal{C} \mid D\varphi \in \mathcal{C} \text{ and } (D\varphi)(0) = F(\varphi(0), \varphi(-1))\}.$$

Thus, the system of differential equations (1) with time delays can be viewed as a transport Eq. (9) with nonlocal boundary conditions (in the domain of definition of the operator D). Furthermore, the solutions of (1) are in one-to-one correspondence with the solutions of the infinite dimensional ordinary differential Eq. (9) and this correspondence is given by

$$u(t)(\theta) = x(t+\theta).$$

This observation originated with Krasovskii and has been crucial in the development of the qualitative theory of differential delay equations.

The linearization of the semiflow $T(t)$ around zero leads to a strongly continuous semigroup $T_0(t)$ with infinitesimal generator $D_0 : \text{Dom}(D_0) \to \mathcal{C}$, which is an unbounded operator defined by

$$D_0\varphi = \frac{d\varphi}{d\theta}.$$

The domain of definition of D_0 is given by

$$\text{Dom}(D_0) = \{\varphi \in \mathcal{C} \mid D_0\varphi \in \mathcal{C} \text{ and } (D_0\varphi)(0) = A\varphi(0) + B\varphi(-1)\}.$$

It is a simple computation to show that the spectrum of D_0 consists of a point spectrum only, which is given by the roots of (7). For more information on this approach of using abstract perturbation theory towards delay equations we refer to [1].

Differential delay equations can exhibit rather complex behaviour, even in scalar equations. To illustrate this fact, we consider the following example

$$\dot{x}(t) = -ax(t) + b\frac{x(t-1)}{1+x(t-1)^n}, \tag{10}$$

where n is a fixed even integer and a and b are positve parameters. This equation was introduced in [5] as a model to describe different periodic diseases.

We have seen that the initial value problem for (10) is well defined if one specifies a continuous function on the interval $[-1, 0]$. The semiflow

$$T(t)\varphi = x_t(\,\cdot\,; \varphi)$$

associated with (10) is point dissipative. This means that there is a bounded set $B \subset \mathcal{C}$ such that, for any state φ, the solution $x(\,\cdot\,;\varphi)$ of (10) satisfies

$$T(t)\varphi \in B \qquad \text{for } t \geq t_0(\varphi).$$

General theory implies that point dissipativeness of the semiflow yields the existence of a global attractor $\mathcal{A} \subset \mathcal{C}$ for (10). In this setting a global attractor $\mathcal{A} \subset \mathcal{C}$ is a compact set, invariant under the action of the semiflow

$$T(t)\mathcal{A} = \mathcal{A}, \qquad \text{for } t \geq 0,$$

which attracts open neighborhoods $U \subset \mathcal{C}$ of $\mathcal{A}$, that is,

$$\text{dist}\,(T(t)U, \mathcal{A}) \to 0 \quad \text{as } t \to \infty.$$

Numerical results show that for n even with $n \geq 8$ and a and b chosen appropriately, Eq. (10) exhibit chaotic motion on the global attractor $\mathcal{A}$ [5,6].

Let us consider the case that $a = 5$, $b = 10$ and $n = 8$ in more detail. It is clear that in this case the steady state solutions are $x = 0$ and $x = 1$. To study the stability properties of the steady states we have to linearize around them. Setting

$$f(x) = 10\frac{x}{1 + x^8},$$

Eq. (10) becomes

$$\dot{x}(t) = -5x(t) + f(x(t-1)). \tag{11}$$

The derivative of f is given by

$$f'(x) = 10\frac{1 - 7x^8}{(1 + x^8)^2}$$

and the linearizations around $x = 0$ and $x = 1$ become, respectively,

$$\dot{x}(t) = -5x(t) + 10x(t-1) \tag{12}$$

and

$$\dot{x}(t) = -5x(t) - 15x(t-1). \tag{13}$$

To verify whether these linearizations are exponentially stable we need a general result about the location of the roots of characteristic equations. A classical result states that the roots of the equation $(\lambda + c_1)e^\lambda + c_2 = 0$ all have negative real part if and only if

$$\begin{cases} c_1 > -1 \\ c_1 + c_2 > 0 \\ c_2 < \zeta \sin \zeta - c_1 \cos \zeta \end{cases}$$

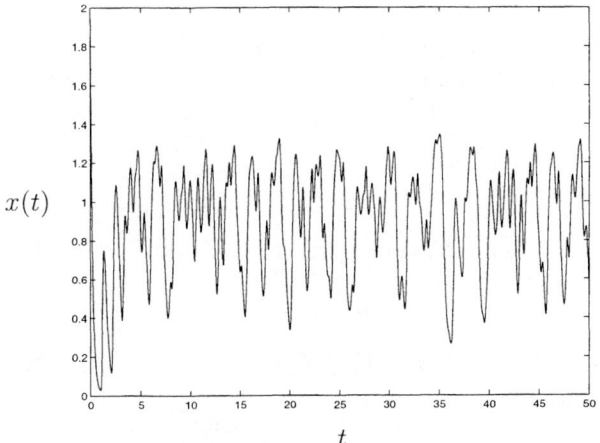

FIGURE 2. Chaotic behaviour of the solutions of (14) for the initial data $\varphi(\theta) = 2, -1 \leq \theta \leq 0$.

where ζ is the root of $\zeta = -c_1 \tan \zeta$, $0 < \zeta < \pi$, if $c_1 \neq 0$ and $\zeta = \pi/2$ if $c_1 = 0$.

The characteristic equation associated with (12) is

$$(\lambda + 5)e^\lambda - 10 = 0.$$

Thus $c_1 = 5$ and $c_2 = -10$, and consequently $x = 0$ is an unstable steady state of (11).

The characteristic equation associated with (13) is

$$(\lambda + 5)e^\lambda + 15 = 0.$$

Thus $c_1 = 5$ and $c_2 = 15$ and it remains to check whether

$$15 < \zeta \sin \zeta - 10 \cos \zeta,$$

where ζ is the root of $\zeta = -5 \tan \zeta$, $0 < \zeta < \pi$. A numerical computation shows that this inequality fails and, consequently, $x = 1$ is also an unstable steady state of (11).

The steady states being unstable complicates the dynamics. To illustrate this, we plot the solution of

$$\frac{dx}{dt}(t) = -5x(t) + 10\frac{x(t-1)}{1+x(t-1)^8}, \qquad x_0 = \varphi \qquad (14)$$

with initial data $\varphi(\theta) = 2$ for $-1 \leq t \leq 0$ in Fig. 2. The time series of x versus t shows the chaotic dynamics in the system. Also note that the solution intersects the steady state $x = 1$. This is no contradiction to the uniqueness of solutions because the states of the solutions are different.

We finish this section with an interesting fact. If we add small rapidly oscillating perturbations in the feedback loop, that is, if we consider the equation

$$\frac{dx}{dt}(t) = -5x(t) + f(x(t-1) + d\cos(t/\epsilon)), \qquad x_0 = \varphi$$

with $\epsilon > 0$ sufficiently small, then we can eliminate the complicated motion on the attractor completely [4].

BEHAVIOUR NEAR EQUILIBRIUM POINTS

In this section, we discuss the behavior of solutions of nonlinear differential delay equations near equilibrium points and periodic orbits.

Let Ω be a neighborhood of zero in $\mathcal{C}$ and let $C_b^1(\Omega, \mathbb{R}^n) \subset C(\Omega, \mathbb{R}^n)$ be the subset of functions from Ω into $\mathbb{R}^n$ that have bounded continuous derivatives with respect to $\varphi \in \Omega$. The results which we present in this section are true for infinite dimensional vector fields $F \in C_b^1(\Omega, \mathbb{R}^n)$ and equations of the form

$$\frac{dx}{dt}(t) = F(x_t).$$

However, to simplify the presentation we only consider nonlinear systems with one time delay τ which, by rescaling time, we can assume to be equal to one

$$\frac{dx}{dt}(t) = F(x(t), x(t-1)), \qquad (15)$$

where $F : \mathbb{R}^n \times \mathbb{R}^n \to \mathbb{R}^n$ is a continuously differentiable function.

Hyperbolic equilibrium points

If $x(t) = x_0$ is a steady state solution of (15), then $F(x_0, x_0) = 0$ and x_0 is an equilibrium point. Without loss of generality we can assume that $x_0 = 0$. Suppose that $F(0, 0) = 0$. In this case $x(t) = 0$ is a steady state of (15). If we linearize (15) around zero, we obtain the linear differential delay equation

$$\frac{dx}{dt}(t) = Ax(t) + Bx(t-1), \qquad (16)$$

where $A = D_1 F(0, 0)$ and $B = D_2 F(0, 0)$. We say that 0 is a *hyperbolic* equilibrium point of (15) if the roots of the characteristic equation

$$\det \Delta(\lambda) = 0 \qquad \text{with} \quad \Delta(\lambda) = \lambda I - A - e^{-\lambda} B, \qquad (17)$$

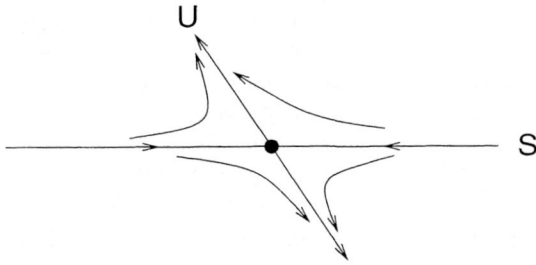

FIGURE 3. The behaviour of solutions of (16) near a saddle point.

have nonzero real parts.

If 0 is a hyperbolic equilibrium point of (15) and $\Lambda \subset \mathbb{C}$ denotes the set of roots of (17) with positive real part, then the phase space $\mathcal{C}$ can be decomposed by Λ as

$$\mathcal{C} = U \oplus S$$

into unstable and stable directions U and S, respectively. It follows from Lemma 1 that U is finite dimensional. Furthermore, the dimension of U is equal to the index of the equilibrium point 0, i.e. the number of roots (counting multiplicity) in Λ.

The decomposition of $\mathcal{C}$ as $U \oplus S$ defines two projection operators

$$\pi_U : \mathcal{C} \to U, \ \pi_U U = U, \qquad \pi_S : \mathcal{C} \to S, \ \pi_S S = S, \qquad \pi_S = I - \pi_U.$$

If Φ is a basis for U and Ψ is a basis for U^T, $(\Psi, \Phi) = 1$, then the projection is given by $\pi_U \varphi = \Phi(\Psi, \varphi)$. Let $\varphi \in \mathcal{C}$ be written as

$$\varphi = \varphi^U + \varphi^S, \qquad \varphi^U \in U, \ \varphi^S \in S.$$

If $T(t)$ is the semigroup generated by the linear Eq. (16), then U and S are invariant under $T(t)$ and $T(t)$ is defined on U for all $t \in \mathbb{R}$.

From the general theory for delay equations [4] it follows that there are positive constants M and α such that for $\varphi \in \mathcal{C}$

$$\begin{cases} \|T(t)\varphi^U\| \leq Me^{\alpha t}\|\varphi^U\|, & t \leq 0, \\ \|T(t)\varphi^S\| \leq Me^{-\alpha t}\|\varphi^S\|, & t \geq 0. \end{cases} \quad (18)$$

Relations (18) imply that the origin of the linearized system (16) is a saddle point with the orbits in $\mathcal{C}$ behaving as shown in Fig. 3.

The set U is uniquely characterized as the set of initial values of those solutions of (16) that exist and remain bounded for $t \leq 0$. In fact, estimates (18) imply that these solutions approach zero exponentially as $t \to -\infty$. The set S is characterized as the set of initial values of those solutions of (16) that exist and remain bounded for $t \geq 0$ and these solutions approach zero exponentially as $t \to \infty$.

It is natural to ask if the solutions of the nonlinear Eq. (15) have the same qualitative behavior near $x = 0$ as the solutions of the linearized Eq. (16). For infinite dimensional dynamical systems the meaning of qualitative behavior must be defined very carefully. The natural definition for finite dimensional systems (see the chapter by *Krauskopf*) is too strong [4]. On the other hand, it is true that important properties of the trajectories are preserved. More specifically, the set of initial values of those solutions of Equation (15) that exist and remain in a neighborhood of $x = 0$ for $t \leq 0$ can be mapped onto each other in a neighborhood in U of zero and these solutions approach zero exponentially as $t \to -\infty$. The same result holds for $t \geq 0$ and S. Any other solution must leave a neighborhood of zero with increasing t and, if it exists for $t \leq -r$, it must also leave a neighborhood of zero with decreasing t.

To be more precise, we let $x(t, \varphi)$ denote the solution of (15) with initial value φ at $t = 0$, and for a given neighborhood V of 0 in $\mathcal{C}$, we define the local stable manifold and the local unstable manifold of the equilibrium point 0 of (15) as, respectively,

$$W^s(0, V) = \{\varphi \in \mathcal{C} \mid x_t(\,\cdot\,, \varphi) \in V \text{ for } t \geq 0\},$$
$$W^u(0, V) = \{\varphi \in \mathcal{C} \mid x_t(\,\cdot\,, \varphi) \in V \text{ for } t \leq 0\}.$$

In order to formulate the main result we first need some definitions.

We say that $W^u(0, V)$ is a C^1-*graph* over $\pi_U \mathcal{C}$ if there is a neighborhood $\widetilde{V}$ of 0 in $\pi_U \mathcal{C}$ and a continuously differentiable function $H : \widetilde{V} \to \mathcal{C}$ such that

$$W^u(0, V) = \{\psi \in \mathcal{C} \mid \psi = H(\varphi),\ \varphi \in \widetilde{V}\}.$$

The set $W^u(0, V)$ is said to be *tangent to* $\pi_U \mathcal{C}$ *at* 0 if $|\pi_S \psi|/|\pi_U \psi| \to 0$ as $\psi \to 0$ in $W^u(0, V)$. Similar definitions hold for $W^s(0, V)$.

A basic result [4] on stable and unstable manifolds of equilibrium points is the following.

Theorem 2 *If 0 is a hyperbolic equilibrium point of Equation (15) and $F \in C^1(\mathbb{R}^n \times \mathbb{R}^n, \mathbb{R}^n)$, then there is a neighborhood V of 0 in $\mathcal{C}$ such that $W^u(0, V)$ is a C^1-graph over $\pi_U \mathcal{C}$ that is tangent to $\pi_U \mathcal{C}$ at 0. In the same way, $W^s(0, V)$ is a C^1-graph over $\pi_S \mathcal{C}$ that is tangent to $\pi_S \mathcal{C}$. Furthermore, the dimension of $W^u(0, V)$ is equal to the index of 0.*

Analogously, we can analyse the behaviour of the nonlinear semiflow near a nonhyperbolic equilibrium point, i.e. when the characteristic Eq. (17) has roots on the imaginary axis. This leads to a center manifold theory for differential delay equations [1]. Since the number of roots of the characteristic Eq. (17) on the imaginary axis is necessarily finite (compare Lemma 1) the center manifold is finite dimensional. As a consequence, results from bifurcation from equilibria (saddle-node, Hopf and Takens-Bogdanov bifurcation) can be "lifted" from the theory of ordinary differential equations to the theory of differential delay equations, see Chapter X of [1].

Hyperbolic periodic sets

In this section, we study the neighborhood of a period orbit γ of the nonlinear Eq. (15). Since a periodic orbit is a C^1-manifold, we can define a transversal cross section Σ at a point $p \in \gamma$ and, therefore, obtain a Poincaré return map on Σ. Since the solutions $T(t)\varphi$ of differential delay equations become smoother in t as t increases, the Poincaré map is differentiable if the vector field F is differentiable. Therefore, it suffices to develop the invariant manifold theory near a fixed point of the Poincaré map and this can be done in a manner analogous to the theory near equilibrium points in the previous section.

If $\gamma = \{p(t) \mid t \in \mathbb{R}\}$ is an ω-periodic orbit, then the linear variational equation of (15) about $p(t)$ is defined by

$$\frac{dx}{dt}(t) = A(t)x(t) + B(t)x(t-1), \qquad (19)$$

where $A(t) = D_1 F(p(t), p(t-1))$ and $B(t) = D_2 F(p(t), p(t-1))$. Equation (19) is a linear differential delay equation with coefficients periodic in t of period ω. Note that $x(t) = p(t)$ is always a solution of (19).

To use the linear theory for periodic delay equations, see Chapter 8 of [4], we first introduce the following definitions.

As before, the method of steps yields the existence of a unique solution for $t \geq s$ given an initial condition $x_s = \varphi \in \mathcal{C}$. The corresponding solution map $T(t, s) : \mathcal{C} \to \mathcal{C}$, $t \geq s$, defined by

$$T(t, s)\varphi = x_t(s; \varphi), \qquad t \geq s,$$

is an evolutionary system, i.e. $T(s, s) = I$ and $T(t, s)T(s, \tau) = T(t, \tau)$, $t \geq s \geq \tau$. Furthermore, periodicity of (19) implies

$$T(t + \omega, s) = T(t, s)T(s + \omega, s) \qquad \text{for } t \geq s.$$

Define the monodromy map $U : \mathcal{C} \to \mathcal{C}$ by

$$U\varphi = T(\omega, 0)\varphi. \qquad (20)$$

It is known that U^m is a compact operator, where m is such that $m\omega \geq 1$. Therefore, the theory of compact operators implies that the spectrum of U is a compact, at most countable, set with 0 as its only possible accumulation point. Furthermore, any nonzero point in the spectrum of U belongs to the point spectrum of U, i.e.

there is a $\varphi \neq 0$ in $\mathcal{C}$ such that $U\varphi = \mu\varphi$. Since $x(t) = p(t)$ is a solution of (19), it follows that $\mu = 1$ always belongs to the point spectrum of U. Any $\mu \in \sigma_p(U)$ is called a *Floquet multiplier* of (19) and λ for which $\mu = e^{\lambda\omega}$ is called a *characteristic exponent* of (19).

Next we extend the above definitions to periodic orbits of (15). The Floquet multipliers of a periodic orbit γ are the Floquet multipliers of the linear variational equation (19) that are nonequal to 1. However, if 1 is not a simple multiplier of the variational Eq. (19), then 1 is also a Floquet multiplier of the periodic orbit γ.

A periodic orbit γ is *hyperbolic* if each Floquet multiplier of γ has modulus different from 1. The *index* $i(\gamma)$ of a hyperbolic orbit γ is the number (counting with multiplicity) of Floquet multipliers with moduli > 1.

Let $x(t, \varphi)$ be the solution of (15) with initial value φ at $t = 0$, i.e. $x_0 = \varphi$. For a given neighborhood $V \subset \mathcal{C}$ of the periodic orbit γ, we define the local stable and the local unstable manifold of the periodic orbit γ of (15) as, respectively,

$$W^s(\gamma, V) = \{\varphi \in \mathcal{C} \mid x_t(\,\cdot\,, \varphi) \in V \text{ for } t \geq 0\}, \qquad (21)$$
$$W^u(\gamma, V) = \{\varphi \in \mathcal{C} \mid x_t(\,\cdot\,, \varphi) \in V \text{ for } t \leq 0\}. \qquad (22)$$

We now have the following theorem [4].

Theorem 3 *If the nonlinearity F in (15) is a C^k-function, $k \geq 1$, and γ is a periodic orbit of (15), then there is a neighborhood V of γ such that $W^s(\gamma, V)$ and $W^u(\gamma, V)$ are C^k-submanifolds of $\mathcal{C}$ with $\dim W^u(\gamma, V) = i(\gamma) + 1$ and $\operatorname{codim} W^s(\gamma, V) = i(\gamma)$.*

In the next section we illustrate how to apply the abstract theory to analyse systems of differential delay equations that arise in the study of lasers subject to optical feedback.

LASERS WITH OPTICAL FEEDBACK

In this section, we consider a laser subject to optical feedback, modeled in a semiclassical approach by single-mode rate equations. After an introduction, we shall apply the thory of differential delay equations that we just introduced to the Lang–Kobayashi equations [7].

Introduction

In numerous applications of lasers it is a fact that optical feedback is unavoidable. In particular, semiconductor lasers are very susceptible to external influences owing to the low reflectivity of their cleaved facets [7]. For this reason, lasers with feedback are a topic of much research, both experimental and theoretical [8–11]; see also the chapters by *Erneux, Fischer et al., Gavrielides* and *Lenstra & Yousefi*.

In this section we consider a single-mode laser subject to optical feedback. In the semiclassical approach, the dynamical interaction between the single mode of the electric field and the population inversion in the laser medium is described by the classical Maxwell's equations and the quantum mechanical Bloch equations. This approach results in a model with three dynamical equations for the electric field, the medium polarization, and the laser inversion. For many laser materials, the dynamics of the medium polarization can be adiabatically eliminated [12,13]. After applying the slowly-varying envelope and rotating-wave approximations one obtains first-order ordinary differential equations for the *complex* electric field amplitude E and for the inversion N. The complex amplitude E is related to the true electric field $\mathcal{E}$ by $\mathcal{E} = 1/2(E\exp(i\omega_0 t) + c.c.)$, where ω_0 is the center frequency of the single mode under consideration.

The abstract first-order system of differential equations for the electric field amplitude E and the inversion N is given by [14]

$$\frac{dE}{dt} = 1/2\big[\mathcal{G}(N, |E|^2) - \Gamma_0(t)\big] E, \tag{23}$$

$$\frac{dN}{dt} = -T_1(N)^{-1} N - \mathrm{Re}\big[\mathcal{G}(N, |E|^2)\big] |E|^2 + J(t), \tag{24}$$

where J is the generation rate of inversion through a pump mechanism, $\mathcal{G}$ is the complex optical gain, Γ_0 is the loss rate, and T_1 is the inversion lifetime.

For a single mode laser, described by Eqs. (23)–(24), optical feedback has only effect on the equation for the electric field (23). When the external resonator is much longer than the laser itself, the effect of feedback can be included in the field equation by means of a difference scheme [15]:

$$\frac{dE}{dt} = 1/2\big[\mathcal{G}(N(t), |E(t)|^2) - \Gamma_0(t)\big] E(t) + \sum_{m=1}^{M} F_m\big(E(t - m\tau)\big), \tag{25}$$

$$\frac{dN}{dt} = -T_1(N)^{-1} N - \mathrm{Re}\big[\mathcal{G}(N, |E|^2)\big] |E|^2 + J(t). \tag{26}$$

The optical feedbacks are given by

$$F_m(x) = \frac{(1 - r_1^2)}{\tau_{in} r_1^2} (-r_1 r_3)^m e^{-im\omega_0 \tau} x, \tag{27}$$

where r_1 is the amplitude reflection coefficient at the opaque side of the facet facing the external cavity, and r_3 the amplitude reflection coefficient of the external resonator. Furthermore, τ_{in} is the round-trip time of the light inside the laser cavity, τ is the round-trip time of the light in the external cavity, and m is an integer denoting multiple round-trips, where up to M round-trips are taken into account.

Note that Eqs. (25)–(26) are of a very general form, allowing for different expressions for the gain and for any number M of external round-trips. A famous example of Eqs. (25)–(26) are the Lang–Kobayashi equations which we shall discuss next.

Equations (25)–(26) form a system of differential delay equations. The state at time t is given by

$$(E, N)_t := \{(E(t-\theta), N(t-\theta)) \mid \theta \in [-M\tau, 0]\}. \tag{28}$$

We have seen, using the method of steps, that the trajectory

$$(E(t), N(t)), \quad t_1 \le t \le t_2,$$

is completely determined by the state of (E, N) at time t_1, i.e. $(E, N)_{t_1}$.

We have also seen that the geometric approach towards differential delay equations is to associate a semiflow $T(t)$ of operators with the differential equation defined by translation along the solution. The map $T(t)$ maps the continuous function $(E, N)_0 = (\varphi, \psi)$, $\varphi, \psi \in C[-M\tau, 0]$, into the state at time t,

$$T(t)(\varphi, \psi) = (E, N)_t,$$

where $(E, N)_t$ is the solution on $[-M\tau + t, t]$ shifted back to the initial interval $[-M\tau, 0]$.

Physically one usually looks at trajectories $(E(t), N(t))$, $t_1 \le t \le t_2$, in (E, N)-space, but one should keep in mind that the (E, N)-space is not the phase space of the system and that the trajectories of different solutions can actually intersect.

Note that the symmetry group $S^1 = \{c \in \mathbb{C} \mid |c| = 1\}$ acts on a trajectory by

$$c \circ (E, N)_{t_1}^{t_2} = (cE, N)_{t_1}^{t_2} \quad \text{for all } c \in S^1$$

and $(cE, N)_{t_1}^{t_2}$ is also a trajectory of (25)–(26); see [16]. The intersection of the group orbit with the two-dimensional half-plane

$$\Sigma = \{(E, N) \mid \operatorname{Im} E = 0 \text{ and } \operatorname{Re} E \ge 0\}$$

is called the *trace* of the trajectory.

The Lang-Kobayashi Equations

An important class of lasers are semiconductor lasers. They are extremely sensitive to small amounts of optical feedback, owing to the fact that their facet reflectivity is very low [17,18]. Due to the importance of semiconductor lasers in technological applications [19], feedback-induced effects have been studied extensively.

The paper by Lang and Kobayashi [7] in 1980 has become a milestone in this research, and for an overview of research in this field we refer to [14,15,20] and the chapters by *Erneux, Fischer et al., Gavrielides* and *Lenstra & Yousefi*.

In [7] a model was presented, now usually referred to as the Lang-Kobayashi equations, to study the effects of a single-mode semiconductor laser subject to optical feedback. The feedback is assumed to be so weak that only one external round-trip needs to be taken into account. Suitably scaled and normalized, the equations can be written as

$$\frac{dE}{dt} = 1/2(1 + i\alpha)N(t)E(t) + \kappa e^{-i\omega_0 \tau} E(t - \tau), \tag{29}$$

$$\frac{dN}{dt} = -\gamma N(t) - A\bigl[1 + N(t)\bigr]\bigl(|E(t)|^2 - 1\bigr), \tag{30}$$

where time is measured in units Γ_0^{-1}, α is known as the linewidth enhancement factor, γ is the dressed inversion decay rate, A is a factor involving both gain coefficient and pump parameter, and $\kappa = (r_1^2 - 1)r_3/\tau_{in} r_1$ is the feedback rate. The dynamical gain saturation is neglected and the optical gain is linearized at the threshold of the solitary laser. It should be noted that N now is the inversion with respect to its threshold value N_0 and that the equations are a special case of Eqs. (25)–(26).

The presence of the external mirror at distance $L = c\tau/2$ of the laser creates an external cavity whose intrinsic resonances allow the laser to operate at various compound-cavity modes, also called continuous wave solutions or CW-states. (A different name is external cavity modes or ECMs.)

CW-States and their Stability

A CW-state of the Lang–Kobayashi equations is a solution with constant intensity and inversion, and a phase which depends linearly on time. In other words, a solution of the form

$$E(t) = E_s(t) = R_s e^{i\omega_s t} \quad \text{and} \quad N(t) \equiv N_s, \tag{31}$$

for fixed values of R_s, N_s and ω_s. Note that a CW-state is rotationally symmetric and the trace of the CW-state is simply a single point (R_s, N_s) in the (R, N)-plane (which we can identify with Σ). More general, if we call a trajectory $(E(t), N(t))$ periodic if $R(0) = R(p)$ and $N(0) = N(p)$, where $p > 0$ is the smallest number with this property (CW-states formally correspond to period zero). The trace of the periodic trajectory $(E(t), N(t))$ in Σ is a closed curve and the group orbit $(cE, N)_0^p$, $c \in S^1$, is a torus in (E, N)-space, see Fig. 4.

Because a CW-state is a rotation around the N-axis in (E, N)-space, we can define a Poincaré return map $P : \Sigma \to \Sigma$ near a CW-state. For a complete analysis of the stability of a CW-state it is necessary to determine the spectrum of the linearization of P around (R_s, N_s). The linearization of P is given by the monodromy operator U associated with the linear variational equation around a CW-state as introduced earlier.

To compute the CW-states and the variational equation around a CW-state, we first write the Lang–Kobayashi equations in polar coordinates. If

$$E(t) = R(t)e^{i\varphi(t)}$$

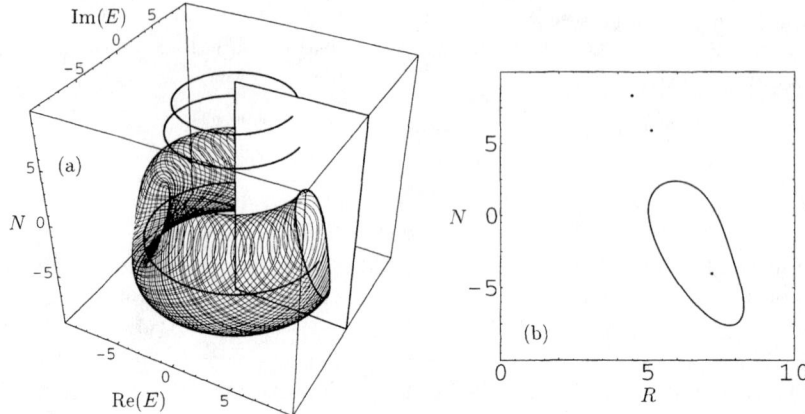

FIGURE 4. Three CW-states and a periodic trajectory. The three CW-states are periodic orbits, while the orbit $(E(t), N(t))$ is a torus (left). The traces of the CW-states are points and the trace of the orbit $(E(t), N(t))$ is a closed curve (right). Reprinted from *Optics Communications* **177**, B. Krauskopf, G.H.M. van Tartwijk, and G.R. Gray, "Symmetry properties of lasers subject to optical fedback", Pages No. 347–353, © 2000, with permission from Elsevier Science.

then the Lang-Kobayashi equations become

$$\frac{dR}{dt}(t) = \frac{1}{2}N(t)R(t) + \kappa R(t-\tau)\cos(\varphi(t) - \varphi(t-\tau) + \omega_0\tau) \tag{32}$$

$$\frac{d\varphi}{dt}(t) = \frac{1}{2}\alpha N(t) + \kappa \frac{R(t-\tau)}{R(t)}\sin(\varphi(t) - \varphi(t-\tau) + \omega_0\tau) \tag{33}$$

$$\frac{dN}{dt}(t) = -\gamma N(t) - A[1 + N(t)](R(t)^2 - 1). \tag{34}$$

Next we substitute a solution of the form

$$R(t) \equiv R_s, \quad \varphi(t) = (\omega_s - \omega_0)t \quad \text{and} \quad N(t) \equiv N_s$$

in equations (32)–(34). This yields

$$\begin{cases} 0 &= \frac{1}{2}N_s R_s + \kappa R_s \cos\omega_s\tau \\ \omega_s - \omega_0 &= \frac{1}{2}\alpha N_s + \kappa \sin\omega_s\tau \\ 0 &= -\gamma N_s - A[1 + N_s](R_s^2 - 1) \end{cases} \tag{35}$$

Since $R_s \neq 0$, these relations can be further simplified to

$$\begin{cases} N_s & = -2\kappa R_s \cos \omega_s \tau \\ \omega_s - \omega_0 & = -\alpha\kappa[\cos \omega_s \tau - 1/\alpha \sin \omega_s \tau] \\ R_s^2 & = 1 + \frac{\gamma N_s}{A(1+N_s)}. \end{cases} \qquad (36)$$

In our computations the parameters α, γ, A and τ are fixed. Given the feedback strength κ, we can use the second equation of (36) to determine the finitely many solutions for ω_s. Once ω_s is known we can use the first and last equation of (36) to solve first for R_s and then for N_s. It is clear that as κ grows, the number of CW-states increases, and for large κ, this number approximates $\kappa\tau\sqrt{1+\alpha^2}/2\pi$; see [14].

Next we compute the linear variational equation around a CW-state. In order to do this we first compute the Jacobian of the vector field that describes the Lang-Kobayashi equation in polar coordinates. If we define

$$F(x, y, z, v, w) = \begin{bmatrix} xz/2 + \kappa v \cos(y - w + \omega_0\tau) \\ \alpha z/2 + \kappa\frac{v}{x}\sin(y - w + \omega_0\tau) \\ -\gamma z - A[1+z](x^2 - 1) \end{bmatrix}, \qquad (37)$$

then the Lang-Kobayashi equation in polar coordinates is given by

$$\frac{d}{dt}\begin{pmatrix} R(t) \\ \varphi(t) \\ N(t) \end{pmatrix} = F(R(t), \varphi(t), N(t), R(t-\tau), \varphi(t-\tau)).$$

The linear variational equation around a CW-state becomes

$$\frac{d}{dt}\begin{pmatrix} R(t) \\ \varphi(t) \\ N(t) \end{pmatrix} = DF(R_s, (\omega_s - \omega_0)t, N_s, R_s, (\omega_s - \omega_0)(t-\tau))\begin{pmatrix} R(t) \\ \varphi(t) \\ N(t) \\ R(t-\tau) \\ \varphi(t-\tau) \end{pmatrix},$$

where DF denotes the Jacobian of F in (37) and is given by

$$DF(x, y, z, v, w) =$$

$$\begin{bmatrix} z/2 & -vs(y-w) & x/2 & c(y-w) & vs(y-w) \\ -\frac{v}{x^2}s(y-w) & \frac{v}{x}c(y-w) & \alpha/2 & \frac{s(y-w)}{x} & -\frac{v}{x}c(y-w) \\ -2A[1+z]x & 0 & -\gamma - A(x^2-1) & 0 & 0 \end{bmatrix},$$

where

$$s(u) = \kappa \sin(u + \omega_0\tau),$$
$$c(u) = \kappa \cos(u + \omega_0\tau).$$

Thus the linear variational equation around a CW-state becomes

$$\frac{dR}{dt}(t) = \frac{1}{2}N_s[R(t) - R(t-\tau)] - \kappa R_s \sin(\omega_s\tau)[\varphi(t) - \varphi(t-\tau)] + \frac{1}{2}R_sN(t)$$
$$\frac{d\varphi}{dt}(t) = -\frac{\kappa}{R_s}\sin(\omega_s\tau)[R(t) - R(t-\tau)] + \kappa\cos(\omega_s\tau)[\varphi(t) - \varphi(t-\tau)] + \frac{1}{2}\alpha N(t)$$
$$\frac{dN}{dt}(t) = -2A[1+N_s]R_sR(t) - \frac{\gamma}{1+N_s}N(t).$$

In general, the linear variational equation around a periodic orbit has periodic coefficients. However, due to the linear form of $\varphi(t)$ for a CW-state, a direct result of the S^1-symmetry [16], the variational equation for CW-states of the Lang-Kobayashi equations is autonomous and therefore has the same structure as a linearization around an equilibrium.

This allows us to study the local stability properties of a CW-state, by studying the roots of the characteristic equation $\det \Delta(z)$, where

$$\Delta(z) = \begin{bmatrix} \lambda - 1/2N_s(1-e^{-\lambda\tau}) & \kappa R_s\sin(\omega_s\tau)(1-e^{-\lambda\tau}) & -1/2R_s \\ \kappa R_s^{-1}\sin(\omega_s\tau)(1-e^{-\lambda\tau}) & \lambda - \kappa\cos(\omega_s\tau)(1-e^{-\lambda\tau}) & -1/2\alpha \\ 2A[1+N_s]R_s & 0 & \lambda + \gamma[1+N_s]^{-1} \end{bmatrix}.$$

A straightforward computation using (36) to simplify the expression for $\det \Delta(z)$ yields

$$\det \Delta(z) = \lambda^3 + \gamma(1+N_s)^{-1}\lambda^2 + A(1+N_s)R_s^2\lambda$$
$$-\kappa^2(1-e^{-\lambda\tau})^2(\lambda + \gamma(1+N_s))$$
$$-\kappa(1-e^{-\lambda\tau})(\alpha\sin(\omega_s\tau) + \cos(\omega_s\tau))A(1+N_s)R_s^2.$$

In the bifurcation analysis the parameters for α, γ, A and τ are fixed and κ is varied. (Recall that R_s, ω_s and N_s vary because they are functions of κ through relations (36)). We are now in a position to use the bifurcation theory for differential delay equations both analytically (see Chapter X-XI of [1]) and numerically [21].

The resulting linear stability analysis should reveal the same results as found in the literature with an indirect approach [22]. Roughly half of the CW-states are saddles (and therefore unstable), while the other half are first stable, but undergo Hopf bifurcations as κ is increased. The saddles have been shown to play a crucial role in the explanation of the phenomenon of Low–Frequency Fluctuations (LFF), see [11,23], while the Hopf-unstable states are responsible for the various routes to chaos that have been found in this system [24–26]. Finding a complete explanation of how LFF develops from simple dynamics when κ is increased is one of the key problems to date.

CONCLUSIONS

We gave an introduction to the mathematical theory of delay equations and showed that differential delay equations can be viewed as transport equations with nonlocal boundary conditions. We have explained that the stability properties of equilibria and periodic orbits can be deduced from properties of the roots of a characteristic equation. We have further discussed the invariant manifold theory to analyse the behaviour of the differential delay equation near hyperbolic equilibria and periodic orbits.

The mathematical theory was then applied to a lasers subject to optical feedback, modeled in a semiclassical approach by single-mode rate equations, in particular, the Lang–Kobayashi equations. We have derived the algebraic relations that determine CW-states of the Lang–Kobayashi equations and computed their linear variational equation. Finally, we arrived at an analytic function of which the roots have to be studied in order to understand the bifurcation diagram and the local behaviour near CW-states.

ACKNOWLEDGEMENTS

The research of S.M. Verduyn Lunel is supported by the Nederlandse Organisatie voor Wetenschappelijk Onderzoek, NWO 600-61-410.

REFERENCES

1. Diekmann, O., van Gils, S.A., Verduyn Lunel, S.M., and Walther, H.O., *Delay Equations: Functional-, Complex-, and Nonlinear Analysis*, Applied Mathematical Sciences **110**, Springer-Verlag, 1995.
2. Mallet-Paret, J., Nussbaum, R., *Ann. Mat. pura appl.* **CXLV**, 33 (1986).
3. Mallet-Paret, J., Nussbaum, R., *SIAM J. Math. Anal.* **20**, 249 (1989).
4. Hale, J.K., and Verduyn Lunel, S.M., *Introduction to Functional Differential Equations*, Applied Mathematical Sciences **99**, Springer-Verlag, 1993.
5. Glass, L., and Mackey, M.C., *Ann. N.Y. Acad. Sci.* **316**, 214 (1979).
6. Hale, J.K, and Sternberg, N., *J. Comp. Phys.* **77**, 221 (1988).
7. Lang R., and Kobayashi, K., *IEEE J. Quantum Electron.* **QE-16**, 347 (1980).
8. Ikeda, K., Otsuka, K., and Matsumoto, K., *Progr. of Theor. Phys. Suppl.* **99**, 295 (1989).
9. Otsuka, K., *Phys. Rev. Lett.* **65**, 329 (1990).
10. Fischer, I., Hess, O., Elsäßer, W., and Göbel, E.O., *Phys. Rev. Lett.* **73**, 2188 (1994).
11. Fischer, I., van Tartwijk, G.H.M., Levine, A.M., Elsäßer, W., Göbel, E.O., and Lenstra, D., *Phys. Rev. Lett.* **76**, 220 (1996).
12. Weiss, C.O., and Vilaseca, R., *Dynamics of Lasers*, Weinheim, 1991.
13. Haken, H., *Light Vol. 2*, North-Holland, 1985.
14. van Tartwijk, G.H.M., and Agrawal, G.P., *Progr. Quantum Electron.* **22**, 43 (1998).

15. van Tartwijk, G.H.M., and Lenstra, D., *Quantum Semiclass. Opt.* **7**, 87 (1995).
16. Krauskopf, B., van Tartwijk, G.H.M., and Gray, G.R., *Opt. Comm.* **177**, 347 (2000).
17. Agrawal, G.P., and Dutta, N.K., *Long-wavelength Semiconductor Lasers*, 2nd ed. Van Nostrand Reinhold, 1993.
18. Petermann, K., *Laser Diode Modulation and Noise*, Kluwer Academic, 1988.
19. Agrawal, G.P. (editor), *Semiconductor Lasers, Past, Present, and Future*, AIP Press, 1995.
20. Masoller, C., and Abraham, N.B., *Phys. Rev. A* **57**, 1313 (1998).
21. Engelborghs, K., *Numerical Bifurcation Analysis of Delay Differential Equations*, PhD thesis, Catholic University of Leuven, 2000.
22. Acket, G.A., Lenstra, D., den Boef, A.J., and Verbeek, B.H., *IEEE J. Quantum Electron.* **QE-20**, 1163 (1984).
23. Sano, T., *Phys. Rev. A* **50**, 2719 (1994).
24. Lenstra, D., Verbeek, B.H., and den Boef, A.J., *IEEE J. Quantum Electron.* **QE-21**, 674 (1985).
25. Mørk, J., Mark, J., and Tromborg, B., *Phys. Rev. Lett.* **65**, 1999 (1990).
26. Mørk, J., Tromborg, B., and Mark, J., *IEEE J. Quantum Electron.* **28**, 93 (1992).

Theory of Delayed Optical Feedback in Lasers

Daan Lenstra and Mirvais Yousefi

Division of Physics and Astronomy & LaserCentreVU
Vrije Universiteit, De Boelelaan 1081
1081 HV Amsterdam, The Netherlands

Abstract. A tutorial introduction to the dynamical behavior of lasers is presented. Simple physical pictures explain the essential ingredients that determine the dynamical properties. The relevant dynamical equations are derived for a laser with delayed external optical feedback, possibly spectrally filtered before re-entry into the laser. Focusing on semiconductor lasers the steady-state solutions are derived and discussed as to their stability, whereafter the various different types of dynamics predicted by the model are presented, highlighting Low-Frequency Fluctuations.

BASIC LASER CONCEPTS

A laser is a device that through a process of stimulated emission produces light of high spectral purity, high intensity and high directionality. The first-mentioned quality, high spectral purity, implies the well-known coherence of laser light, which is measured in terms of length typically on a scale ranging from cm's to km's or more. In terms of bandwidth it is typically 10 GHz or below, or in terms of wavelength, 0.3 nm or less. It is of course the combination of these three qualities that characterizes laser light and it is the process of stimulated emission that makes this combination possible.

Stimulated emission is the reverse process of absorption, as is illustrated in Fig. 1. Here, in case of absorption, light described by an N photon state (since photons are bosons, N identical photons can occupy the same quantum state) is incident on a two-level atom in the ground state. After the interaction, the atom is in the excited state, while the light is in an $N - 1$ photon state. In the time-reversed replica of this process the atom, initially in the excited state, adds to the optical field a clone of the incident identical photons. Provided that sufficiently many excited atoms are available, a snowball effect will occur resulting in amplification and phase preservation of the original incoming optical field.

The relevant ingredients that define a laser device are the amplifying medium where the stimulated emission takes place and the resonator which serves as a

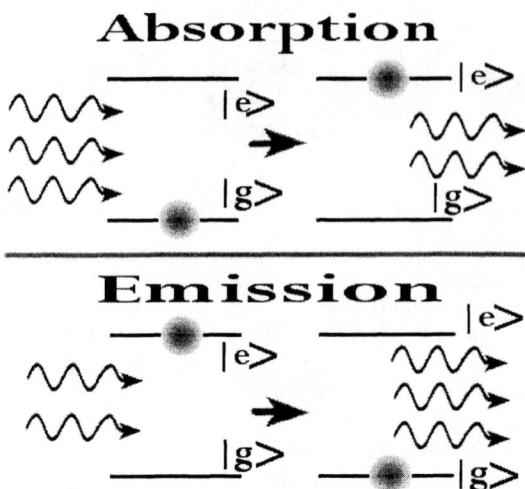

FIGURE 1. Sketch of absorption (top) and emission (bottom) process. In this example $N = 3$; see text.

frequency selective element. The amplifying properties of the active medium are characterized by the gain, a complex-valued quantity, where the real part describes the amplification of the amplitude while the imaginary part affects the phase of a traversing electromagnetic field. The resonator provides a mechanism of direct feedback through multiple transits of the light before coupling out of the device, thus sustaining frequencies of operation for which constructive roundtrip interference occurs and suppressing all other frequencies. Relevant time scales involved are the field decay rate or inverse photon life time, i.e. the rate at which photons will escape from the device, and the roundtrip time which defines the frequencies of operation.

Taking a closer look at the gain as provided by a stimulated emission medium, the real part of the gain as function of field frequency will generally look as sketched in Fig. 2(a), with the imaginary part as in Fig. 2(b). The real and imaginary parts are related through the Kramers-Kronig causality relations [1]. This usually implies that the frequency where the gain is maximal (denoted by ω_0 in Fig. 2(a)) more or less coincides with the zero crossing of the imaginary part, meaning that if the laser chooses to operate close to maximum gain, as it usually will, the operation frequency will not change significantly due to imaginary gain induced refractive index changes. In some gain media, however, the asymmetry of the gain profile brings about that the gain maximum is not at the zero dispersion point, in which case substantial frequency shifts and self-phase modulation effects like chirping may occur. The latter is especially the case in semiconductor lasers, as will be discussed

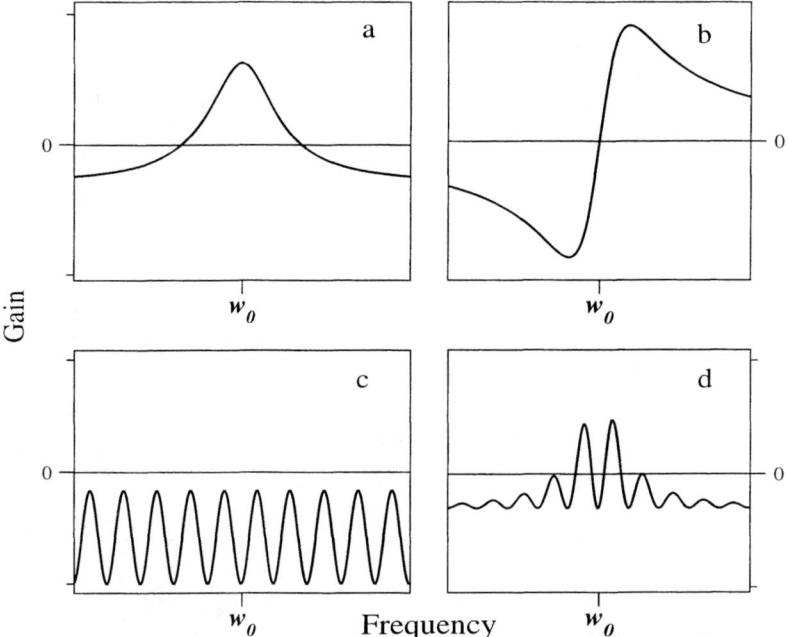

FIGURE 2. Sketch of the real part of the gain (a), the imaginary part of the gain (b), the laser cavity mode structure (c) and the final combined mode structure (d)

in more detail later on.

It should be realized that the gain function depends proportionally on how hard one pumps the medium, i.e. on the amount of sustainable inverted atoms in the medium. Furthermore, the roundtrip resonance condition implies that the effective gain will be a convoluted version of the curve in Fig. 2(a) with the roundtrip resonance transmission function sketched in Fig. 2(c), resulting in a net gain function sketched in Fig. 2(d). This illustrates that, depending on the width of the gain profile compared to the frequency spacing of roundtrip resonances, a certain number of frequencies of operation would be possible (two in the example of Fig. 2(d)), although in general the laser will just operate in the mode whose frequency has the highest gain.

LASER DYNAMICS

Quite generally, the dynamical state of a free-running laser is described by three physical time dependent quantities, the optical field strength $\mathcal{E}$, the polarization field strength $\mathcal{P}$ and the inversion $\mathcal{N}$. The polarization $\mathcal{P}$ is the induced dipole moment density in the medium, which expresses the gain and dispersion properties of the medium at inversion $\mathcal{N}$. It should be realized that these quantities are also dependent on space coordinates. However, in order not to complicate this presentation of concepts we will ignore the spatio-temporal aspects here completely and fully concentrate on the time dependence.

Representing the optical fields as

$$\mathcal{E}(t) = E(t)e^{i\omega_0 t} + \text{c.c.},\tag{1a}$$
$$\mathcal{P}(t) = P(t)e^{i\omega_0 t} + \text{c.c.},\tag{1b}$$

where ω_0 denotes a fixed optical frequency close to the operation frequency and E, P are weakly time-dependent complex amplitudes, the laser equations of motion can generally be cast in the form [2,3]:

$$\left(\tfrac{d}{dt} + \lambda_E - i\omega_E\right) E = -iP,\tag{2a}$$
$$\left(\tfrac{d}{dt} + \lambda_P - i\omega_P\right) P = iNE,\tag{2b}$$
$$\left(\tfrac{d}{dt} + \lambda_N\right) N = J - i(EP^* - E^*P).\tag{2c}$$

The equations for E and P correspond to damped harmonic oscillators, each having its own resonance frequency (with respect to ω_0) ω_E and ω_P and damping rates λ_E and λ_P. The electric field oscillator is driven by the polarization, while the polarization oscillator in turn is driven by the product of inversion and field. The inversion is created by a certain external pump process, expressed as a pumping rate J. The inversion tends to decay under the influence of spontaneous processes, like spontaneous emission and nonradiative transitions. This is described by the decay rate λ_N. The last two terms in the right hand side of (2c) describe the stimulated emission or absorption. Equation (2a) describes the field dynamics and Eqs. (2b,2c) the gain dynamics.

Since E and P are complex valued quantities, the system Eqs. (2) consists of 5 coupled nonlinear first-order differential equations. As long as propagation effects within the active medium can be neglected this set of coupled equations (2) should give an adequate description, at least qualitatively. On the other hand, if propagation effects are to be included one ends up with partial differential equations, as is in fact done in the chapters by *Hess*, *McInerney at al.* and *Moloney*.

It turns out that the dynamical behavior predicted by (2) is crucially dependent on the relative magnitude of the three inverse decay time scales λ_E, λ_P and λ_N. If the photon decay rate is much smaller than both the polarization and inversion decay rates, i.e. $\lambda_P, \lambda_N \gg \lambda_E$, the fast dynamical response of N and P will make them adiabatic followers of E, so that they can be eliminated. In that case the system can be adequately described by the single nonlinear first-order differential equation for E

$$\left(\tfrac{d}{dt} + \lambda_E - i\omega_E\right) E = \tfrac{1}{2}(1+i\alpha)\frac{JE}{I_{\text{sat}}+|E|^2},\tag{3}$$

where I_{sat} is a saturation intensity whose value is proportional to the product $\lambda_P \lambda_N (1+\alpha^2)$ and $\alpha = \frac{\omega_P}{\lambda_P}$ is a detuning parameter which quantifies the frequency shift and self-phase modulation effects mentioned in the first section. Lasers that fall into this category are referred to as class A lasers. Most gas lasers, notably He-Ne lasers, belong to this class.

Similarly, class B lasers are lasers for which $\lambda_P \gg \lambda_N, \lambda_E$. For this class P is the adiabatic follower and can be eliminated from (2), resulting in the equations

$$\left(\tfrac{d}{dt} + \lambda_N\right) N = J - \xi N |E|^2 , \tag{4a}$$

$$\left(\tfrac{d}{dt} + \lambda_E - i\omega_E\right) E = \tfrac{1}{2}\xi(1+i\alpha)NE , \tag{4b}$$

where ξ is a gain coefficient proportional to $\frac{\lambda_P}{(\lambda_P^2 + \omega_P^2)}$. Semiconductor lasers, solid state lasers and CO_2 lasers fall into class B. In the last sections we will focus entirely on semiconductor lasers and discuss different dynamical behaviors, some of which are representative for all class B lasers, although the typical absolute time scales will vary from one laser to the other.

Lasers that are not class A or B are automatically class C. The classification of lasers into three groups makes sense from a dynamical point of view. The dynamical complexity increases from A to C. Class A lasers exhibit stable behavior. Depending on whether the pumping strength J is below or above a certain threshold value determined by the decay rate λ_E, the power $I = EE^*$ reaches the steady-state value of 0 or a nonzero value, respectively. In the latter case, there are two steady-state equilibria, the off-state with zero power and the on-state with nonzero power. Class B lasers exhibit richer dynamical behavior due to an intrinsic resonance which leads to oscillatory damped evolution to the steady-state, as will be discussed in more detail below. Since the class-B equations are basically a two-dimensional model, class-B lasers can not exhibit torus bifurcations or chaos at least when isolated from external perturbations. Any form of external driving, however, will increase the dimensionality of the system to at least three, in which case instabilities associated with the above-mentioned bifurcations can readily occur. One such external driving mechanism is delayed optical feedback, the central theme of this chapter.

Class C lasers are described by a system of ordinary differential equations with dimension 3 or more. This system may show very rich nonlinear dynamics, including period doubling and torus bifurcations into chaos all by itself even without external influences.

As mentioned, class B lasers exhibit an intrinsic damped oscillation. Physically it corresponds to a damped oscillatory exchange of excitation between the intensity of the optical field and the inversion. This can be made clear by differentiating Eq. (4a) with respect to time and substituting the equation for $\tfrac{d}{dt}I$ where $I = EE^*$ derived from (4b), which yields

$$\ddot{n} + 2\lambda_R \dot{n} + \omega_R^2 n = \dot{J} - \xi^2 I n^2 , \tag{5a}$$

$$\lambda_R = \tfrac{1}{2}(\lambda_N + \xi I) , \tag{5b}$$

$$\omega_R^2 = 2\xi\lambda_E I , \tag{5c}$$

where $n \equiv N - 2\lambda_E/\xi$ denotes the deviation of N from its equilibrium value.

We recognize an equation for a damped (an)harmonic oscillator, possibly driven when $\dot{J} \neq 0$, where both the damping constant and the resonance frequency are intensity dependent. It is not surprising from the physical point of view that when class B lasers are subject to external influences (feedback, modulation, injection) the relaxation oscillation can be sustained with finite amplitude. Indeed, in many cases such as discussed in the last two sections of this chapter, or for instance in the chapters by *Erneux*, *Gavrieledes* and *Krauskopf*, Hopf bifurcations are discussed which do correspond to the sustained excitation of the relaxation oscillation. In combination with the self-phase or frequency modulation effect as described by α in (4b), relaxation oscillations lead to chirp and, in case of external driving, to enhanced tendency for instabilities, or phrased more positively, to very rich nonlinear dynamics.

We can take a closer look at the phase modulation effect by re-expressing Eq. (4b) in terms of the intensity and phase of the electric field,

$$E = \sqrt{I} e^{i\phi}. \tag{6}$$

This yields

$$\dot{I} = \xi n I, \tag{7a}$$

$$\dot{\phi} = \omega_E + \alpha \lambda_E \frac{1}{2} \xi \alpha n, \tag{7b}$$

$$n = N - 2\frac{\lambda_E}{\xi}. \tag{7c}$$

It follows straightforwardly from these equations that any deviation from steady-state ($n = 0$) causes a frequency shift $\xi \alpha n / 2$. For $\alpha \neq 0$, this implies chirping of the laser when modulating the pump strength, frequency broadening due to pick-up of noise and, as said before, destabilization in general. In most types of lasers the α factor is rather small. For the simple Lorentzian gain profile as assumed in (2b) α is just the relative detuning parameter ω_P/λ_P. In those cases values of $|\alpha|$ exceeding 0.1 hardly occur. Semiconductor lasers are a remarkable exception in that the gain profile is by far not Lorentzian, but has the typical asymmetric profile as can be found, for instance, in Ref. [4]. This generally leads to much larger values of α ranging from 1 to 15.

DELAYED OPTICAL FEEDBACK

In this section we will derive the modified rate equation that describes the laser when part of its output intensity is coupled back into the laser cavity after some external delay time τ. The relevant configuration is sketched in Fig. 3. Here, the laser is assumed to have two mirrors, one with reflection coefficient r_0, while the outcoupling mirror has reflection coefficient r_1. The external mirror has reflection

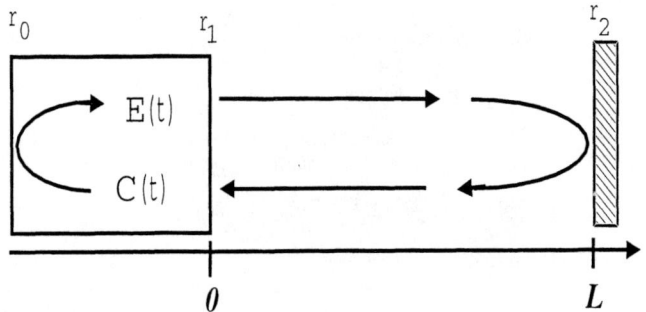

FIGURE 3. Sketch of laser diode with external optical feedback.

coefficient r_2. The field amplitude of the right-travelling field just before leaving the laser is $E(t)$, while the left-travelling wave just after reflection from r_1 is $C(t)$.

Balancing the fields E and C, taking into account the direct reflection r_1, as well as two transmissions through r_1 and one external reflection r_2 with total delay time τ, we readily find

$$C(t) = r_1 E(t) + (1 - r_1^2) r_2 E(t - \tau) e^{-i\omega_0 \tau}. \tag{8}$$

On the other hand, the field E can be related to C through one internal transit, that is,

$$E(t + \tau_{\text{in}}) = r_0 G C(t), \tag{9}$$

where G is the round trip gain. Combining (8) with (9) we can write

$$\dot{E}(t) = \frac{E(t + \tau_{\text{in}}) - E(t)}{\tau_{\text{in}}} = \frac{r_0 r_1 G - 1}{\tau_{\text{in}}} E(t) + \frac{r_0 r_1 G (1 - r_1^2) r_2}{r_1 \tau_{\text{in}}} E(t - \tau) e^{-i\omega_0 \tau}. \tag{10}$$

Now realizing that the first term in the right-hand side of (10) gives just the solitary laser evolution and that in the second term the combination $r_0 r_1 G$ will be very close to unity, at least when the laser operates in a single-longitudinal mode, we can write to good approximation

$$\dot{E}(t) = \left.\frac{dE(t)}{dt}\right|_{\text{SolitaryLaser}} + \gamma E(t - \tau) e^{-i\omega_0 \tau}, \tag{11}$$

where

$$\gamma \equiv \frac{(1 - r_1^2) r_2}{r_1 \tau_{\text{in}}} \tag{12}$$

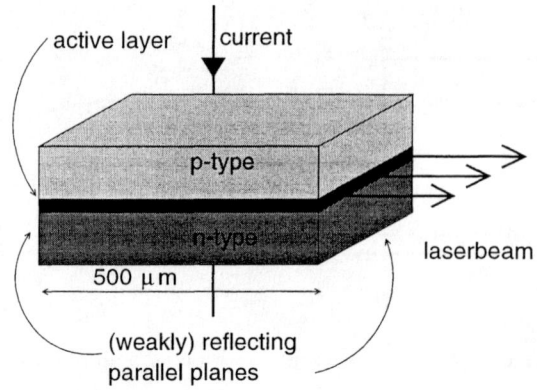

FIGURE 4. Sketch of an Edge Emitting diode Laser (EEL).

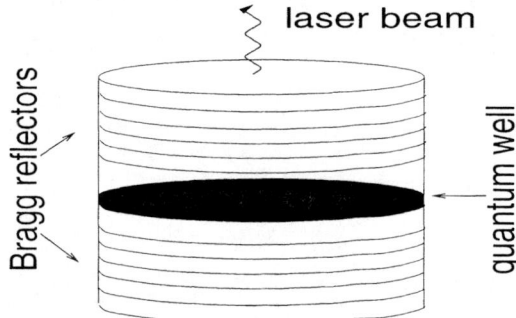

FIGURE 5. Sketch of a Vertical Cavity Surface Emitting Laser diode (VCSEL).

is the strength of the feedback expressed as a rate. Clearly, the feedback introduces two new time scales into the problem: the external delay time τ and the inverse feedback rate γ^{-1}. The delay actually makes the system infinite dimensional. Instead of specifying a finite number of initial values, one has to specify a continous function on the time interval of length τ in order to uniquely define the solution of the delay-differential equation (11); see also the chapter by Verduyn Lunel & Krauskopf. Equation (11) is generally valid for any laser with external optical feedback provided

i) multiple reflections in the external system can be neglected, i.e. $r_1 r_2 \lesssim 0.1$;

ii) there is no significant change of E during one internal round trip, i.e. no longitudinal multimode operation.

One can easily extend (11) to include multiple reflections in the external system as well, see for instance [5]. The single-mode requirement, however, is more of a fundamental restriction as to the validity of (11). In view of the above-given derivation it is clear that through the roundtrip gain G an additional time dependence enters the problem in such a way that the feedback rate is enhanced during periods of high gain and reduced during high loss.

In many situations the external feedback light has undergone some spectral filtering before it re-enters the laser cavity. This is for instance the case when a grating is used as an external reflector, or when a Fabry-Perot cavity is placed in the external system. In those cases the feedback term in the rate equation will be more complicated than in (11) and we then formulate the problem by adding an additional rate equation which accounts for the time evolution of the field in the external system. In cases when the spectral filtering in the external system can be approximated by a Lorentzian, we obtain [6]

$$\dot{E}(t) = \left.\frac{dE(t)}{dt}\right|_{\text{SolitaryLaser}} + \gamma F(t), \tag{13}$$

$$\dot{F}(t) = \lambda E(t-\tau)e^{-i\omega_0 \tau} + (i\omega_m - \lambda)F(t), \tag{14}$$

where γ is given by (12) and where (14) describes a damped oscillator with damping rate λ and detuning frequency (with respect to solitary laser frequency) ω_m driven by the field emitted by the laser at time τ earlier. A more detailed derivation, including an expression for a general type of (linear) filtering can be found in [6].

SEMICONDUCTOR LASERS WITH FEEDBACK

From here on we will focus entirely on semiconductor lasers and formulate the rate equations for them. As said before, semiconductor lasers are class B lasers and exhibit interesting nonlinear dynamical operation only after applying external disturbances. Until not so long ago this "interesting" dynamical behavior was considered as instabilities, a nuisance that should be avoided by proper design. This is precisely the kind of approach followed in the chapter by *McInerney et al.* Other approaches try to take advantage of the peculiar nonlinear dynamical behavior. We mention the possible application of chaotic laser light for chaotic communication and encryption [see the chapters by *Mirasso* and *Roy*] or the exploitation of the various multistabilities for optical switches. In the present section we will analyze the nonlinear dynamics of a semiconductor laser with delayed optical feedback from a purely curiosity driven perspective.

Semiconductor lasers are found basically in two different designs, both of which are fabricated using multi layer epitaxial growth techniques. In the more conventional, so-called Edge-Emitting semiconductor diode Laser (EEL) no special reflecting elements at the facets of the device are necessary. The light travels in the lateral direction, see Fig. 4, following an active layer where electrons and holes can recombine (bulk or quantum well active region). Since the net gain that can be realized is on the order of 25 cm^{-1}, and the refractive index of the material is 3.6, a chip of length 0.3 mm already provides a gain of 5.5 per roundtrip. Since the Fresnel reflectivity associated with the crystal-to-air transition equals 0.32, no special reflecting elements at the facets of the device are necessary. With an internal roundtrip time of 7 ps, the feedback rate (12) can be expressed as

$$\gamma_{EEL} = 180 r_2 \text{ GHz} \qquad (15)$$

with r_2 the external reflection coefficient.

In the other semiconductor laser design the laser light travels in the direction perpendicular to the layers. Since an active layer is traversed only once per single pass, the gain per roundtrip is a factor 1000 smaller than for an EEL. Therefore, one definitely needs highly reflecting mirrors on top and bottom in order to have the light make enough roundtrips before leaving the cavity. These highly reflecting mirrors are realized by growing stacks of dielectric multi layers which define Bragg reflectors on top and bottom. These lasers are referred to as Vertical Cavity Surface Emitting Lasers (VCSEL); see Fig. 5. With a reflectivity of 0.998 and an internal roundtrip time of 10 fs, we estimate the feedback rate for these lasers as

$$\gamma_{VCSEL} = 200 r_2 \text{ GHz}, \qquad (16)$$

that is, not very different from the EEL case.

The two different types of lasers are described by the same rate equations, at least in the approximation of vanishing transverse effects and neglect of polarization effects. The first mentioned approximation becomes incorrect for VCSELS with larger width, since then typical broad-area instabilities like dynamic-filamentation and self-focusing will occur (this topic is in fact studied in the chapters by *Hess, McInerney at al., Moloney,* and *Rorison*), but for sufficiently narrow VCSELS it should be correct. The second approximation means that we disregard the vectorial nature of the optical field, which is correct in devices where the light has one well-defined polarization. This is usually true for EELs where the combination of wave guiding and gain favors TE polarization. For VCSELs it may be less

TABLE 1. Representitive values of the parameters for COF and FOF.

Symbol	Value	Description
α	$1 \leftrightarrow 15$	Linewidth enhancement factor
γ	$0 \leftrightarrow 200$ GHz	Feedback rate
ξ	$2 \cdot 10^{-4}\ s^{-1}$	Differential gain
τ	$1 \leftrightarrow 50\ ns$	External delay time
ω_0	$\sim 10^{15}\ Hz$	Solitary laser frequency
Γ_0	$4 \cdot 10^{11}\ s^{-1}$	Inverse photon life time
T_1	$14 \cdot 10^{-10}\ s$	Carrier life time.
I_0	$\sim 10^5$	Solitary laser photon number; $I_0 = (J - J_{Thr})/\Gamma_0$
ω_m	$-10 \leftrightarrow 10\ GHz$	Detuning between the filter and the solitary laser frequency
λ	$0 \leftrightarrow 100\ GHz$	Filter width (HWHM)

often true, especially if the configuration symmetry gives rise to near degeneracy of the polarization (see the chapter by *Hess*). In any case, the rate equations for semiconductor diode lasers with external (filtered) optical feedback are formulated as

$$\dot{E}(t) = \frac{1}{2}(1 + i\alpha)\xi n(t) E(t) + \gamma F(t), \tag{17}$$

$$\dot{n}(t) = J - J_{\text{Thr}} - \frac{n(t)}{T_1} - (\Gamma_0 + \xi n(t))|E(t)|^2, \tag{18}$$

$$\dot{F}(t) = \lambda E(t - \tau) e^{-i\omega_0 \tau} + (i\omega_m - \lambda) F(t). \tag{19}$$

Here, the role of the inversion is taken by the electron-hole pair number $n(t)$ measured with respect to the laser threshold value. Furthermore, J is the electric pump current and J_{Thr} its value at threshold; T_1 the life time for spontaneous recombination of electrons and holes, Γ_0 the photon escape rate from the cavity, and ω_0 is the solitary laser (i.e. the free-running laser) angular frequency. In Table 1 the semiconductor laser parameters are summarized and their typical values given.

The rate equations are formulated for the case of a Lorentzian filter in the external system. Conventional feedback, that is, without filtering, is obtained as the limiting case $\lambda \to \infty$. In that case we can write the solution of (19) as

$$F(t) = E(t - \tau) e^{-i\omega_0 \tau} \tag{20}$$

and Eqs. (17) and (18) assume with (20) the form known as Lang-Kobayashi equations for semiconductor lasers with conventional feedback [7].

It should be noted that with external power reflectivities as small as 0.01% (i.e. $r_2 = 0.01$) the feedback rate according to (15) and (16) already amounts to $2\ GHz$. This should be compared with the relaxation oscillation frequency ω_R and damping rate λ_R, which, for the laser described by (17) and (18), are given by

$$\omega_R = \sqrt{\xi \Gamma_0 I}, \tag{21}$$

$$\lambda_R = \frac{1}{2}\left(\frac{1}{T_1} + \xi I\right), \tag{22}$$

where $I = |E|^2$ is the intensity of the laser light measured in photon number. With the values given in Table 1 we obtain relaxation oscillation frequencies and damping rates typically in the GHz range. This is why a power reflectivity of less than 0.01%, although corresponding to weak optical feedback, is already powerful enough to sustain a permanent relaxation oscillation. Since this undamping of the relaxation oscillation is in many cases the first step towards destabilization, the above order-of-magnitude estimate explains the unusually high sensitivity of semiconductor lasers to external feedback.

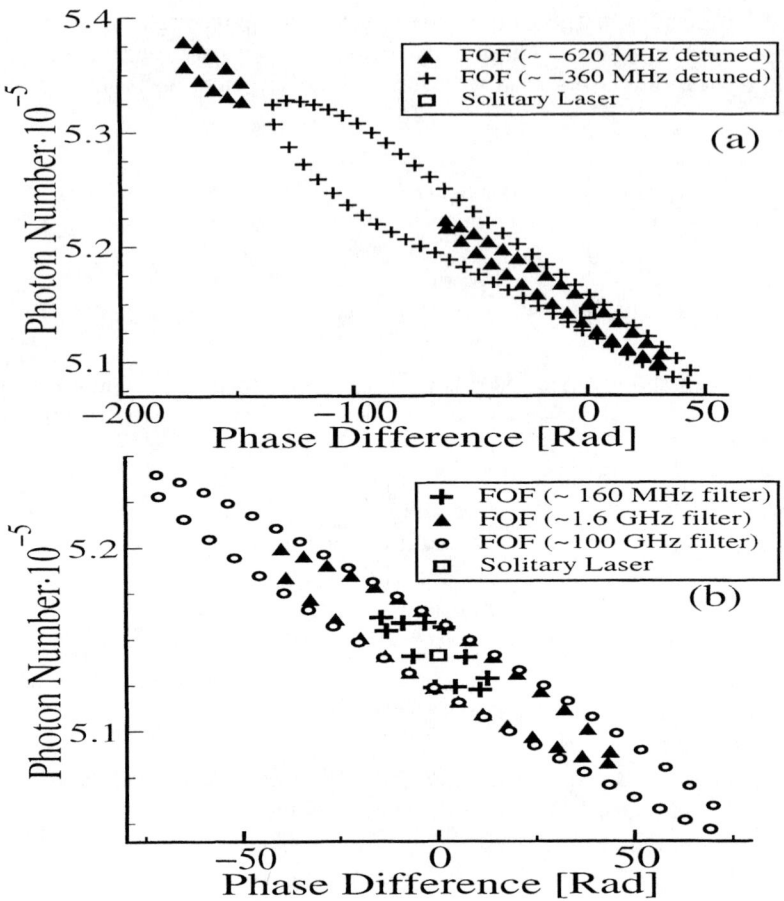

FIGURE 6. Fixed points (η_s, I_s) in case of filtered feedback. Panel (a) shows the effect of varying the detuning. The islands (triangles) are formed when the filter is detuned too far away from the solitary laser point. The crosses show intermediate detuning. Panel (b) shows the fixed points for a few different filter widths at zero detuning. When the filter width is $\sim 100\ GHz$ the fixed points coincide with those of conventional optical feedback, and they have been included for reference.

STEADY STATE ANALYSIS

We will now find the steady-state continuous wave type solutions to Eqs. (17), (18) and (19). These correspond to the states of highly coherent, single-frequency operation of the laser. We therefore seek solutions of the form

$$E(t) = E_s e^{-i\omega_s t}, \tag{23a}$$
$$F(t) = F_s e^{-i(\omega_s t + \phi_s)}, \tag{23b}$$
$$n(t) = n_s, \tag{23c}$$

where E_s, F_s, n_s, ω_s and ϕ_s are all real constants independent of time ($E_s, F_s > 0$). We will refer to these solutions as fixed points, because the phase, intensity and inversion have fixed values in spite of the trivial dynamics involved by the exponential time factor. Introducing the quantity $\eta \equiv \omega_s \tau$ (η can be seen as an external round trip phase difference), the fixed points correspond to the solutions of (see [6] for a precise derivation)

$$\eta = -\frac{C}{\sqrt{1 + \left(\frac{\omega_m - \eta/\tau}{\lambda}\right)^2}} \sin\left[\eta + \omega_0 \tau + \arctan(\alpha) - \arctan\left(\frac{\omega_m - \eta/\tau}{\lambda}\right)\right]. \tag{24}$$

The transcendental equation (24) has at least one solution, but depending on the magnitude of the effective feedback strength $C = \gamma \tau \sqrt{1 + \alpha^2}$ and the actual filter parameters λ, ω_m, multiple solutions may be expected. For each η solving (24) the corresponding values for the amplitudes, phase and inversion are given by

$$I_s = |E_s|^2 = \frac{T_1(J - J_{\text{Thr}}) - n_s}{T_1(\Gamma_0 + \xi n_s)}, \tag{25}$$

$$n_s = \frac{2\gamma}{\xi \sqrt{1 + \left(\frac{\omega_m - \eta/\tau}{\lambda}\right)^2}} \cos\left[\eta + \omega_0 \tau - \arctan\left(\frac{\omega_m - \eta/\tau}{\lambda}\right)\right], \tag{26}$$

$$F_s = \frac{E_s}{\sqrt{1 + \left(\frac{\omega_m - \eta/\tau}{\lambda}\right)^2}}, \tag{27}$$

$$\phi = -\eta - \omega_0 \tau + \arctan\left(\frac{\omega_m - \eta/\tau}{\lambda}\right). \tag{28}$$

A most relevant question is whether or not a fixed point solution corresponds to a stable state of operation. On increasing the feedback rate γ, the fixed points are usually created in pairs in saddle-node bifurcations. This already suggests that half of the fixed points are saddle points and, hence, do not correspond to stable states of operation. Physically, these points correspond to maximally *destructive* interference between the light inside the laser and the externally delayed light. The other points correspond to maximally *constructive* interference and, hence, can be identified as the external modes. Similarly, the saddle points can be identified as the external anti-modes.

Figure 6 shows the modes and the anti-modes for some different values of the filter parameters λ and ω_m. In Fig. 6(a) the detuning ω_m has been varied and the filterwidth λ has been fixed to 1 GHz while $C = 178$. When the detuning is 0 the ellipse resembles that of conventional optical feedback (COF); see Fig. 7. Moving

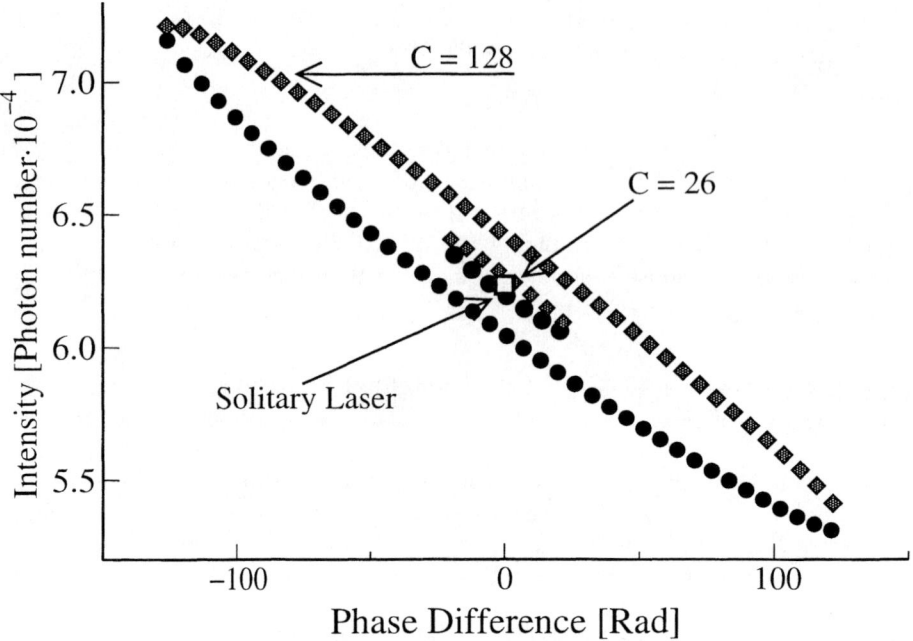

FIGURE 7. The fixed points of the COF system (29)-(31). The fixed points on the edges of the ellipse are created in saddle-node bifurcations. Compare the $C = 26$ with the larger C case. The solitary laser point has been indicated for reference.

the detuning in either direction changes the shape of the ellipse and imposes the filter profile on top of the COF-ellipse. This results in the suppression of some of the modes. Moving the filter even further away splits the ellipse in two parts. Only the fixed points at the center of the filter and a few around the solitary laser frequency survive, the rest are suppressed by the filter. The fixed points of the system are re-arranged in phase space in this manner. Figure 6(b) shows the effects of varying the filter width. Here the filter detuning has been fixed to 0 and $C = 72$. For a very broad filter the COF-ellipse is recovered. On decreasing the filter width the fixed points at the edge of the filter are annihilated in saddle-node bifurcations, resulting in a decreased number of fixed point solutions for the system. The effective feedback to the system can be controlled in this manner.

It turns out, for instance as a result of integration of the full equations (17-19), that even the fixed point solutions corresponding to modes may in fact not be stable. Instead, a large flora of dynamical behaviors is observed ranging from simple limit cycles up to wild chaotic fluctuations. In order to discuss these complicated dynamical behaviors we will concentrate first on COF, i.e. without the filtering and, where adequate, add some comments about the particular role of filtering.

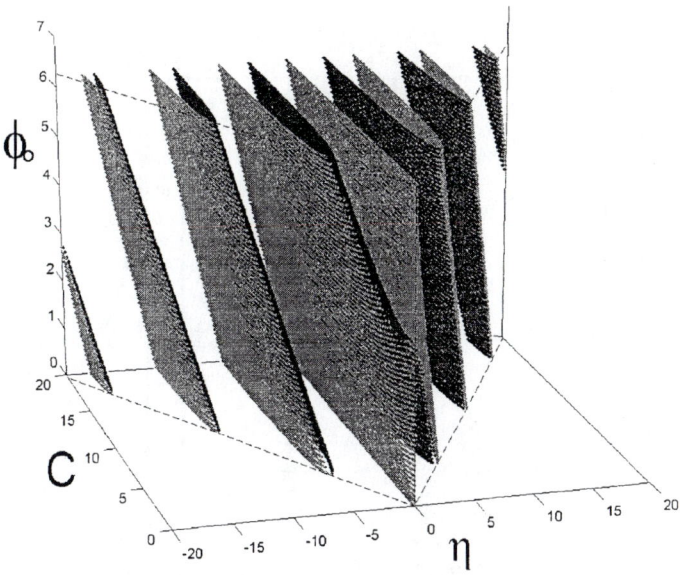

FIGURE 8. The fixed points of the COF system (29)-(31) in (η, C, ϕ_0)-space. The dark sheets represent the modes while the light sheets represent the anti-modes. Note the pitchfork bifurcation at $\phi_0 = \pi$; see also Fig. 9.

The special case of COF can be derived from the general result by taking the limit of an infinitely broad filter, i.e. $\lambda \to \infty$. This yields the fixed points equation

$$\eta = -C \sin\left[\eta + \omega_0 \tau + \arctan(\alpha)\right], \tag{29}$$

$$n_s = \frac{2\gamma}{\xi} \cos\left[\eta + \omega_0 \tau\right], \tag{30}$$

$$I_s = |E_s|^2 = \frac{J - J_{\text{Thr}} + \frac{2\gamma}{\xi T_1} \cos(\omega_0 \tau + \eta)}{\Gamma_0 - 2\gamma \cos(\omega_0 \tau + \eta)}. \tag{31}$$

Here, the parameter C, as before, is the effective feedback strength,

$$C = \gamma \tau \sqrt{1 + \alpha^2}. \tag{32}$$

The number of fixed points is approximately equal to $2C/\pi$. In the (η, n) plane, these points lie on an ellipse. The role of α is to make the ellipse lean in such a way that for positive α (as is commonly the case in semiconductor lasers) the positive η's have larger n, while the negative η's have smaller n. Since smallest n implies

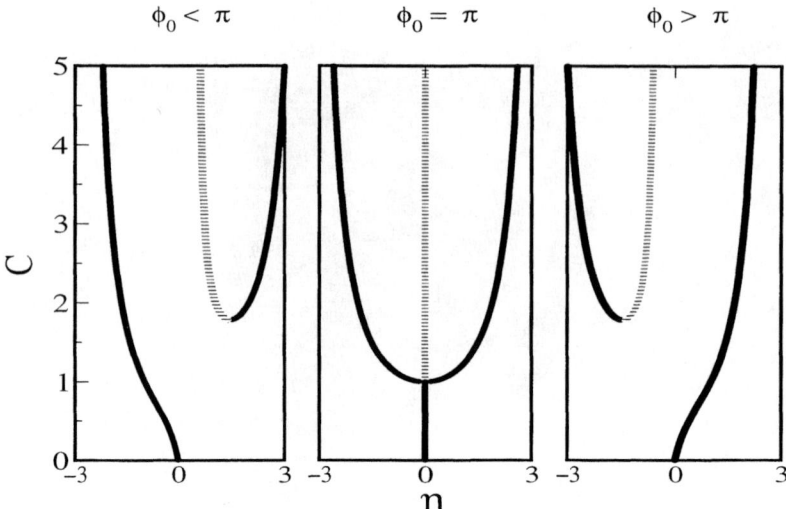

FIGURE 9. The pitchfork bifurcation at $\phi_0 = \pi$.

largest I, the points with maximum intensity are found close to the most negative η's.

Figure 7 shows the fixed points in the (η, I) plane for a case with $\alpha = 5, \tau = 5, \omega_0 \tau = 0$. Here, the modes are indicated by diamonds and the anti-modes by circles. The open box indicates the solitary laser point of operation ($\eta = 0$). The fixed points lie on an ellipse-like curve, which becomes more and more banana-shaped as the solitary laser operates more close to threshold. If α would be 0, the (η, n) ellipse would have its principle axes along η and n; in that case the maximum intensity mode would be close to $\eta = 0$, i.e. close to the solitary laser frequency.

Apart from C, also the phase combination

$$\phi_0 = \omega_0 \tau + \arctan(\alpha) \qquad (33)$$

is a parameter that influences the fixed point solutions of (29). In Fig. 8 the continuum of fixed points in the three-dimensional (η, C, ϕ_0) space is presented for C-values up to 20, the dark sheets represent the modes while the light sheets represent the anti-modes. It is clearly seen that at fixed ϕ_0, on increasing C, the fixed points are created in saddle-node bifurcations. Only in the highly symmetric case $\phi_0 = \pi$, there is a pitch-fork bifurcation at $C = 1$; see Fig. 9.

A direct experimental verification of the existence of external modes is presented in Fig.10, where the measured external feedback light intensity is plotted versus the solitary laser frequency, where the latter is varied by varying the pump current (upper triangular curves). Increasing the pump corresponds to decreasing the frequency. By including a filter with a known profile in the external cavity one

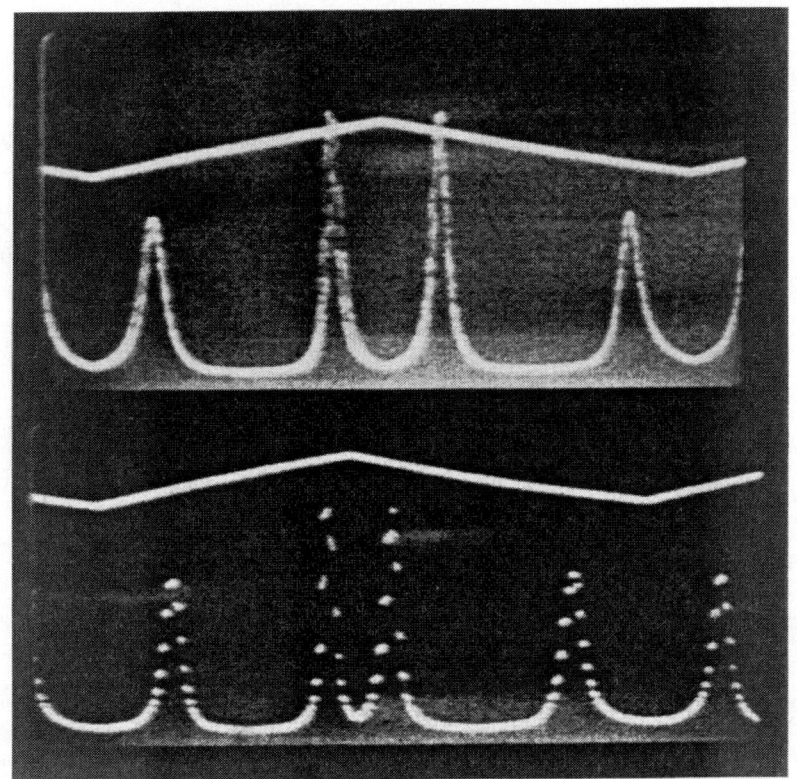

FIGURE 10. Experimental demonstration of the external cavity modes. In the top figure there is no feedback, therefore the upper trace is the filter profile, i.e. the power transmitted through the filter versus frequency. In the bottom figure the feedback strength is $C \simeq 20$ and the external cavity modes are clearly visible. (Figure courtesy of A.P.A. Fischer and O.K. Anderson.)

can directly relate the power through the filter to the frequency of the laser diode and, hence, measure the latter. The top figure shows the case of no feedback and thus the filter profile is shown. In the bottom figure the feedback rate has been increased ($C \sim 20$) and the external cavity modes are visible. The discretization of the filter profile in the bottom figure means that the laser is locking (in frequency) to the external cavity modes for a certain range of the pump current. Therefore, the power transmitted through the filter is (almost) fixed for that current range. At the end of the locking range the laser jumps to the next external cavity mode. This whole procedure results in a discrete filter profile as the output signal. For more details concerning these experiments see Fischer *et al.* [8].

It has been a known experimental fact already for a long time that semiconductor lasers with external feedback of sufficiently high strength, such that C is well

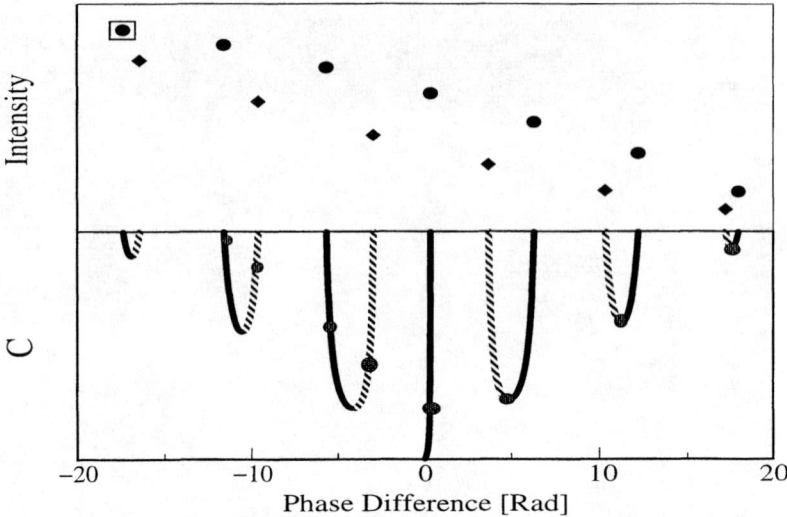

FIGURE 11. Lower part: Creation of fixed points in saddle-node bifurcations. The dark branches represent the modes while the dashed branches represent the anti-modes. The dot on each branch indicates the Hopf bifurcation point (see text). Upper part: Fixed point situation at C=15; parameter values are $\alpha = 3$ and $\tau = 30$ ns. The boxed point is the stable highest intensity mode.

above 1 are hardly ever observed in steady-state operation. Instead, they usually show chaotic-type dynamics with linewidth several orders of magnitude larger than expected for steady-state operation. On the basis of measured spectral properties of the output light, Heil et al. [9] presented an overview of the different types of unstable operation. At high feedback, only for low pump strength stable steady-state operation is observed. In the next subsection this will be discussed further in some detail. Numerical as well as analytical methods have shown that the low-intensity modes, i.e. the ones with positive η (see Fig. 7), are always unstable, straight on from their creation in the saddle-node bifurcations that were mentioned earlier. In contrast, on the high-intensity side the modes are always stable upon creation in a saddle-node bifurcation [10]. On increasing the feedback level further, however, a Hopf bifurcation instability occurs, which marks the feedback strength at which a sustained relaxation oscillation of typically a few GHz starts modulating the output intensity. If the feedback strength is increased further, a period-doubling scenario into chaos may be observed [11], but details depend very much on parameter values; see also the chapter by *Gavrielides*. In the meantime, a new stable mode will have been created at the upper left end of the banana with still higher intensity than the former one. Thus, the maximum gain mode is always stable.

In Fig. 11 the situation is summarized in terms of the creation of fixed points, i.e. modes and anti-modes, and the position where the Hopf bifurcation occurs, as a function of increasing feedback level. This particular picture refers to some given fixed value of ϕ_0. On can easily imagine by combining Figs. 8 and 11 how the situation will somewhat change when ϕ_0 is varied.

DYNAMICS

If a semiconductor diode laser is pumped close to but still below its free-running threshold, it will not be lasing. Subjection of the laser to external optical feedback will produce an effective threshold reduction that can be as large as 10% or even more. In fact, from Eq. (31) we obtain a maximum threshold shift of

$$\Delta J = \frac{2\gamma}{\xi T_1}. \qquad (34)$$

Note that measuring the threshold reduction is one way of calibrating the feedback rate. Another and independent way would be to observe the $C = 1$ point of operation. In any case, the very presence of feedback can raise the laser substantially above its threshold, without having to pump harder.

One procedure for observation of fixed point operation would be to fix the feedback rate and then increase the pump current from below the effective threshold, through the free-running laser threshold to values far above threshold. In this manner one has a good chance of observing the system in fixed point operation. This can be explained best by referring to the banana-type curve of Fig. 7, where one should realize that varying the pump strength means shifting the curve in the vertical direction. There will also be a shift in solitary laser operation frequency, but that has been scaled away by adopting the solitary laser frequency as reference frequency; see Eq. (1). Therefore, no horizontal shift will occur in Fig. 7. For $J < J_{\text{Thr}} - \Delta J$, there will be no fixed points with positive intensity. The laser will be in the off state, which is represented in Fig. 7 by the line $I = 0$ (in the degenerate case $I = 0$ the phase is completely undetermined). For J just above $J_{\text{Thr}} - \Delta J$ one mode is available and this mode is stable. In principle this mode remains stable while the pump current is increased. However, as the banana is lifted higher above the $I = 0$ line, more modes become available, most of which will not be stable due to the above-mentioned scenario. In fact, coexisting with the stable high-intensity mode is a strange attractor, also referred to as Sisyphus attractor [12], in which various limit cycles around individual (unstable) modes have merged together. The corresponding dynamics is known as low-frequency fluctuations (LFF). Some general features of LFF will be discussed below, after we have completed the overview of what happens on increasing the pump current.

In the coexistence region, both stable maximum gain mode operation and LFF can be observed. Experimentally, also spontaneous transitions between the two types of dynamical behavior have been observed [9]; see also the extensive discussion in the chapter by *Fischer et al.* on this topic. For even higher pump currents such

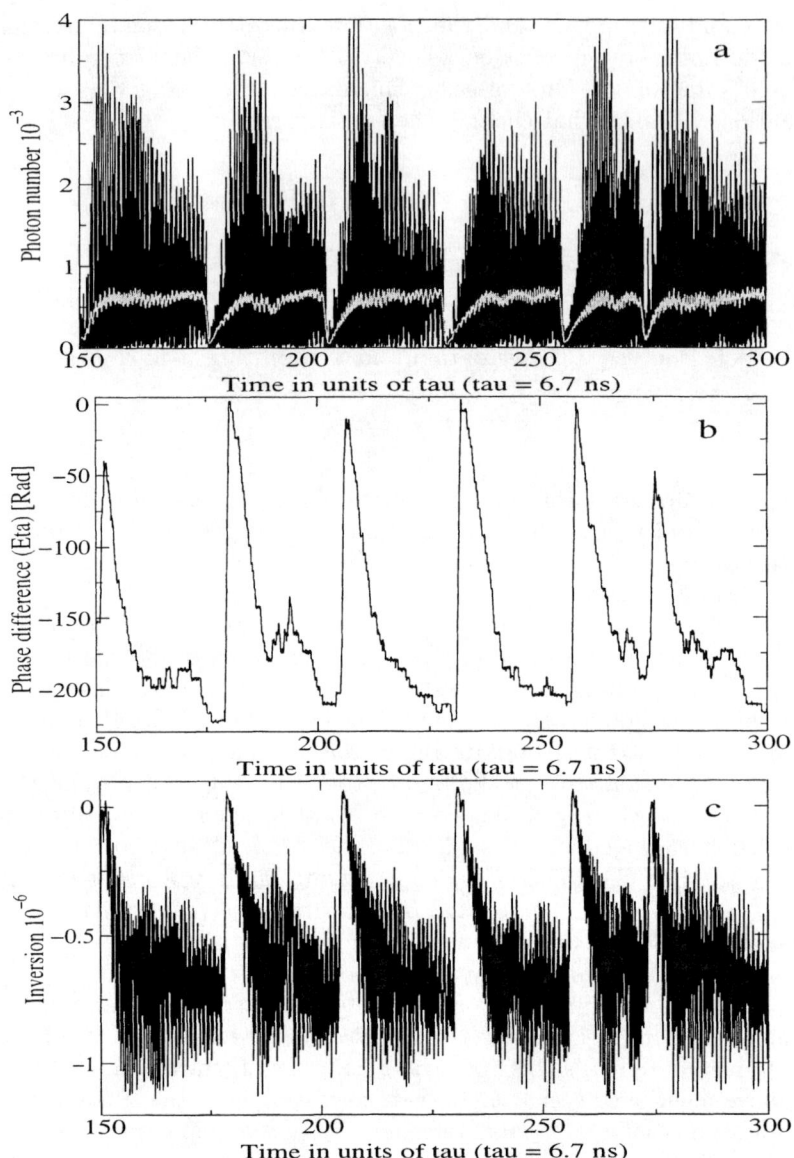

FIGURE 12. Time traces of the intensity (a), the roundtrip phase difference (b), and the inversion (c). In the top trace, the intensity avaraged over ∼ 1 ns is indicated by the light curve

that the laser operates well above the free-running threshold, stable maximum gain mode operation has not been observed anymore, although it is still predicted by

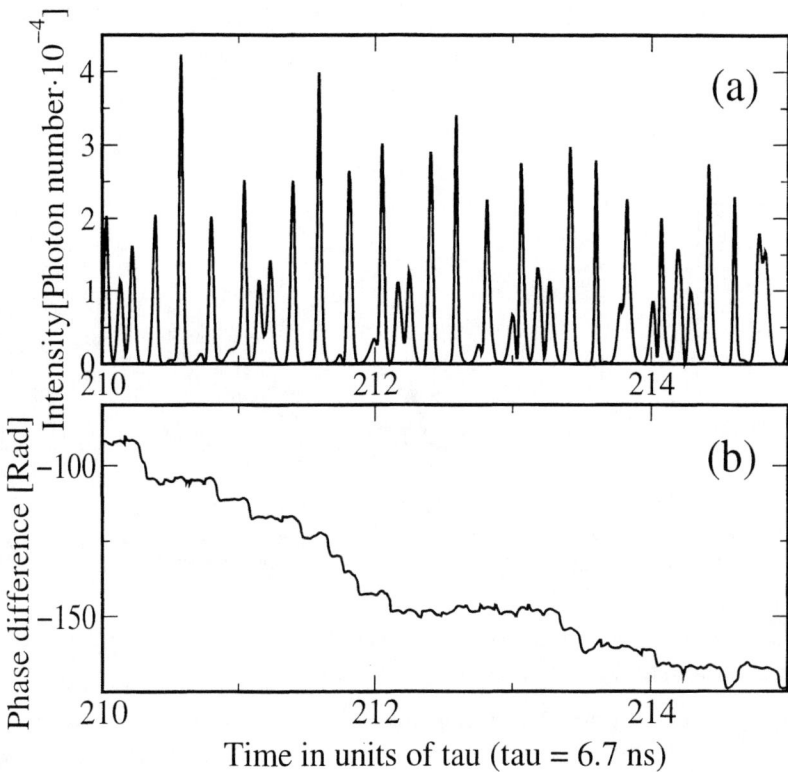

FIGURE 13. Blow-up of the time traces in Fig. 12, showing the intensity (a) and the phase difference η (b).

theory. In stead, a chaotic type dynamics is observed which differs from LFF in some respects, but which is predicted by theory as well. This latter type of behavior is known under the name Coherence Collapse (CC).

If at moderate to strong amounts of feedback, i.e. $C > 20$ and threshold reduction $> 2\%$, the full equations (17), (18) with (20) are numerically integrated with all other parameters as in Table 1 and $\alpha = 5, \tau = 6.7\ ns$, while the laser is pumped at its solitary threshold, one will obtain a typical result as depicted in Figs. 12(a)–(c) . The intensity versus time is plotted in Fig. 12(a). Irregular cycles are seen, the duration of which are distributed around an average duration of $\sim 20\tau \simeq 170 ns$. Since this time scale is longer than any relevant intrinsic semiconductor laser time scale, this dynamical behavior is called low-frequency fluctuations (LFF). The corresponding inversion n and roundtrip phase difference η are plotted in Fig.12(b) and (c). Comparing the plots of I and η, we summarize the time evolution during one cycle as follows:

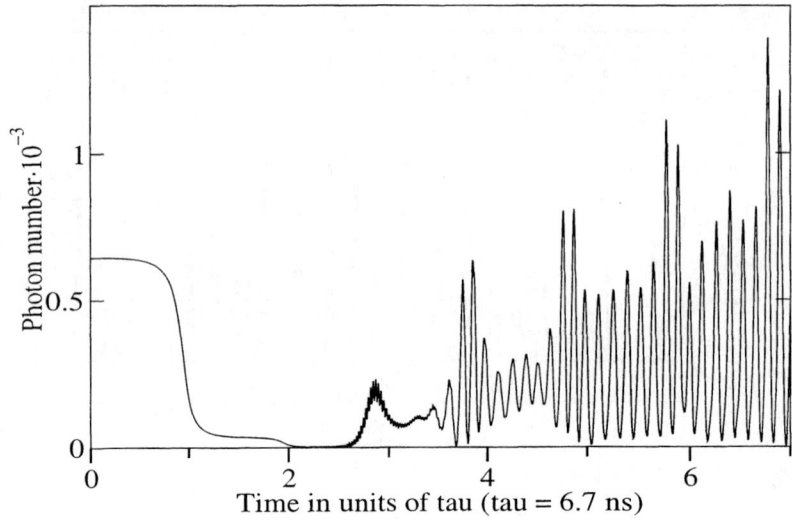

FIGURE 14. A drop-out event. The simulation was started very close to a saddle point.

1. The light is emitted in short pulses, where each pulse has a width of $50-100ps$; see the blow-up in Fig. 13(a).

2. The intensity averaged over $1ns$ gradually increases during 5 to 15 roundtrips to a more or less constant value.

3. The end of a cycle is characterized by a "drop out", an abrupt ending of the pulse train, i.e. a sudden drop of the mean intensity to a significantly lower value.

4. The gradual increase of mean intensity is accompanied by an almost monotoneous decrease of η to large negative values.

5. The intensity drop is accompanied by an equally sudden "jump up" of η.

6. η is quantized in integer multiples of 2π, at least to a good approximation, as can be observed from the blow-up in Fig. 13(b). It can also be seen that jumps of η occur precisely in between the intensity pulses, i.e. at very low instantaneous intensities.

The strange attractor with the above described features was first identified by Sano [13]. Soon after that, the attractor was studied in great detail and given the name Sisyphus attractor [12]. An earlier study of LFF was carried out by Mørk *et al.* already in 1989. They developed a global theory of LFF in terms of spectral injection and the creation of a dynamical bistability [14]. During intensity build-up the frequency spectrum broadens and shifts to negative values, while at the same

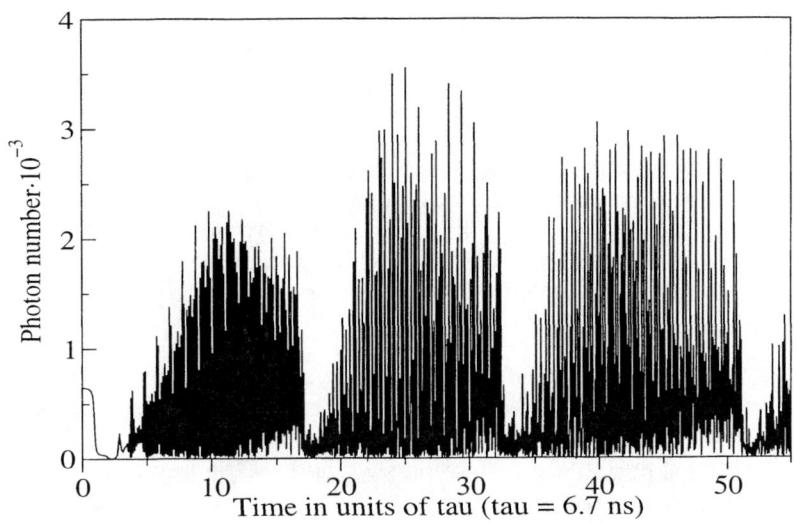

FIGURE 15. A drop-out event continues into Sisyphus cycles.

time a coexisting state is created which corresponds more or less to the free-running laser. At a certain point a spontaneous transition occurs (the "drop out") from the frequency-broadened high-intensity state back to the free running laser state. This theory has recently been revived and was extended to include a return map analysis yielding correct global description of the attractor, including the typical time scales [15].

It was observed from numerical simulations in [12] that the drop-out at the end of an LFF cycle always seemed to occur after a close approach of a saddle point (antimode). If in a simulation one deliberately starts the system during the time interval $[-\tau, 0]$ very close to a saddle point, one observes typically a time trace as depicted in Fig. 14. The intensity is repelled, first exponentially slow, as expected, from the saddle point, but then shows a sudden drop down to a very small value, after which an irregular series of pulses marks the gradual recovery of the mean intensity. The pulses are deformed relaxation oscillations. Fig. 15 shows the continuation of the simulation after the drop out shown in Fig. 14. Clearly, the Sisyphus cycles which are so characteristic for LFF are immediately visible.

The Sisyphus attractor seems to be the interesting result of a compromise between several conflicting requirements for the laser when operating with external feedback:

1. In the presence of external feedback the laser maximizes its intensity by maintaining a multiple of 2π roundtrip phase difference. This seems to be a very strong requirement.

2. The laser maximizes its intensity also by profiting from the effective threshold reduction. This implies operation in the maximum gain mode, for which the operation frequency, or equivalently η, must shift to high negative values. (This is the α-induced effect.)

3. The α-induced shift towards maximum intensity mode is incompatible with the multiple of 2π phase locking condition mentioned in 1, except during time intervals of vanishing or very small intensity. By producing pulses with deep modulations in between, the phase locking condition is released during short time intervals, but long enough for the phase to jump to a next higher negative multiple of 2π.

4. Thus, the step-wise α-induced drift of the frequency will bring the system closer and closer to the maximum gain mode, but at the same time the distance in η between mode and anti-mode becomes smaller and smaller, somehow inevitably leading to a too close passage of a saddle point after which a drop out as in Fig. 14 takes place.

It has been shown by Gavrielides et al. [16] that under certain very special conditions the Sysiphus cycles may become periodic due to synchronization between the round trip period, the relaxation oscillation period and the length of the cycle and this was also observed in experiments; see the chapter by *Gavrielides* for more details. One may conclude from this that the role of noise, which is always present in the system due to spontaneous emission, is of minor significance in shaping the Sisyphus attractor. We strongly believe that new investigations will provide new insights and clues.

REFERENCES

1. Jackson, J.D., *Classical electrodynamics*, second edition, Wiley, 1975, p.311.
2. Iga, K., *Fundamentals of laser optics*, Plenum Press, 1999, Sec. 10.2.
3. Pantell, R.H., and Puthoff, H.E., *Fundamentals of Quantum Electronics*, Wiley, 1969.
4. Chow, W., Koch, S.W., Sargent III, M., *Semiconductor-Laser Physics*, Springer-Verlag, 1994.
5. Van Tartwijk, G.H.M., and Lenstra, D., *IEEE Quantum Semiclassical Optics* **7**, 87 (1995).
6. Yousefi, M., and Lenstra, D., *IEEE J. Quantum Electron.* **53**, 970 (1999).
7. Lang, R., and Kobayashi, K., *IEEE J. Quantum Electron.* **16**, 347 (1980).
8. Fischer, A.P.A., Andersen, O.K., Yousefi, M., Stolte, S., and Lenstra, D., *IEEE J. Quantum Electron.* **36**, 375 (2000).
9. Heil, T., Fischer, I., and Elsäßer, W., *Phys. Rev. A.* **58**, R2622 (1998); *Phys. Rev. A.* **60**, 634 (1999).
10. Levine, A.M, Van Tartwijk, G.H.M., Lenstra, D., and Erneux, T., *Phys. Rev. A.* **52**, R3436 (1995).

11. Hohl, A., and Gavrielides, A., *Phys. Rev. Lett.* **82**, 1148 (1999).
12. Van Tartwijk, G.H.M., Levine, A.M., and Lenstra, D., *IEEE J. Sel. Topics QE* **1**, 466 (1995).
13. Sano, T., *Phys. Rev. A.* **50**, 2719 (1994).
14. Mørk, J., Tromborg, B., and Christiansen, P.L., *IEEE J. Quantum Electron.* **24**, 123 (1988).
15. Mørk J., Sabbatier H., Sørensen M.P., Tromborg B., *Optics Communications* **171**, 93 (1999).
16. Gavrielides, A., Newell, T.C., Kovanis, V., Harrison, R.G., Swanston, N., Dejin Yu, and Weiping Lu, *Phys. Rev. A.* **60**, 1577 (1999).

Applications of Semiconductor Lasers to Secure Communications

Claudio R. Mirasso

Departament de Física, Universitat de les Illes Balears, E-07071 Palma de Mallorca, Spain

Abstract. We numerically study the synchronization of two chaotic semiconductor lasers in a *master-slave* configuration. To synchronize the lasers a small amount of the output power from the master laser is injected into the slave laser. We show that the output of the master laser can be used as a chaotic carrier to encode a digital message which can be recovered at the receiver. We also check the quality of the synchronization diagram when the two lasers are slightly different.

INTRODUCTION

The world is currently witnessing an ever-growing demand for communications services. This is being supported by a rapid increase in transmission and switching capacity in networks of all kinds. Two major issues in these networks are privacy and security. Although software codification has proven to be useful for data encoding, the continuous increase of computer speed threatens the safety of this procedure. A contribution to solve this problem is by additionally encoding at the physical level using chaotic carriers generated by components operating in the non-linear regime.

Semiconductor lasers are ideal candidates for the realization of these non-linear transmitter and receiver systems because they are already inherently non-linear devices that under various operation conditions exhibit non-linear dynamical behaviour associated with, e.g., fast irregular pulsations of the optical power or wavelength hopping. In the today existing and operating communication schemes these types of behaviour are not desired and sometimes tremendous efforts have to be performed to avoid them. It is our particular aim to explicitly benefit from these behaviours, usually called optical chaos.

The essence of the optical chaotic communication resides in the fact that two spatially separated chaotic lasers, in our case semiconductor diode lasers, can synchronize to each other. Synchronization means that the irregular time evolution of the emitter laser, either in the optical power P or in the wavelength, can be perfectly reproduced by the receiver laser; see Fig. 1.

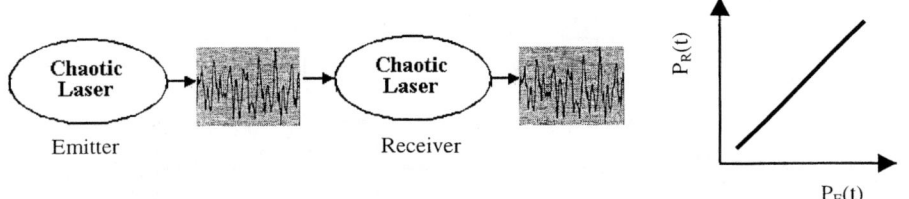

FIGURE 1. Schematic representation of synchronization. The synchronization diagram, i.e. the evolution of output signal of the emitter vs. the output signal of the receiver, is plotted at the right side. The 45° line represents perfect synchronization.

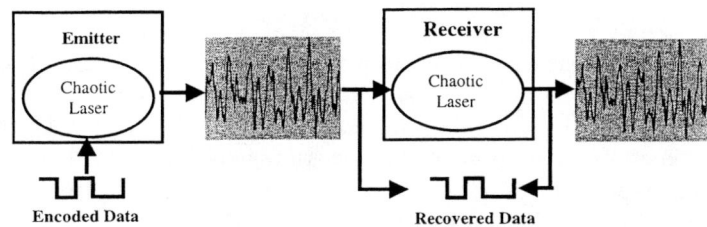

FIGURE 2. Schematic representation of the encoding/decoding process

Once the two lasers have been synchronized by a suitable method, we can use the chaotic output of the transmitter laser as the carrier on which the message is encoded. The other laser, at the receiver, allows the message to be extracted. Decoding is based on the non-linear phenomenon of chaos synchronization between emitter and receiver [1]. The key issue resides in the fact that the receiver synchronizes to the chaotic oscillations of the emitter (the carrier) suppressing the encoded message. Therefore, by comparing the input (carrier + message) and output (carrier only) of the receiver the message can be extracted (see Fig. 2).

The key for effective synchronization (and thus for decoding) lies in the use of very similar components for both chaotic systems, with close matching with respect to parameters and operation conditions. Decoding without the right receiver is very difficult to achieve due to the high frequencies involved and the large number of dynamical degrees of freedom of the chaotic carrier. It is worth noting that chaotic carriers provide in semiconductor lasers a broad spectrum (typically in the GHz range) where the message is hidden. The properties of these carrier signals, and the way the message is encoded, are such that with a linear filtering process it is not possible to extract the message. Also correlators and frequency-domain analysis would fail.

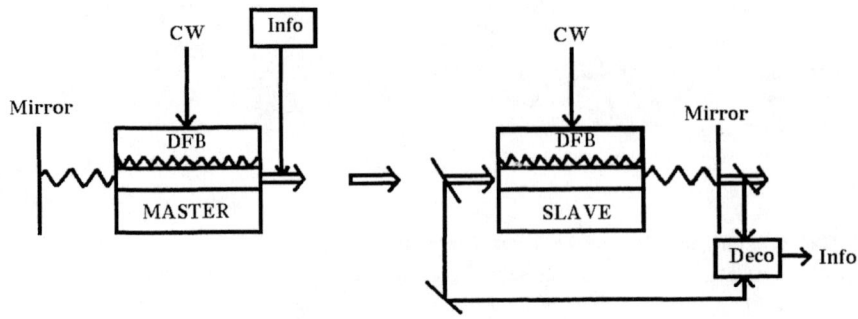

FIGURE 3. Proposed set-up for chaos synchronization and message encoding using semiconductor lasers (similar to the one proposed in Refs. [5,6])

The idea of communicating using transmitters and receivers operating in the nonlinear dynamical regime has recently attracted the attention of many researchers. Different chaotic synchronization schemes and their applications to encoded communications have been proposed. This was first studied using electronic circuits [2] and it has been more recently proposed in solid state [3], fibre ring lasers [4], semiconductor lasers [5,6,8–15] and microchip lasers [16].

Chaotic signals have similar features as those used in broad spectrum communications and related fields where information is hidden within complex codes or noisy signals. Up to the present, the methods for generating chaotic signals have generally used simple electrical circuits. However, there have been two main problems with this approach. Firstly, most of the circuits have been designed in the audio range with message bandwidths limited to about 10 kHz. Although RF circuits may be contemplated, it is difficult to attain the multi-GHz frequencies required in many communication channels. Furthermore, most already installed high speed communication networks are based on optical fibers. Therefore, a codification scheme directly based on an optical carrier is highly desirable. The second more fundamental problem of the chaotic carrier generated electronically is its low-dimensionality, which results in a low level of confidentiality when applied to secure communication transmissions. Both these issues are overcome by the proposed use of high-dimensional chaos generated by lasers subject to feedback, either all optical or electro-optical.

Recent studies have shown the feasibility of synchronizing the so-called hyperchaos, i.e. attractors with a very large number of degrees of freedom, in optical systems. A set of two semiconductor lasers subjected to optical feedback, operating in the chaotic regime, and linked through an optical fiber were numerically studied in Refs. [5,6]. The two lasers were coupled as in Fig. 3, the scheme that will be followed during the rest of the paper. The master laser (ML) operates in the coherence collapse regime due to the optical feedback, i.e. the output power oscillates chaotically; see also the chapter by *Fischer et al.* The message to be encrypted can be added at the output of the ML (chaos masking) or included in the injection current (chaos shift keying). The slave laser (SL) is similar to the ML and is subjected to the same feedback. When the ML and SL are not connected, their output power are uncorrelated and the synchronization diagram (a plot of the ML output vs. the SL output) exhibits a cloud of points, as will be shown later. However, when a small amount of light coming from the ML is injected into the SL the two subsystems are able to synchronize and an almost 45° line in the synchronization diagram is obtained. As it was already mentioned, the SL laser synchronizes to the carrier of the ML by filtering the message. Under such conditions the message can be recovered by processing the input and output of the SL.

Another common way to produce a chaotic output from a semiconductor laser is by optical injection. Optical injection allows for different kinds of routes to chaos and types of instabilities. A carefully study of this can be found in Ref. [7]; see also the chapter by *Krauskopf*. Synchronization of semiconductor lasers with an injected signal operating in the chaotic regime was very recently studied by Chen and Liu [8], although the original idea was proposed by V. Annovazzi and coworkers [9], to encrypt information. They proposed a set-up that is shown in Fig. 4 of Ref. [9]. In the system, the transmitter is switched between two different chaotic orbits by modulating the injection current of the laser by the bit stream to be transmitted. Decoding at the receiver is performed with two replicas of the emitter: the first is biased with a current (J_0) corresponding to the bit "0" and the second at a bias (J_1) corresponding to the bit "1". The received signal is sent to both decoders, however only the one having the pump current corresponding to the transmitted bit synchronizes, as shown in Fig. 5 of Ref. [9].

Encryption of information within a chaotic carrier was experimentally demonstrated by van Wiggeren and Roy [4] using two similar chaotic erbium doped fibre ring lasers (EDFRL). The set-up is shown in Fig. 1 of Ref. [4]. A coupler injects the message to be encrypted into the first EDFRL. The output is extracted by another coupler and directed to the second EDFRL (the receiver). Both EDFRL at the emitter and receiver were fabricated to match as closely as possible. With an appropriate decoding process they were able to encode/decode a digital message at a bit rate up to 250 Mbit/s. Moreover, they showed the possibility of transmitting the message up to 35 km of single mode dispersion shifted optical fibers [4].

Almost at the same time, a different approach to encode information was proposed and experimentally demonstrated by J. P. Goedgebuer and coworkers [11]. They proposed a method for encrypting a signal within the high dimensional chaotic fluctuations of the wavelength from a delayed feedback tunable laser. The novelty of the idea was not only to use for the first time chaos in wavelength, but also a non-linear electro-optical feedback was included to generate the chaos (see Fig. 1 of Ref. [11]). This non-linear electro-optical feedback guarantees a large dimension of the generated chaos (between 100-1000), also called hyperchaos, which is very important for the security of the encrypted message. They also performed message encoding/decoding of a sine wave, although at a quite low rate (KHz).

One year later, Sivaprakasam and Shore demonstrated experimentally the synchronization of two external-cavity semiconductor laser diodes, in the configutration shown in Fig. 3. They found an optimal coupling strength between the ML and the SL for which a good synchronization is obtained (see Fig. 2 and 3 of Ref. [12]). In a later paper they showed the possibility of encoding/decoding a square wave signal, however at a low bit rate (KHz).

The first demonstration of GHZ signal encoding/decoding within a chaotic carrier was performed by I. Fischer and coworkers [13]. They used a similar scheme as the one shown in Fig. 3 but for the slave laser they removed the optical feedback, i.e. the SL operates under continuous wave. Although one could think that this system would not be useful for encoding and decoding information, they showed that the receiver synchronizes to the chaotic output of the transmitter (as shown in Figs. 2 and 3 of Ref. [13]) suppressing the encoded message, as it is clearly shown in Fig. 4 of Ref. [13]. By operating with the input and output of the receiver they were able to reconstruct the encoded signal.

In this work we consider a model for chaos synchronization and digital communications using as a carrier the chaotic output of a single-mode semiconductor laser subject to external optical feedback (Master Laser, ML), as shown in Fig. 3. We encode a digital message within the chaotic output of the master laser by either a time dependent attenuation of the carrier (chaos masking), in a scheme similar to the one proposed in Refs. [5,6], or by slightly modulating the injection current (as in Ref. [9]) (chaos shift keying). After codification, the carrier containing the message is injected into the receiver. For simplicity, we will consider a free space link, although transmission through dispersion shifted optical fibers has been shown to be feasible [6]. The receiver, the Slave Laser (SL), is also a single-mode semiconductor laser with external feedback, with parameters similar to the ML and operating under similar conditions. The two lasers display chaotic behavior with different time traces if they operate separately, i.e. with no-coupling between them. Even for identical initial conditions the presence of spontaneous emission noise will make time traces to depart from each other in time. However, if a small amount of the ML power is injected in the SL they can synchronize. This synchronization will allow us to encrypt messages within the chaotic output of the emitter.

The paper is organized as follows. We first describe the model we use for the transmitter and the receiver. Next we present numerical results for both synchronization and message encoding/decoding. Finally, we give a summary and the conclusions of our work.

THE MODEL

We model the transmitter (ML) and the receiver (SL) by using the rate equations for the complex slowly varying amplitude of the electrical field $E_{t,r}$ and carrier number inside the cavity $N_{t,r}$ [5] (t,r stand for transmitter and receiver respectively). We also include spontaneous emission noise in the field equation. The equations read:

$$\frac{dE_{t,r}(t)}{dt} = (1+j\alpha_{t,r})\left(G_{t,r}(t) - \frac{1}{\tau_{t,r}}\right)\frac{E_{t,r}(t)}{2} + \gamma E_{t,r}(t-\tau)\,e^{j\omega_{t,r}\tau}$$
$$+ \kappa_r E_{ext}(t)e^{-j\Delta\omega\,t} + \sqrt{2\beta N_{t,r}(t)}\,\xi_{t,r}(t), \qquad (1)$$

$$\frac{dN_{t,r}(t)}{dt} = \frac{I}{e} - \frac{1}{\tau_{n_{t,r}}}N_{t,r}(t) - G_{t,r}(t)\,|E_{t,r}(t)|^2, \qquad (2)$$

$$G_{t,r}(t) = \frac{g(N_{t,r} - N_{o_{t,r}})}{(1 + s|E_{t,r}(t)|^2)}. \qquad (3)$$

We consider the transmitter (t) and the receiver (r) to be very similar lasers, so we take, in principle, the same parameters for them: g=1.5 ×10^{-8} ps^{-1} is the gain parameter, s=5×10^{-7} is the gain saturation coefficient, $\alpha = 5$ is the linewidth enhancement factor, $\beta = 1.1 \times 10^{-9}\,ps^{-1}$ is the spontaneous emission rate, $e = 1.602 \times 10^{-19}$ C is the electronic charge, $\tau_n = 2$ ns is the carrier lifetime, $\tau_{t,r} = 2$ ps is the photon lifetime, $N_o = 1.5 \times 10^8$ is the carrier number at transparency, $\gamma = 30$ ns^{-1} is the feedback coefficient, τ is the external cavity roundtrip time and $\Delta\omega$ is the detuning between the optical frequencies of the free-running ML and SL. $I = 44$ mA is the bias current (the threshold current is $I_{th} \approx 14.7$ mA) and $\omega_{t,r} \approx 1.2 \times 10^3$ ps^{-1} is the angular frequency of both ML and SL under continuous wave operation,. The random spontaneous emission process is modeled by a complex Gaussian white noise term $\xi(t)$ of zero mean and correlation $<\xi_a(t)\xi_b^*(t')> = 2\delta_{ab}\delta(t-t')$ where a, b stand for t, r. The term $\kappa_r E_{ext}$ in (1), which takes into account the signal generated by the ML, exists only for the receiver laser. The field E_{ext} is the input signal at the receiver and κ_r is the coupling parameter of the injected field into the SL. Equation (1)-(3) are written in the reference frame where the emission frequency is zero when neglecting spontaneous emission and the laser is biased at threshold.

NUMERICAL RESULTS

Synchronization

A typical time trace and optical spectrum for the ML and SL operating in the chaotic regime are displayed in Fig. 4; see also the chapter by *Fischer et al.* In the time trace it can be clearly seen large intensity fluctuations indicating a well established chaotic regime, and consequently a broad optical spectrum of some tens of GHz width. The parameters used in these simulations, and the ones that we will use for the rest of the paper, are those given in the previous section. In Fig. 4 $\tau = 0.3$ ns and $\kappa_r = 50$ ns^{-1}.

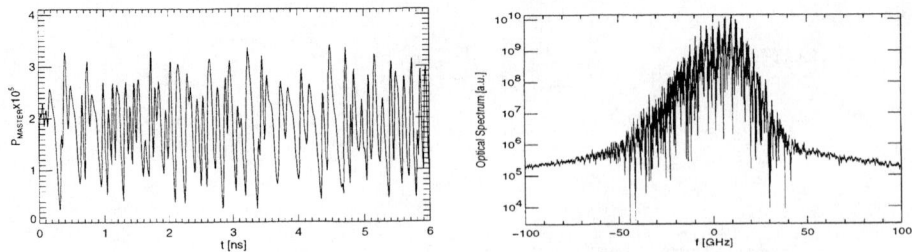

FIGURE 4. Typical time trace for a semiconductor laser subjected to optical feedback operating in the coherence collapse regime (left), and its optical spectrum (right). $\kappa_r = 50$ ns^{-1}, $\Delta\omega = 0$ $\tau = 0.3$ ns and the rest of the parameters are those given in the previous section.

For identical laser parameters and operating conditions the synchronization diagram, i.e. the output power of the SL vs. the output power of the ML, would exhibit an almost 45° straight line. However, a small deviation is obtained due to the fact that the SL has an extra power coming from the light injected from ML. By taking slightly different photons lifetime this difference can be compensated. The quality of the synchronization between the ML and SL laser strongly depends on some parameters as coupling coefficient, noise, feedback strength, etc. The robustness of the synchronization scheme was partially studied in Ref. [6] were the internal parameters of the ML and SL were varied. Through this section we will analyse the effects of some other parameters involved in the scheme in some detail.

We first examine the dependence of the synchronization quality with respect to the coupling coefficient. A simple expression for this coefficient, valid for Fabry-Perot semiconductor lasers, can be given: $\kappa_r = \sqrt{1-R}\, \eta_{ext}/(\tau_L \sqrt{R})$ where R is the facet power reflectivity of the SL, τ_L is the cavity roundtrip time of the light within the laser and η_{ext} accounts for losses different than those introduced by the laser facet. For our calculations we take $\tau_L = 6.6$ ps (which corresponds to a laser length of ~ 250 μm) and $R = 30\%$. The different values of κ_r are obtained by considering different values of $\eta_{ext} < 1$. It is evident that a minimum value for the coupling between the ML and SL is required in order to reach good synchronization. Even for the case of identical lasers and operating conditions a small coupling would not allow for synchronization. The synchronization diagram as a function of the coupling strength (κ_r) is shown in Fig. 5. When the two lasers are decoupled, their trajectories depart each other as time evolves and the synchronization map becomes a cloud of points with no correlation between ML and SL outputs (Fig. 5 a). However, this situation may change if a small amount of the light emitted by the ML is injected into the SL. Even for small amounts of coupling the synchronization diagram presents clear signs of synchronization between ML and SL. It can be clearly seen in the figure that the synchronization improves for increasing κ_r. For our parameters values, $\kappa_r \gtrsim 50$ ns^{-1} starts to give a good synchronization diagram.

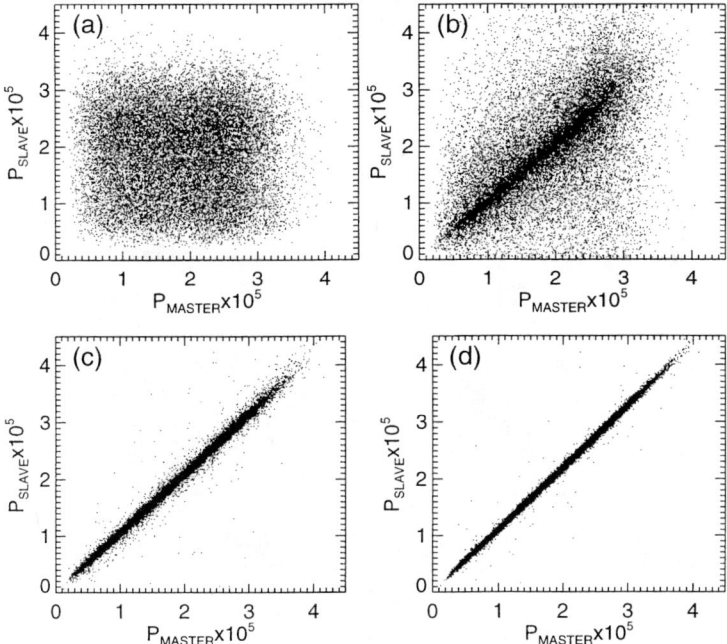

FIGURE 5. Synchronization diagram for different values of κ_r, namely $\kappa_r = 0$ (a), $\kappa_r = 25$ ns^{-1} (b), $\kappa_r = 50$ ns^{-1} (c), and $\kappa_r = 100$ ns^{-1} (d). $\Delta\omega = 0$, $\tau = 0.3$ ns and the rest of parameters are those given in the previous section

In practice, it is not possible to find pairs of identical semiconductor lasers. Variations of internal parameters have been partially studied in [6], as was already mentioned. In the this article, the photon lifetime was identified as the most critical parameter and changes in its value larger than $\sim 5\%$ could yield to a loss of synchronization. Synchronization quality would be also strongly influenced by, e.g., noise effects. Several noise sources would be present in real communication systems, including spontaneous emission noise in diode lasers, current noise, amplified spontaneous emission noise, in case of optical amplification, etc. In Fig. 6 we show how the synchronization degrades with the spontaneous emission factor β of the lasers. It is evident that for low values of β a good synchronization diagram is observed but it strongly degrades for larger values of the noise level, until the signals become almost uncorrelated. These results indicate that noise sources should be very well controlled in real systems.

Another sensible parameter is the detuning. It accounts for the difference between the free-running laser frequencies of the ML and SL. Fig. 7 shows the synchronization diagram as a function of the detuning. This effect was studied in detail

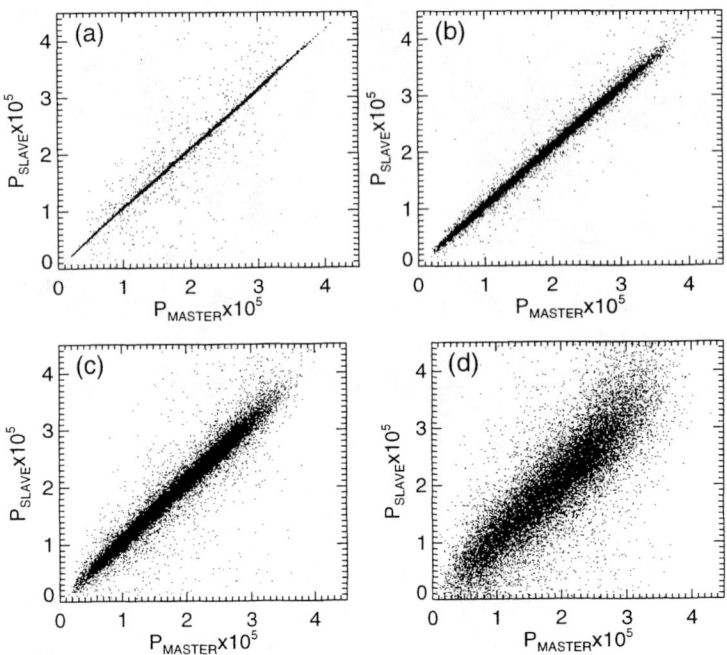

FIGURE 6. Synchronization diagram for different values of β, namely $\beta = 0$ (a), $\beta = 1 \times 10^{-6}$ ns^{-1} (b), $\beta = 1 \times 10^{-5}$ ns^{-1} (c), and $\beta = 1 \times 10^{-4}$ ns^{-1} (d). $\kappa_r = 50$ ns^{-1}, $\Delta\omega = 0$, $\tau = 0.3$ ns and the rest of parameters are those given in the previous.

in Ref. [15] using an iterative model, valid for weak to strong optical feedback, in the case of a Vertical Cavity Surface Emitting Laser (VCSEL). In this reference, the authors showed that there exists a restricted set of detunings, which form a synchronization window, that allows for a good synchronization. This window is roughly some tens of GHz wide and is asymmetric due to the linewidth enhancement factor parameter α. The same effect is observed in our case using the rate equation model. The synchronization window also ranges roughly for 40 GHz, for the parameter values we have chosen, and is asymmetric due to the α parameter in such a way that a mirror image of Fig. 7 is obtained when changing the sign of α. However, it is important to point out that in our case the detuning can be continuously varied within the synchronization window and the synchronization is still maintained. This fact is in contrast to the results presented in Ref. [15] where only specific values for the detuning, and small island around these values, allow for synchronization, even within the synchronization window. The discrepancy is due to the fact that in Ref. [15] the feedback phase $\omega_{t,r}\tau$ was fixed for both ML and SL at a constant value for all the detunings, while in our case this phase changes

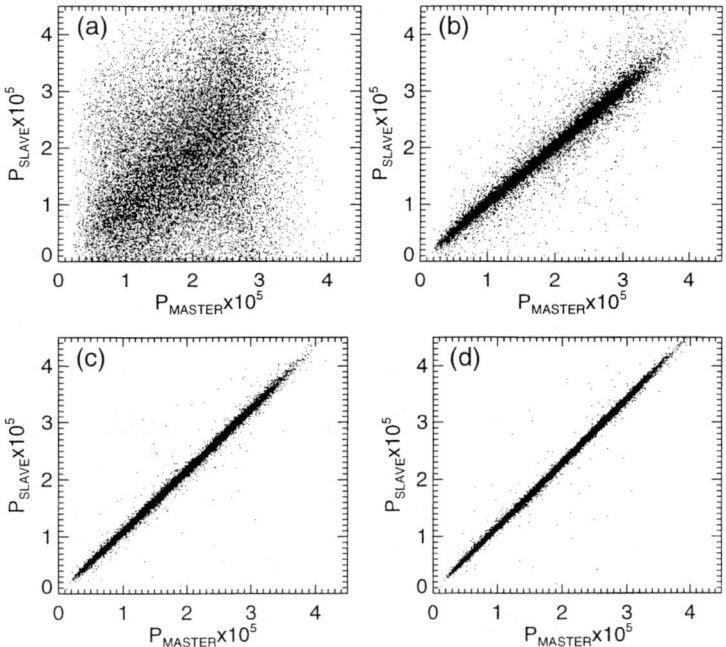

FIGURE 7. Synchronization diagram for different detunings $\Delta\nu = \Delta\omega/(2\pi)$, namely $\Delta\nu = -15$ GHz (a), $\Delta\nu = -5$ GHz (b), $\Delta\nu = 5$ GHz (c), and $\Delta\nu = 15$ GHz (d). $\kappa_r = 50$ ns^{-1}, $\tau\omega = 0.3$ ns and the rest of parameters are those given in the previous section.

continuously with the detuning, accordingly to Eq. (1).

Finally, we analyze how the synchronization diagram changes with the feedback time τ. From the mathematical point the phase space of a delayed system is infinite, and one should give an infinite number of initial conditions for the delayed feedback term when solving equation (1). This is why an attractor of a delay system can be of any, and even infinite, dimension. However, in practice this is not the case and it is well accepted that the number of degree of freedom associated with the ML and SL dynamics, i.e. the dimension of the chaotic attractors, increases with the feedback time τ and with the feedback coefficient γ. This point is important since one would expects that the higher the chaotic attractor dimension the better the confidentiality of the encrypted message. The feedback time τ can be easily varied by changing the position of the external mirrors. We have observed that the synchronization diagram degrades for increasing τ, as shown in Fig. 8. To compensate this loss of synchronization κ_r has to be increased. However, one would expect that the coupling between ML and SL has a maximum value in a real experiment. Once this limiting value is reached, the synchronization would be lost

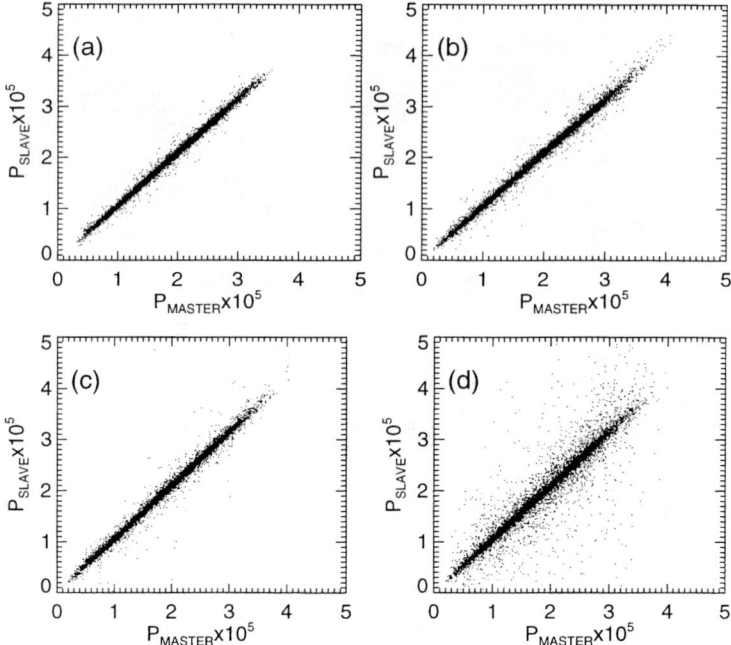

FIGURE 8. Synchronization diagram for $\tau=0.1$ ns (a), $\tau=2$ ns (b), $\tau=10$ ns)c), and $\tau=20$ ns (d). $\kappa_r = 50$ ns^{-1}, $\Delta\omega = 0$ and the rest of parameters are those given in the previous section.

for larger values of τ. A way to solve this problem is by using the set-up described in Refs. [11] and [13] where the SL operates without the external feedback. These two schemes offer the same filtering properties at the receiver and additionally they are more robust from the practical point of view.

Message Encoding/Decoding

In this part we show that, under good synchronization conditions, it is possible to encrypt and decrypt a message encoded within the chaotic carrier. For simplicity, we take identical parameters and operating conditions for both ML and SL, although slightly different lasers give rise to similar results.

As a first example we show in Fig. 9 the output of the ML and SL, and the encoded and decoded message, when a 1 GBit/s and 4 Gbit/s non-return to zero digital message is added at the output of the ML (chaos masking) following the technique used in Refs. [5,6]. When comparing the output of the ML with and without the message it is difficult to observe the fluctuations introduced by the en-

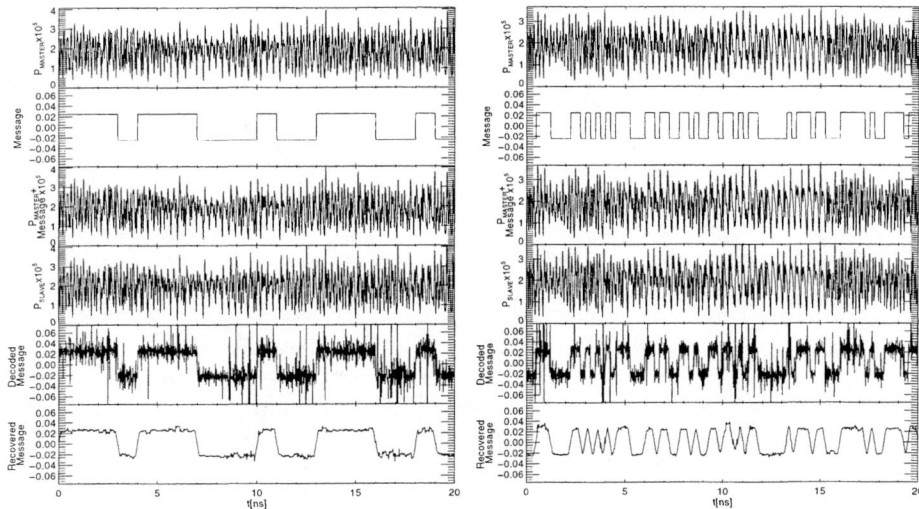

FIGURE 9. From top to bottom there is the output of the ML, the encoded message, the output of the ML with the message, the output of the SL, the decoded message, and the recovered message after filtering. $\tau=0.3$ ns, $\kappa_r = 50$ ns^{-1}, $\Delta\omega = 0$ and the rest of parameters are those given in the previous section. The message modulation rates are 1 Gbit/s (left panel) and 4 Gbit/s (right panel)

coded message. Also, optical spectra look very similar, as was shown in Ref. [6]. In both cases the recovery message exhibits fast oscillations that can be easily reduced by a simple filtering process. After this process, the recovered message compares quite well with the transmitted message. Also the synchronization diagram is good, as it is shown in Fig. 10 (left panel).

On the right panel of Fig. 10 the synchronization diagram can be seen when the message is included by modulating the injection current of the ML while keeping the other parameters fixed (chaos shift keying). Two parallel straight lines can be observed: one corresponds to the synchronization between ML and SL when a "1" bit is injected and the other when a "0" bit is injected. The separation between the two lines depends on the amplitude of the message and can be reduced by modulating the injection current with a smaller amplitude.

In Fig. 11 we show the ML and SL laser output together with the encoded and decoded message when a 1 Gbit/s and 4 Gbit/s non-return to zero digital message is encrypted. The quality of the recovered message after filtering is as good as the one obtained in the case of chaos masking (Figs. 9). However, this encoding technique is much simpler to implement in real systems. Moreover, it can be clearly seen that the message can be recovered very well with only one receiver which indicates that

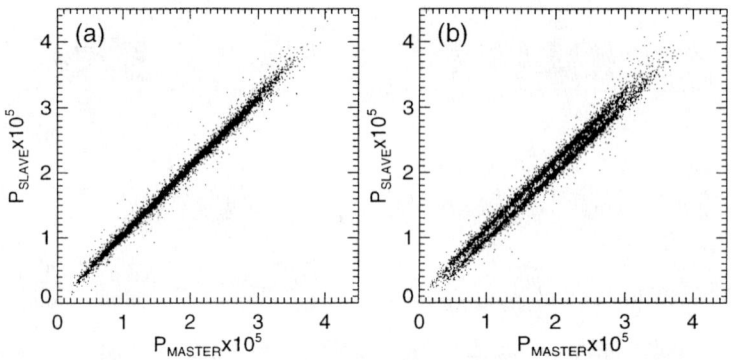

FIGURE 10. Synchronization diagram for the ML vs. SL when the message is added (chaos masking) at the output of the ML (a), and when the message is injected through the injection current (chaos shift keying) (b). $\tau=0.3$ ns, $\kappa_r = 50$ ns^{-1}, $\Delta\omega = 0$ and the rest of parameters are those given in the previous section.

two subsystems, one synchronizing to the "1" bit and the other one synchronizing to the "0" bit, suggested in Ref. [9] are not necessary.

CONCLUSIONS

In conclusion, we have numerically studied the synchronization of two chaotic semiconductor lasers with external optical feedback in a *master-slave* configuration. We have shown that a very good synchronization can be obtained when a minimum value for the coupling between the lasers is reached, even when the lasers are not identical. We have observed that the synchronization quality degrades with the noise, with the detuning between the ML and SL free running frequencies and with the dimension of the chaotic attractor (external-cavity round trip time and feedback coupling). It has been also shown that one the most critical parameter seems to be the detuning, although a good synchronization diagram can be continuously obtained for a detuning range, or detuning window, of tens of GHz. This detuning window is asymmetric due to the α factor.

We have also shown that a digital message modulated at frequencies up to 4 Gbit/s can be encoded in the chaotic signal of the master laser and recovered at the SL. The large and fast fluctuations that appear in the recovered message can be strongly reduced by a simple filtering process. We have shown that both chaos masking and chaos shift keying technique can be implemented and that both yield a similar quality of the recovery message, the chaos shift keying technique being the simplest one to implement in real systems. Moreover, we have also shown that,

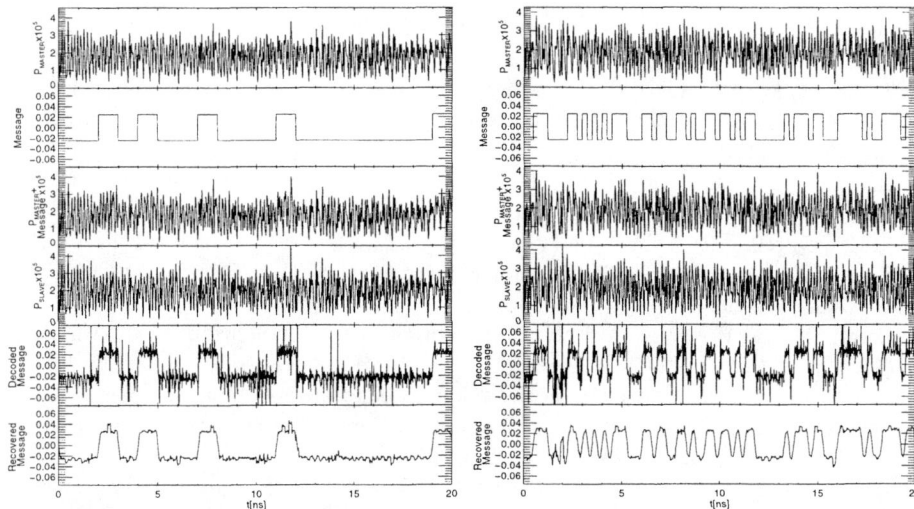

FIGURE 11. From top to bottom there is the output of the ML, the encoded message, the output of the ML with the message, the output of the SL, the decoded message, and the recovered message after filtering. $\tau=0.3$ ns, $\kappa_r = 50$ ns^{-1}, $\Delta\omega = 0$ and the rest of parameters are those given in the previous section. The message modulation rates are 1 Gbit/s (left panel) and 4 Gbit/s (right panel)

for the latter, only one receiver is necessary in contrast to previous studies.

To summarize, we have presented a simple but robust communication system based on chaotic synchronization that supports high modulation frequencies for encoded messages. This technique, understood as a codification at hardware level, could very well complement already existing encoding techniques, namely software codification.

ACKNOWLEDGEMENTS

I would like to specially thank Josep Mulet for helping with the simulations and figures, and to Pere Colet, Luis Pesquera and José Revuelta for fruitful discussions and valuable comments. This work has been supported by the CICYT, Spain, Project TIC99-0645-C05-02 and DGES, Spain, project PB97-0141-C02-01.

REFERENCES

1. Pecora, L.M., and Carroll, T.L., "Synchronization in chaotic systems", Phys. Rev. Lett. **64**, 821-823 (1990); "Driving systems with chaotic signals", *Phys. Rev.* **A 44**, 2374–2383 (1991).
2. Cuomo, K.M., and Oppenheim, A.V., "Circuit implementation of synchronized chaos with applications to communications", *Phys. Rev. Lett.* **71**, 65-68 (1993); Cuomo, K.M., Oppenheim, A.V., and Strogatz, S.H., "Synchronization of Lorenz-based chaotic circuits with applications to communications", *IEEE Trans. Circuits Sys.* **40**, 626–633 (1993).
3. Colet, P., and Roy, R., "Digital communications with synchronized chaotic lasers", *Opt. Lett.* **19**, 2056–2058 (1994).
4. Van Wiggeren, G.D., and Roy, R., "Communication with chaotic lasers", *Science* **279**, 1198-1200 (1998); "Chaotic communications using time delayed optical systems", *Int. J. of Bif. and Chaos* **9**, 2129–2156 (1999).
5. Mirasso, C.R., Colet. P., and García-Fernández, P., "Synchronization of chaotic semiconductor lasers: application to encoded communications", *Phot. Tech. Lett.* **8**, 299–301 (1996).
6. Sánchez-Díaz, A., Mirasso, C., Colet, P. and García-Fernández, P., "Encoded Gbit/s digital communications with synchronized chaotic semiconductor lasers", *IEEE J. Quant. Electron.* **35**, 292–297 (1999).
7. Krauskopf, B., Wieczorek, S.M., and Lenstra, D., "Different types of chaos in optically injected semiconductor lasers", *Appl. Phy. Lett.* **77**(11) 1611–1613 (2000)
8. Chen, H.F., and Liu, J.M., "Open loop chaotic synchronization of injection locked semiconductor lasers with Gigahertz range modulation", *IEEE J. of Quant. Electron.* **36**, 27–34 (2000).
9. Annovazzi-Lodi, V., Donati, S., and Scire, A., "Synchronization of chaotic injected-laser systems and its application to optical cryptography"", *IEEE J. of Quant. Electron.* **32**, 953–959 (1996); "Synchronization of chaotic lasers by optical feedback for cryptographic applications", *IEEE J. of Quant. Electron.* **33**, 1449–1454 (1997).
10. Rahman, L., Li, G., and Tian, F., "Remote synchronization of high frequency chaotic signals in semiconductor lasers for secure communications", *Opt. Comm.* **138**, 91–94 (1997).
11. Goedgebuer, J.P., Larger, L., and Porte, H., "Optical cryptosystem based on synchronization of hyperchaos generated by a delayed feedback tunable laser diode", *Phys. Rev. Lett.* **80**, 2249–2252 (1998); Larger, L., Goedgebuer, J.P., and Delorme, F., "Optical encryption system using hyperchaos generated by an optoelectronic wavelength oscillator", *Phys. Rev.* **E 57**, 6618–6624 (1998).
12. Sivaprakasam, S., and Shore, K.A., "Demonstration of optical synchronization of chaotic external-cavity laser diodes", *Opt. Lett.* **24**, 466–468 (1999); "Message encoding and decoding using chaotic external-cavity diode lasers", *IEEE J. Quant. Electron.* **36**, 35–39 (2000).
13. Fischer, I., Liu, Y., and Davis, P., "Synchronization of chaotic semiconductor laser dynamics on sub-ns timescales and its potential for chaos communication", *Phys. Rev.* **A 62**(1) 011801(R) (2000).

14. Spencer, P., Mirasso, C.R., Colet, P., and Shore, A., "Synchronization of chaotic VCSELs", *IEEE J. Quant. Electron.* **34**, 1673–1679 (1998).
15. Spencer, P., and Mirasso, C.R., "Synchronization of chaotic VCSELs", *IEEE J. Quant. Electron.* **35**, 803–809 (1999).
16. Uchida, A., Shinozuka, M., Ogawa, T., and Kannari, F., "Experiments on chaos synchronization in two separate microchip lasers", *Opt. Lett.* **24**, 890–892 (1999).

Theory and Simulation of Spatially Extended Semiconductor Lasers

Ortwin Hess

Theoretical Quantum Electronics, Institute of Technical Physics, DLR, Pfaffenwaldring 38-40, D-70569 Stuttgart, Germany
and
Department of Physics, University of Munich, Amalienstrasse 54, D-80799 München, Germany

Abstract. A hierarchy of theoretical models ranging from microscopic levels of description to effective rate equations is presented in terms of characteristic physical processes of the active semiconductor medium and the light field occuring in space and time. The applicability of the various levels of approximations of the hierarchy to describe the dynamics of spatially extended edge-emitting and vertical-cavity surface-emitting semiconductor lasers is discussed with characteristic results of numerical simulations. The influence of delayed optical feedback on the spatio-temporal dynamics is described. In particular, we show how suitably tailored delayed coherent feedback can be employed for the control of spatio-temporal dynamics.

INTRODUCTION

Due to their small size and unrivalled efficiency in transforming electrical current into coherent light, semiconductor lasers have in recent years become key elements in daily life, being applied in optical information networks, compact disc players or laser printers. In spite of the tremendous improvements in laser processing technologies, there are still two major situations which continue to be serious sources of instabilities. First, due to its very high gain and outcoupling rate, the semiconductor laser is very sensitive to delayed optical feedback (DOF) caused by distant reflecting surfaces such as an optical fiber. Second, in high-power in-plane edge-emitting lasers (IPEELs) and vertical-cavity surface-emitting semiconductor lasers (VCSELs) the nonlinear interaction of spatial with temporal degrees of freedom leads to chaotic spatio-temporal instabilities. Clearly, for practical reasons it is highly desired to understand the nature of these complex temporal and spatio-temporal processes and, building on this knowledge, to develop schemes which may allow to control and suppress them.

In this chapter, we review the theoretical framework for describing the spatio-temporal dynamics of spatially extended semiconductor lasers. A hierarchy of microscopic and phenomenological models is discussed along with the various approximations which lead from fundamental microscopic descriptions (Maxwell-Bloch equations, explicitly taking into account the quantum-kinetical processes within the active semiconductor laser) to simplified rate equations. On the one hand, we will discuss as examples the characteristic spatio-temporal dynamics of broad-area and vertical-cavity surface-emitting lasers, which is strongly determined by the microscopic spatio-temporal and spatio-spectral processes. With the knowledge of the microscopic processes, we then, on the other hand, exlpoit for the case of arrays of small lasers the possibility to simplify the theory considerably and treat their dynamics on a phenomenological level. The inclusion of spatially resolved delayed optical feedback into the phenomenological theory represents a spatial extension of the famous Lang-Kobayashi description to spatially extended semiconductor lasers. After a brief discussion of the influence of delayed optical feedback on the spatio-temporal dynamics of the twin-stripe semiconductor laser, we highlight a scheme for stabilization of spatio-temporal chaos which is based on suitably tailored delayed optical feedback.

SEMICONDUCTOR LASER THEORY – A HIERARCHY OF MODELS

In the following, we briefly review a hierarchy of theoretical models which has been developed for spatially extended and coupled semiconductor lasers. Details on their derivation, limits of validity and application to various examples can be found in Ref. [1]. Generally, one can group the approaches into several levels related to the relevant time-scales and physical effects taken into account:

- On the most fundamental level of the hierarchy, the nonlinear spatio-temporal dynamics of the microscopic effects within the active semiconductor medium are considered self-consistently with the macroscopic properties of the laser devices. Starting point are the Maxwell-Bloch equations for spatially extended semiconductor lasers [1,2]. On this level of description, the semiconductor band-structure as well as the intrinsic many-body processes are taken into account. In particular, the mechanisms and consequences of spectral as well as spatial holeburning are considered with microscopically computed momentum and density dependent scattering rates representing carrier-carrier and carrier-phonon scattering.

- Disregarding the characteristic band-structure of the semiconductor medium, the two-level approximation effectively tries to model the semiconductor laser dynamics as transitions between two levels while the characteristic influences of the bands are treated in the form of functional expressions for the induced refractive index and gain. It may thus be regarded as an intermediate level between the microscopic semiconductor-laser Maxwell-Bloch approach and the phenomenological description. A consequence of the two-level approximation is that spectral hole-burning is neglected.

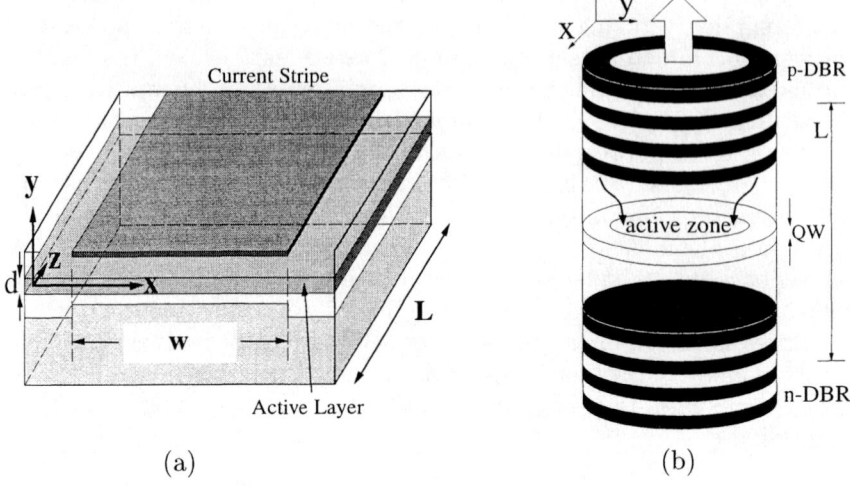

FIGURE 1. Schematics of a broad-area (a) and a vertical-cavity surface-emitting semiconductor laser (b).

- As a further step of simplification, the dynamics of the polarization is adiabatically eliminated such that one only has to consider the wave equations for the field(s) together with a diffusion equation for the charge carrier density. One should keep in mind that a major consequence of an adiabatic elimination of the polarization is to limit the reliability of the resulting models to time-scales larger than the characteristic spectral carrier-relaxation and dipole-dephasing processes which range from about 50 fs up to about 100 ps.

- If we further disregard the phase-information carried in the optical fields we may formulate rate equations for the charge carrier and photon densities. The rate equations may approximately model the linear and slow response of semiconductor lasers.

Next to the classification with respect to the relevant time scales, various levels of description exist regarding the explicit treatment of spatial degrees of freedom. Figure 1 schematically depicts the typical geometries of an edge-emitting and a vertical-cavity surface-emitting laser, respectively.

In the edge-emitting laser in Fig. 1(a) carriers are injected via the contacts on the top and light travels in the longitudinal (z)-direction inside the cavity formed by the two mirrors at the front and back. To assure sufficient gain during the counterpropagation (induced by reflection at the mirrors) the resonator length L measures at least 100 times the wavelength (e.g., for GaAs with $\lambda \approx 815$ nm the resonator length typically ranges from 300 μm to 2000 μm). The width of the active layer in the vertical (y) direction is as a result of the semiconductor epitaxial layer structure, and to assure waveguiding it is extremely small (about 0.1 μm). This is in strong contrast to the transverse (x)-direction which may be considerably larger (about 3–5 μm for a single-mode laser and 50–200 μm in case of a multi-mode high-power broad-area laser). From the viewpoint of spatio-temporal dynamics an edge-emitting semiconductor laser combines one dynamically relevant transverse direction with light-field counterpropagation.

In the vertical-cavity surface-emitting lasers in Fig. 1(b), the geometry of the cavity is completely different; see also the chapter by *Rorison*. Most notably, the length of the resonator now only measures about one wavelength. Consequently, only a single longitudinal mode will be dynamically relevant and propagation effects may be disregarded. At the same time, both transverse (x and y) directions are equally large (typically 3–30 μm). To assure sufficient gain the mirror reflectivities have to be sufficiently high and are technologically realized by dielectric multi-layers. For the VCSEL we thus will have to consider two transverse dimensions.

For a representation of spatio-temporal dynamics there are two important approximations – the plane-wave and the mean-field approximation – which may directly be applied to semiconductor lasers with typical in-plane edge-emitting or vertical-cavity surface-emitting geometries. If an edge-emitting laser is transversely small enough one may frequently disregard transverse modes and only consider the dynamics in the propagation direction (*plane-wave approximation*). On the other hand, in the VCSEL we may approximate the propagation effects by assuming a longitudinally averaged mean-field (*mean-field approximation*). For small VCSELs we may think of combining the plane-wave with the mean-field approximation resulting in a complete neglect of spatial degrees of freedom.

The way from Maxwell's equations and the quantum kinetic description of the semiconductor to the rate equations in the semiconductor laser is mapped out in the diagram in Fig. 2. The successive approximations which lead from model to model (in the boxes) are indicated in the ovals (dark shading). In each model-box, the relevant degrees of freedom (Dim.) and the independent variables (Var.) of the model are listed. Recall that E and P represent complex variables and consequently include the information of the amplitude and the phase. The effect of counterpropagation of the optical fields in the laser cavity is implicitly included in every model with a z-dependence (substitute: $E \to E^{\pm}$ and $P \to P^{\pm}$). The abbreviations MF and PW refer to the mean-field and the plane-wave approximation, respectively (discussed for the case of semiconductor lasers in [1]).

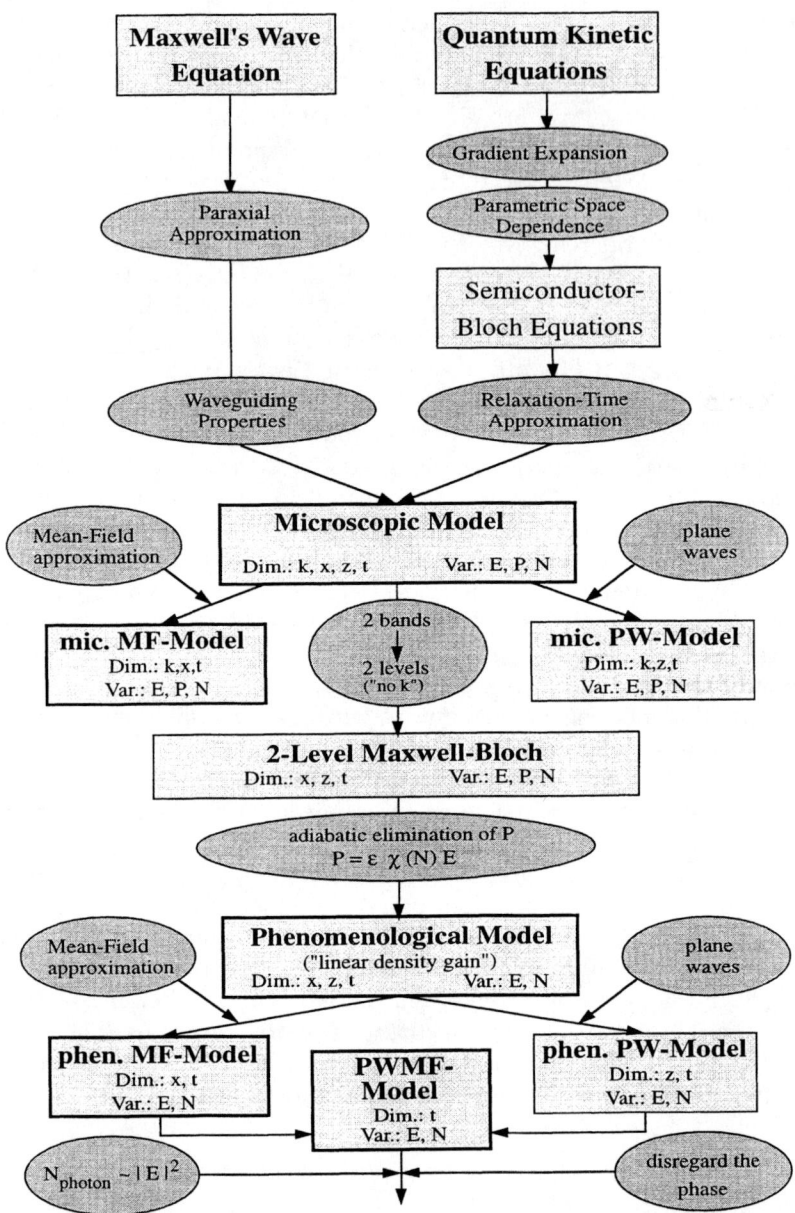

FIGURE 2. From Maxwell's wave equation and the quantum kinetic description of the semiconductor to the rate equations – A hierarchy of theoretical models for the semiconductor laser.

DYNAMIC FILAMENTATION IN BROAD-AREA SEMICONDUCTOR LASERS

Due to the width w of the transverse stripe contact being typically one order of magnitude larger than in the case of a single-stripe laser, broad-area laser devices can emit considerably high output power and are frequently used to pump solid-state lasers or as powerful light sources in optical communication systems. Often, in order to facilitate production-steps, they are manufactured as a pure gain-guided device and the lower cladding lacks any imposed transverse waveguiding structure. As neither optical confinement nor the confinement of the injected charge carriers diffusing in the transverse x-direction are provided by the device structure, the possibility of self-focusing (filamentation) and multi-transverse mode operation exists in broad-area lasers.

As the filamentation effects are a genuine and veritable consequence of the semiconductor nature of the active area, various efforts have been undertaken in order to include in a model spatial and temporal variations as well as characteristic semiconductor laser properties. For that purpose, approximate treatments of the semiconductor medium have been derived from microscopic semiconductor Bloch equations [3–5]. In an alternative approach based on effective Bloch equations for semiconductor lasers and amplifiers, the carrier-density dependence of the gain and refractive index and their respective dispersions are efficiently approximated by a superposition of several Lorentzians [6]. In microscopic simulations on the basis of Maxwell-Bloch equations for spatially inhomogeneous semiconductor lasers [2,7,8], the full space and momentum dependence of the charge carrier distributions and the polarization has been included. With the direct consideration of the microscopic spatio-temporal semiconductor dynamics, good quantitative agreement of the simulation results with streak-camera measurements of the spatio-temporal near-field intensity dynamics of a broad-area semiconductor laser has been obtained [9]. In addition to the macroscopic spatio-temporal intensity dynamics, information on the complex internal interplay of the spatio-temporal light-field dynamics with the active semiconductor medium can been revealed, demonstrating, in particular, the presence of dynamic spatio-spectral hole burning and spatio-temporal carrier-carrier as well as carrier-phonon scattering processes [2,8] and nonequilibrium spatio-spectral dynamics [10].

We now discuss this complex interplay revealed by the microscopic simulations performed on the basis of Maxwell-Bloch equations [2,10]. Figure 3 displays a representative example of the spatio-temporal dynamics and the temporally averaged spatial profile of the near-field intensity (left column), the far-field intensity (middle column), and the corresponding density (right column) of charge carriers of the broad-area laser during its initial 5 ns after start-up. The electrical pump current ($J = 1.5 J_{thr}$) is applied at $t = 0$. In this free-running condition (i.e., without any controlling force, e.g., by delayed optical feedback or optical injection), the formation of filaments and the onset of their migration are clearly visible in Fig. 3 (left

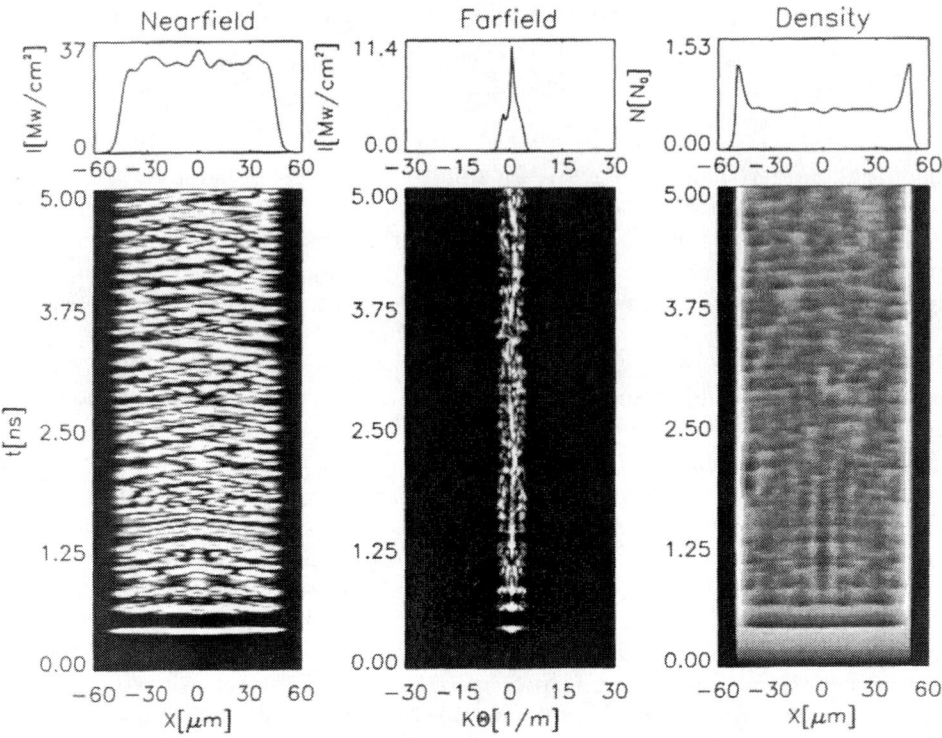

FIGURE 3. Representative start-up spatio-temporal dynamics of a free-running broad-area semiconductor laser. Shown are the temporal evolvement of the near-field intensity and time-averaged near-field intensity profile (left column), the corresponding dynamics of the Fraunhofer-far-field with time-averaged far-field profile, with optical wavenumber K and far-field angle θ (middle column), and the spatio-temporal distribution of the charge carrier density and temporally averaged profile (right column).

column). The processes which lead to this peculiar filamentation behavior are of a microscopic nature and depend in the broad-area laser only to a certain degree on the particular lateral boundary conditions, e.g., the imposed waveguiding structure. Rather, they are are a direct consequence of a concert of processes. Due to the fast microscopic Coulomb-scattering processes, the relaxation times of the carrier density is larger than that of the optical field and the interband-polarization. This leads to a localization of filaments of high intensity in wave-guiding channels formed by the carrier density: the carrier density is locally depleted by a filament of high intensity. An optical filament is thus located in a region of low gain (low carrier-density) and relatively high refractive index. By the process of gain-guiding, the filament thus provides itself with the dielectric waveguide which is necessary for its support during propagation in the laser cavity. Due to the length of the internal cavity of the broad-area laser, however, the filaments are longitudinally inhomogeneous. The result is a wave-like reflection of the filaments. Due to the spatially non-uniform transport of carriers, the filament-front becomes longitudinally and laterally more and more inhomogeneous. At the same time, with uniform injection of charge carriers the local carrier density outside the filament is not being depleted by stimulated emission and, consequently, rises quickly to levels above the threshold charge carrier density. A new optical filament is thus created. Every new filament interacts with the previous filament via the medium nonlinearly, thereby destabilizing it. The result is a vividly irregular and fundamentally chaotic longitudinal and transverse interaction of optical filaments. Also, the gain-guiding processes inside the laser cavity ensure that, by sustaining relatively stable high values of the density at the edge of the laser stripe, the optical field has created its own optical waveguide, i.e., an effective waveguide is formed. The combination of a transverse modulational instability with the propagation of the filaments as well as the carrier transport then leads to symmetry-breaking and the start of the filament-migration on a time-scale of about 500 ps [9]. As a consequence, the temporally averaged profiles of the near-field intensity and the carrier-density appear asymmetric while the far-field shows a strong angular spread next to the asymmetry. The spatio-temporal far-field trace reveals that this is due to an irregular sub-ns beam-fanning.

SPATIO-TEMPORAL POLARIZATION DYNAMICS IN VERTICAL-CAVITY SURFACE-EMITTING LASERS

Vertical Cavity Surface Emitting Lasers (VCSEL) have recently attracted intensive experimental and theoretical effort. One of the major advantages VCSELs have over conventional in-plane edge-emitting semiconductor lasers is the highly symmetric coin-shaped geometry around the axis of laser light emission. In large VCSELs, in addition to the spatio-temporal effects, this highly symmetric cavity is the reason for the fact that the polarization of light is not determined by any strong anisotropy of the device architecture, as is the case in edge emitting lasers. (We used this fact in the case of the broad-area lasers implicitly to uniquely consider only one single polarization direction.) Instead, due to the transverse symmetry of the VCSEL cavity and the active region, the polarization is highly sensitive to more subtle effects, such as small anisotropies in the crystal structure, the microscopic carrier systems, or strain and optical anisotropies in the mirrors [11–15].

At the same time, VCSELs are very susceptible to transverse filamentation effects, which now occur in two transverse dimensions, while in the case of broad-area edge-emitting lasers the filaments appeared as quasi-one dimensional structures. Indeed, in the VCSELs, ultra high frequency oscillations with frequencies up to 240 GHz have recently been observed and attributed to dynamical varia-

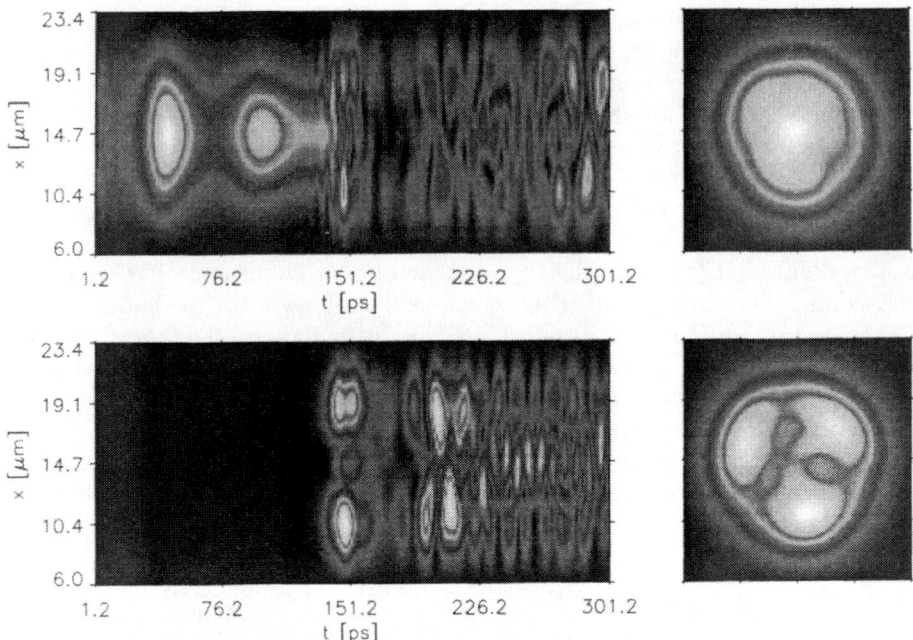

FIGURE 4. Excitation of spatio-temporal polarization dynamics in an isotropic VCSEL. Shown is the temporal evolvement of a center-slice of the near-field intensity (left column) in the x-polarization (top) and the y-polarization (bottom). The two circular intensity-pictures show the two-dimensional temporally averaged near-field intensity characteristics in the x-polarization (top) and the y-polarization (bottom).

tions of multiple transverse modes [16]. There, the reasons for the simultaneous amplification of various transverse modes was associated with mode competition and frequency beating for lower (10–30 GHz) and ultra-high (~ 200 GHz) oscillation frequencies, respectively. Other phenomena related to multi-transverse mode behavior are the onset of self-pulsations [17] or an extremely narrow emission line-splitting [18]. In the time averaged regime, near-field optical measurements reveal the simultaneously observed spatial and spectral variations of the transverse laser modes in VCSELs [19].

In our theoretical description of spatially extended VCSELs we thus have to consider both, spatio-temporal and polarization effects.

In 1995, San Miguel, Feng and Moloney introduced a rate equation model for quantum well VCSELs, based on the observation that the electron-hole pairs in the bands closest to the band gap can be separated into electron-hole pairs emitting only right circular polarized light and electron-hole pairs emitting only left circular polarized light [20]. Effectively, this corresponds to two independent reservoirs of

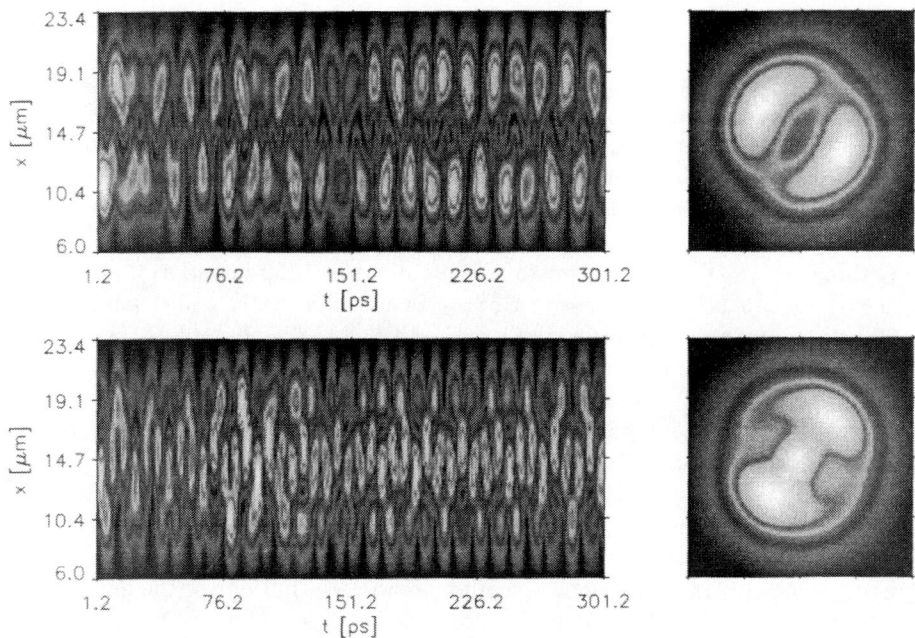

FIGURE 5. Ultrafast spatio-temporal polarization dynamics continuing the traces from Fig. 4, and presented in the same format.

electron hole pairs, each characterized by its own electron-hole pair density, coupled only by a phenomenological spin relaxation process. Instead of simply extending this – spatially homogeneous model – to represent the spatial extension of VCSELs we now discuss results obtained on the basis of a more general and microscopic 4-band Maxwell-Bloch description which takes into account the spin-resolved spatio-spectral carrier dynamics of the carriers along with the spatio-temporal dynamics of the polarized optical fields. Details of the model and the consequences will be presented elsewhere [21]. Although we refrain from an explicit discussion of spin effects, it is important to note that in this model the spin-relaxation of the carriers will be described microscopically, and it occurs via Coulomb-scattering processes.

To unravel some of the closely tied transverse and polarization effects in VCSELs we consider an ideal circular symmetric VCSEL (diameter $d = 15\mu m$) without any externally imposed anisotropy (resulting, e.g., from a non-circular mirror). When highlighting the start-up dynamics we assume for specifity that our VCSEL starts with the preferred light-field polarization direction oriented parallel to the x-direction. Figure 4 shows the time traces of the first 300 ps of the x-polarized (upper row) and y-polarized (lower row) light-field intensity as slices obtained by cutting though the VCSEL-center. The circular intensity patterns in the right column are the corresponding time-averaged intensity profiles. During the initial 150 ps the VCSEL displays the well-known relaxation-oscillations. The characteristic formation of a gain-guide leads to continuous beam-narrowing until the field becomes thin enough such that at the sides higher-order modes gain sufficient room to appear. Up to this point, this single transverse mode is uniquely polarized in the x-direction. With the appearance of the higher-order transverse modes the dynamics changes profoundly: the other polarization direction is being excited by spectrally selective depletion and spin-relaxation due to carrier-carrier scattering. From then on, both polarizations are present with different spatio-temporal dynamics. This can be directly seen in Fig. 5, which illustrates the dynamics during the 300 ps time-frame following that of Fig. 4. As before, the upper row shows the x-polarization and the lower one the corresponding y-polarization. Both polarizations spatio-temporally avoid each other and also differ in their characteristic time scales governed by the microscopic spatio-spectral spin-dynamics. In comparison to the x-polarization which favors a twin-peaked intensity pattern, the y-polarization oscillates quicker between a single peak in the center of the VCSEL and a twin-peaked structure at the rim of the laser.

While the study of spatio-temporal and polarization dynamics certainly will require more detailed investigation, we could already see that one of the major reasons for the excitation of a second polarization mode is the appearance of multiple transverse modes and that the effects responsible for this behavior lie in the microscopic spatio-spectral and spin-relaxations of the carriers.

DELAYED OPTICAL FEEDBACK AND CONTROL OF SPATIO-TEMPORAL DYNAMICS

Up to now, the destabilizing mechanism in the broad-area and vertical-cavity surface-emitting lasers is closely related to the nonlinear interaction of multiple transverse modes. Similar to the situation for broad-area lasers, both temporally and spatially irregular behavior corresponding to deterministic spatio-temporal chaos is observed in coupled multi-stripe laser arrays [1,22]. As this constitutes a severe problem when good beam quality at high intensities is desired, there have recently been initiatives to apply concepts from the field of chaos-control for stabilization. To stabilize spatio-temporal instabilities and chaos a number of schemes using delayed optical feedback [23] and stabilization of travelling waves by means of delayed optical feedback and spatial filtering in a broad-area laser model [24,25] were demonstrated in recent theoretical works. While theoretical investigations of delay-induced temporal instabilities in single-stripe lasers using the Lang-Kobayashi rate equations has been a subject of intense research until today, the investigation of spatio-temporal phenomena occurring in multi-stripe laser arrays [22] and broad-area lasers [9,26] has evolved to an issue of great interest in the last few years. In recent theoretical works [27], both destabilizing mechanisms were combined in investigations on the influence of delayed optical feedback on spatio-temporal dynamics.

In the following, we apply delayed optical feedback to achieve suppression of spatio-temporal instabilities. Thus, we can stabilize stationary operation in the fundamental transverse mode from an originally spatio-temporally chaotic state. We first give a brief survey of the model equations and next discuss a complex eigenmode analysis via the Karhunen-Loève algorithm, which yields a decomposition of the complex optical field into a set of orthonormal transverse modes and the individual oscillation frequencies of these modes. We then show that this permits setting up optical feedback conditions which lead to steady-state operation in the fundamental transverse mode by the suppression of higher-order transverse modes.

Transverse Lang-Kobayashi Model

To represent the transverse light-field diffraction and carrier diffusion effects responsible for the coupling of lasers in arrays we extend the plane-wave Lang-Kobayashi (LK) model [28] (see also the chapters by *Gavrielides* and *Lenstra & Yousefi*) to include transverse effects (TLK) [27]. Due to the structural similarity of these model equations with the LK-model, while still being comparatively simple in structure (in fact, it is the most simple transverse extension of the LK equations), we will explicitly discuss this model here. Generally, the transverse dependence is caused, on the one hand, by transverse coupling via optical diffraction (diffraction coefficient D_p) and charge carrier diffusion (D_f) and, on the other hand, by the transverse inhomogeneity of stripe-geometry lasers. We thus obtain the following set of nonlinear coupled partial differential equations (PDEs) for the complex optical field $E(x,t)$ and the charge carrier density $N(x,t)$:

$$\frac{n_l}{c}\partial_t E = i D_p \partial_x^2 E - [\gamma_E + i\eta(x)] E + \Gamma(x) [g(N) + ik_0 \delta n(N)] E \qquad (1)$$
$$+ \frac{1}{2L}\gamma_R e^{i\Phi} E(x, t-\tau),$$

$$\partial_t N = \Lambda(x) + D_f \partial_x^2 N - \gamma_{nr} N - \frac{2\epsilon_0 c}{\hbar \omega_0 n_l} g(N)|E|^2. \qquad (2)$$

The transversely varying parameters $\Lambda(x)$, $\eta(x)$ and $\Gamma(x)$ describe current injection via stripe electrodes of a given width w and spacing s, index-guiding through transverse index steps located below the stripe electrodes, and the transversely varying confinement-factor, respectively [27]. Further parameters are the nonradiative decay rate γ_{nr} of the carrier density, the refractive index of the active layer n_l, and the carrier frequency and vacuum wave number, ω_0 and k_0, respectively. The variation of the optical gain and the refractive index of the active medium with the carrier density is approximated by the phenomenological linear gain-function $g(N) = a(N - N_0)$ (where a is linear gain coefficient, and N_0 is carrier density at transparency), and $\delta n = -\alpha a N/k_0$, respectively. The linewidth enhancement factor α is fixed at $\alpha = 2$.

The distributed mirror loss is represented by the damping constant $\gamma_E = -\frac{\log\sqrt{R_1 R_2}}{2L}$, where R_1 and R_2 are the power reflectivities of the front and rear facet, respectively. Delayed optical feedback is represented by the feedback parameters γ_R (feedback strength), τ (delay time), and Φ (feedback phase), respectively. The values of the relevant parameters are given in Table 1. Equations (1)-(2) are solved using a Hopscotch method, assuming absorbing boundary conditions at the transverse edges [29].

TABLE 1. Parameters of the multi-stripe semiconductor laser arrays.

L	$250\,\mu m$	cavity length
w	$5.0\,\mu m$	stripe width
s	$6.0\,\mu m$	stripe separation
d	$0.15\,\mu m$	thickness of active layer
R_1	0.32	power reflectivity of the front facet
R_2	0.99	power reflectivity of the rear facet
λ	$815\,nm$	laser wavelength
n_l	3.59	refractive index of active layer
n_c	3.32	refractive index of cladding layer
a	$1.5 \times 10^{-16}\,cm^2$	linear gain coefficient
b	$1.0 \times 10^2\,cm^{-1}$	linear loss coefficient
D_p	$18 \times 10^{-9}\,m$	diffraction coefficient
γ_{nr}	$2 \times 10^8\,s^{-1}$	nonradiative recombination coefficient
Γ	0.5014	confinement factor below stripes
	0.5149	confinement factor between stripes
α_w	$30\,cm^{-1}$	surface absorption constant
α_{sr}	$10^8\,cm\,s^{-1}$	surface recombination constant

Eigenmode analysis

In order to characterize the spatio-temporal complexity in the laser output, it is highly desirable to find out how many optical modes are involved and how they are spatially structured. To this end, eigenmode analysis via Karhunen-Loève decomposition (KLD) has been successfully applied [30,31]. Given a time series of a spatially extended system (from experiment or numerical simulation), this method provides a decomposition into an orthonormal set of eigenmodes. In its original form, this algorithm computes eigenmodes for real input data. In optics, one therefore generally uses data of the output intensity. On this basis, however, the structure of the complex field modes and their respective frequencies are not directly available. As it turns out, for efficient stabilization schemes, information on both, spatial modes and frequency distribution, is crucial. For convenience we briefly add the analysis presented in [32] where the traditional eigenmode analysis has been extended to complex input data pertaining to the spatially resolved optical field.

The complex eigenmodes $\mathbf{p}_n$ corresponding to transverse modes $p_n(x)$ of the optical field are obtained by solving the eigenvalue problem of the Hermitian covariance matrix:

$$\mathbf{C}\mathbf{p}_n = \lambda_n \mathbf{p}_n,$$

$$\text{where} \quad C_{jk} = \frac{1}{T}\int_0^T E^*(x_j,t)E(x_k,t)\,dt \quad j,k = 1\ldots N_x$$

and T and N_x denote the length of the time series and the number of transverse grid points, respectively. The real eigenvalues λ_n yield the relative importance of the complex eigenmodes $\mathbf{p}_n$, which form an orthonormal set, meaning that

$$\sum_{k=1}^{N_x} p_m^*(x_k)p_n(x_k) = \delta_{mn}. \tag{3}$$

The time-varying modal amplitudes are also complex quantities. They govern the dynamics of the individual eigenmodes and are obtained according to

$$a_n(t) = \sum_{k=1}^{N_x} p_n^*(x_k)E(x_k,t). \tag{4}$$

It is the frequency spectra of these modal time series $a_n(t)$ which yield the oscillation frequencies of the complex eigenmodes and will later help us in establishing suitable conditions for achieving stabilizing feedback. ¿From the eigenmodes and expansion coefficients, the original complex optical field can be reconstructed by

$$E(x_k,t) = \sum_n a_n(t)p_n(x_k). \tag{5}$$

The same procedure can be carried out for the carrier density $N(x,t)$. Since $N(x,t)$ is real, its eigenmodes q_n and modal amplitudes $b_n(t)$ are real quantities. The reconstruction of the carrier density is obtained by

$$N(x_k,t) = \bar{N}(x_k) + \sum_n b_n(t)q_n(x_k) \tag{6}$$

where $\bar{N}$ is the time-averaged carrier density profile.

The eigenmodes obtained from complex KLD can be used as an orthonormal basis for mode projection, in analogy to the Galerkin procedure. In this way, the set of PDEs Eqs. (1)-(2) can be reduced to a set of ordinary differential equations (ODEs) [32].

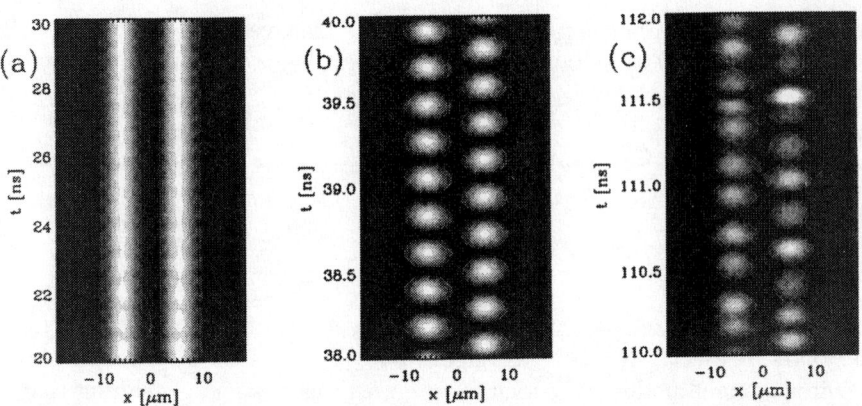

FIGURE 6. Spatio-temporal dynamics of the intensity $I(x,t) = (1-R_1)\epsilon_0 c/n_l |E(x,t)|^2$ for the twin-stripe laser in the regime of coherent continuous wave light emission ($J = 30\,mA$) (a), in the periodic regime ($J = 45\,mA$) (b), and in the chaotic regime ($J = 70\,mA$) (c), for an intermediate diffusion of $D_f = 4 \cdot 10^{-4} \frac{m^2}{s}$. Light shading corresponds to high intensity values.

Stabilization of chaotic laser arrays

Depending on the separation between the laser stripes, semiconductor laser arrays are strongly, moderately or weakly coupled [1,33,34]. In the regime of strong coupling (small separation between the lasers) the dynamical behavior is strongly dependent on the amount of external pumping via the electrical current. With increasing pump current, spatio-temporal instabilities arise in multi-stripe lasers due to the nonlinear interaction of multiple transverse modes. Not surprisingly, the higher the number of transversely coupled laser stripes, the larger the spatio-temporal complexity. The basic effects, however, can already be demonstrated with the simplest laser array, the twin-stripe laser. Therefore, we start our discussion with this simple array. We also show results for a configurations of five transversely coupled laser stripes.

For low values of the pump current, the twin-stripe laser operates in a steady state of coherent light emission with one single double-lobed transverse mode as shown in Fig. 6 (a). However, when we increase the pump current above a critical threshold value J_c, an instability sets in that leads to spontaneous periodic intensity pulsations shown in Fig. 6 (b). Several dynamical regimes with alternating pulsations in the two stripes are observed [32]. Further increase of the current drives the laser into a both spatially and temporally irregular regime as is shown in Fig. 6(c). It is found that the value of J_c strongly varies with the amount of charge carrier diffusion, which is quite significant in semiconductor lasers.

Eigenmode analysis reveals that the spatio-temporal dynamics is governed by the two eigenmodes shown in Fig. 7. The first antisymmetric mode, which will be referred to as the fundamental mode, represents a mode where the two adjacent laser

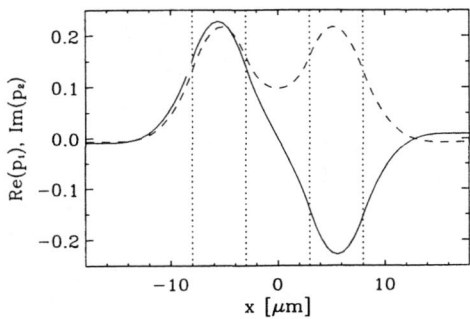

FIGURE 7. Eigenmodes $p_1(x)$ and $p_2(x)$ for the twin-stripe laser. Shown are the real part of the first eigenmode (solid line) and the imaginary part of the second eigenmode (dashed line). The vertical lines indicate the locations of the laser stripes.

stripes oscillate with opposite phase. This mode represents the "natural" operation condition of stable transversely coupled laser arrays. Due to the antisymmetric shape of the fundamental mode, its modulus is vanishing in the center of the gap separating the laser stripes. The second eigenmode is symmetric with respect to the two lasers. The time-dependent expansion coefficients associated with it represent in-phase oscillation of the stripes. In contrast to the anti-symmetric mode, its modulus is clearly non-vanishing in the center of the gap. Therefore, the symmetric eigenmode can profit from the charge carriers diffusing towards the gap more than the fundamental mode. This effect becomes more and more important, as the pump current increases until, at the critical current value, the symmetric mode becomes involved in the dynamics. In the above-mentioned periodic regime, it coexists with the fundamental mode, leading to transverse mode-beating. This beating generates the periodic intensity pulsations, whose frequency equals the difference between the mode frequencies. For high currents there is mode competition rather than coexistence, which leads to chaotic behavior. The frequencies $\Omega_{1,2}$ pertaining to the first and second eigenmodes are obtained via the corresponding frequency spectra of the time series $a_{1,2}(t)$ in Fig. 8. In the periodic regime, the frequencies emerge as sharp lines, while in the chaotic regime the peaks are considerably broadened.

Using the frequencies $\Omega_{1,2}$ of the transverse optical eigenmodes, we are able to establish a destructive interference condition in the symmetric mode, in order to re-obtain steady-state operation in the fundamental transverse mode. In a steady state condition, we can easily realize the effects of the delay term. Imagine that the complex Eq. (1) for $E = R \cdot e^{i\theta}$ is written as two real equations for the amplitude R and the phase $\theta = \omega t$. Then the delay term produces a contribution $\gamma_R R(x) \cos(\Phi - \omega\tau)$ in the amplitude equation, while a term $\gamma_R \sin(\Phi - \omega\tau)$ arises in the equation for the phase. This means that there is a contribution to the gain of strength γ_R times a cosine dependence on the overall phase $\Psi = \Phi - \omega\tau$ which governs the interference condition. The contribution in the phase equation amounts to a

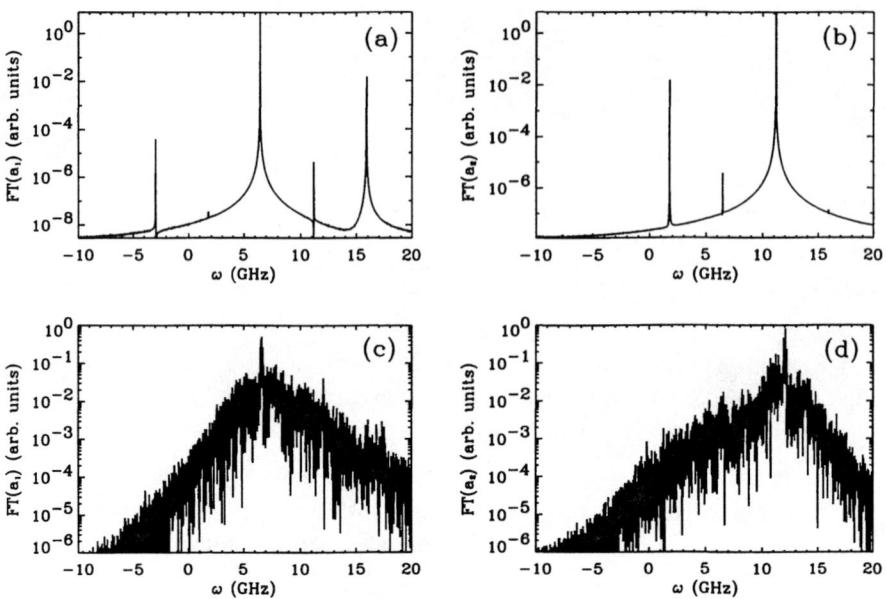

FIGURE 8. Frequency spectra of the time-varying modal amplitudes $a_1(t)$ and $a_2(t)$ of the two eigenmodes in the periodic regime ($J = 45\,mA$), (a,b), and in the chaotic regime ($J = 70\,mA$) (c,d).

frequency shift depending on γ_R and Ψ.

The contribution of delayed optical feedback to the gain can be graphically interpreted as the presence of a cosine-shaped filter in the frequency domain which, for appropriately chosen feedback parameters τ and Φ, can cause destructive interference for light of a particular frequency. This effect can be used to provide damping in a transverse mode oscillating at that frequency.

Let ω_1 be the frequency of the fundamental mode and ω_2 that of the mode which we aim to suppress. The two parameters τ and Φ permit to set up the two interference conditions:

$$\Phi - \omega_2 \tau = \left(n + \frac{1}{2}\right), \tag{7}$$

$$\Phi - \omega_1 \tau = m \quad n, m = 0, \pm 1, \pm 2 \ldots \tag{8}$$

(Here and in what follows, all phases are written in units of 2π). Equation (7) represents destructive interference in the transverse mode to be suppressed, while Eq. (8) describes constructive interference in the fundamental transverse mode, which seems to be a natural choice. Then τ and Φ are determined by the following equations as

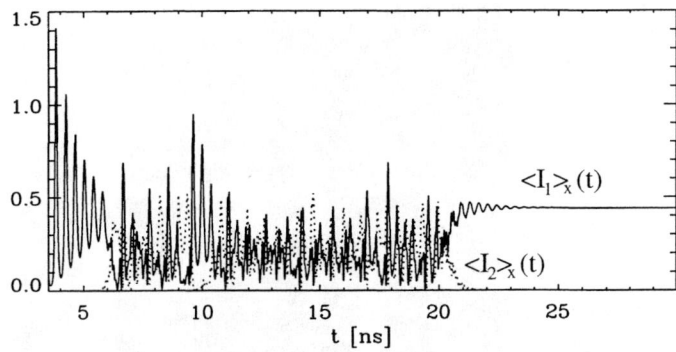

FIGURE 9. Time series of the transversely averaged intensities $\langle I_1 \rangle_x(t)$, $\langle I_2 \rangle_x(t)$ pertaining to the two eigenmodes (solid line: fundamental mode, dashed line: symmetric mode). Stabilization of the antisymmetric fundamental transverse mode is achieved after switching on time delay at $t_0 = 20\,ns$. For $t < t_0$, $\gamma_R = 4 \cdot 10^{-2}$, $\tau = \Phi = 0$ and for $t > t_0$, $\gamma_R = 4 \cdot 10^{-2}$, $\tau = 0.0862\,ns$, $\Phi = 0.543$.

$$\tau = \frac{1}{2(\omega_2 - \omega_1)} \quad (9)$$

$$\Phi = \omega_1 \tau \quad (10)$$

where the integers n and m were chosen such that the two frequencies for which the interference conditions are provided correspond to adjacent extrema of the cosine filter.

For a chaotic twin-stripe laser with $J = 80\,\text{mA}$, $D_f = 4.0 \cdot 10^{-4} \frac{m^2}{s}$ (which is well within the chaotic range) we obtain the modal frequencies $\Omega_1 = 6.3\,\text{GHz}$ and $\Omega_2 = 12.1\,\text{GHz}$. Identifying $\Omega_{1,2}$ with $\omega_{1,2}$ from Eqs. (9)-(10) then yield the values $\tau = 0.0862\,\text{ns}$ and $\Phi = 0.543$. With these parameters we indeed obtain stable cw operation in the fundamental antisymmetric mode. The stabilizing effect of delayed optical feedback is demonstrated in Fig. 9. For $t < t_0 = 20\,ns$, $\gamma_R = 4 \cdot 10^{-2}$, $\tau = \phi = 0$, and for $t > t_0$, the paramters τ and Φ are set to the above values. After the chaotic interval, the symmetric eigenmode is quickly damped to zero, while the antisymmetric eigenmode approaches cw emission via damped relaxation oscillations. When stabilization is achieved and constructive interference is exact, the original system of equations (without delayed feedback) is effectively recovered. In this case, only the mode exists for which $\Psi = 0$. Thus, the delay term in the phase equation is zero and the delay term in the amplitude equation reduces to a slightly enhanced reflectivity (γ_R). In this sense, our procedure bears some analogy to genuine schemes of controlling chaos where, upon successful control, the controlling force vanishes [35,36].

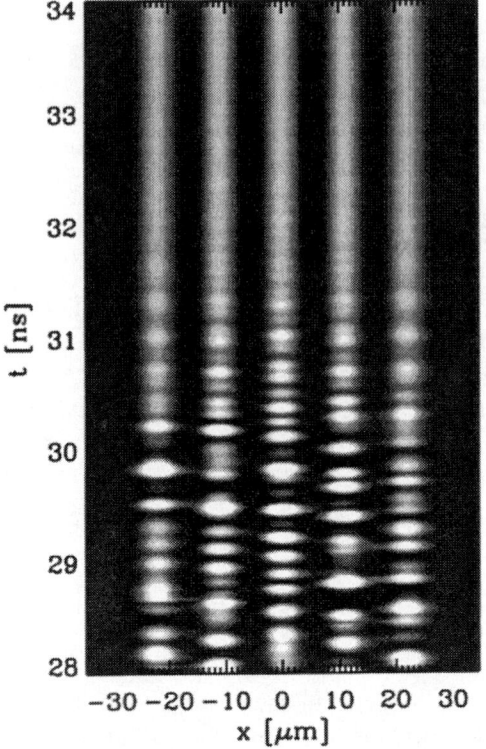

FIGURE 10. Spatio-temporal dynamics of the five-stripe laser for $J = 80\,mA$, $\gamma_R = 3 \cdot 10^{-2}$, $\tau = 0.0877\,ns$ and $\Phi = 0.06$ where the time delay and feedback phase are switched from zero to the above values at $t = 30\,ns$.

Eigenmode analysis of multi-stripe lasers yields that the maximum number of relevant eigenmodes in the complex spatio-temporal regimes is equal to the number of stripes. Thus, in principle, several transverse modes have to be suppressed in order to stabilize laser arrays with more than two stripes. In the case of the five-stripe array, where four relevant transverse modes exist besides the fundamental mode, stabilization of the fundamental mode can still be achieved. An example where the five-stripe laser is brought from a state of spatio-temporal chaos to cw operation is shown in Fig. 10. However, the criteria in respect to the way how to choose ω_2 are not as obvious as in the case of the twin-stripe laser. The mode to be suppressed varies with pump current and diffusion strength, and in some cases the best results are obtained when ω_2 lies between two transverse mode frequencies. The fact that stabilization is still possible could be plausible, if the successive transverse modes are nonlinearly coupled in a cascade-like fashion. Then, damping of particular modes could cause the suppression of the whole cascade of modes.

CONCLUSIONS

In this chapter we sketched the hierarchy of theoretical models for spatially extended semiconductor lasers. The hierarchy ranges from microscopic semiconductor laser Maxwell-Bloch models which take into account the spatio-temporal dynamics of the charge carriers and interband polarization to a phenomenological and linear representation of the gain and induced refractive index change. Moreover, typical approximative treatments – the plane wave and mean-field approximations – were discussed with respect to their application to modeling edge-emitting or vertical-cavity surface-emitting semiconductor lasers. Representative results from microscopic simulations showed the spatio-temporal dynamics of broad-area lasers and the connection of spatio-spectral and polarization dynamics in VCSELs. The application of delayed optical feedback on spatially extended and coupled laser arrays was discussed on the basis of the transverse Lang-Kobayashi-equations. With the characteristic spectra pertaining to the stabilizing and destabilizing modes of the laser arrays (which have been obtained by complex eigenmode-analysis) a control-scheme was described which allows to suppress the spatio-temporal instabilities.

REFERENCES

1. Hess, O., and Kuhn, T., em Prog. Quant. Electr. **20**, 85 (1996).
2. Hess, O., and Kuhn, T., *Phys. Rev.* **A 54**, 3347 (1996).
3. Bowden, C.M., and Agrawal, G.P., *Phys. Rev.* **A 51**, 4132 (1995).
4. Yao, J., Agrawal, G.P., Gallion, P., and Bowden, C.M., *Opt. Comm.* **119**, 246 (1995).
5. Balle, S., *Opt. Comm.* **119**, 227 (1995).
6. Ning, C.Z., Indik, R.A., and Moloney, J.V., *IEEE J. Quant. Electr.* **33**, 1543 (1997).
7. Hess, O., Koch, S.W., and Moloney, J.V., *IEEE J. Quant. Electr.* **QE-31**, 35 (1995).
8. Hess, O., and Kuhn, T., *Phys. Rev.* **A 54**, 3360 (1996).
9. Fischer, I., Hess, O., Elsäßer, W., and Göbel, E., *Europhys. Lett.* **35**, 579 (1996).
10. Gehrig, E., and Hess, O., *Phys. Rev.* **A 57**, 2150 (1998).
11. Chang-Hasnain, C.J., Harbison, J.P., Hasnain, G., Lehmen, A.C.V., Florez, L.T., and Stoffel, N.G., *IEEE J. Quant. Electr.* **27**, 1402 (1991).
12. Koyama, F., Morito, K., and Iga, K., *IEEE J. Quant. Electr.* **QE-27**, 1410 (1991).
13. Vakhshoori, D. *Appl. Phys. Lett.* **65**, 259 (1995).
14. Choquette, K. D., Schneider, J.P., Lear, K.L., and Leibenguth, R.E., *IEEE J. Selec. Topics* **1**, 661 (1995).
15. van Doorn, A.K.J., van Exter, M.P., and Woerdman, J.P., *Appl. Phys. Lett.* **69**, 1041 (1996).
16. Buccafusca, O., Chilla, J.L.A., Rocca, J.J., Wilmsen, C., Feld, S., and Leibenguth, R., *Appl. Phys. Lett.* **67**, 185 (1995).
17. Nugent, D. .H., Plumb, R.G.S., Fischer, M.A., and Davies, D.A.O., *Electron. Lett.* **31**, 43 (1995).
18. Epler, J.E., Gehrsitz, S., Gulden, K.H., Moser, M., Sigg, H.C., and Lehmann, H.W., *Appl. Phys. Lett.* **69**, 2312 (1996).

19. Hörsch, I., Kusche, R., Marti, O., Weigl, B., and Ebeling, K.J.. *Appl. Phys. Lett.* **79**, 3831 (1996).
20. San Miguel, M., Feng, Q., and Moloney, J.V., *Phys. Rev.* **A 52**, 1728 (1995).
21. Hamm, J., and Hess, O., submitted (2000).
22. Adachihara, H., Hess, O., Indik, R.A., and Moloney, J.V., *J. Opt. Soc. Am. B* **10**, 496 (1993).
23. Simmendinger, C., and Hess, O., *Physics Lett. A* **216**, 97 (1996).
24. Bleich, M.E., Hochheiser, D., Moloney, J.V., and Socolar, J.E.S., *Phys. Rev.* **E 55**, 2119 (1997).
25. Simmendinger, C., Münkel, M., and Hess, O., *Chaos, Solitons & Fractals* **10**, 851 (1999).
26. Hess, O., and Schöll, E., *Physica D* **70**, 165 (1994).
27. Münkel, M., Kaiser, F., and Hess, O., *Physics Lett. A* **222**, 67 (1996).
28. Lang. R., and Kobayashi, K., *IEEE J. Quant. Electr.* **QE-16**, 347 (1980).
29. Hess, O., *Spatio-Temporal Dynamics of Semiconductor Lasers*, Wissenschaft und Technik Verlag 1993.
30. Hess, O., *Chaos, Solitons & Fractals* **4**, 1597 (1994).
31. Merbach, D., Hess, O., Herzel, H., and Schöll, E., *Phys. Rev.* **E 52**, 1571 (1995).
32. Münkel, M., Kaiser, F., and Hess, O., *Phys. Rev.* **E 56**, 3868 (1997).
33. Winful, H.G., and Wang, S.S., *Appl. Phys. Lett.* **53**, 1894 (1988).
34. Wang, S.S., and Winful, H.G., *Appl. Phys. Lett.* **52**, 1774 (1988).
35. Auerbach, D., Grebogi, C., Ott, E., and Yorke, J., *Phys. Rev. Lett.* **69**, 3479 (1992).
36. Pyragas, K., *Physics Lett. A* **170**, 421 (1992).

Semiconductor Laser Device Modeling

J.V. Moloney

Arizona Center for Mathematical Sciences
Department of Mathematics
University of Arizona
Tucson, Arizona 85721

Abstract. The semiconductor laser offers enormous advantages as a practical device in many technological applications because of its very large gain per unit length. This translates into compact size, making it an ideal system for integration into complex integrated opto-electronic systems networks. However, this advantage comes at a price. Achieving large gain in small volumes introduces significant thermal and electrical transport issues. The enormous gain in bandwidth of the laser and the complicated dependence of gain and refractive index on carrier density, plasma and lattice temperature, makes it particularly challenging for study as a nonlinear dynamical system. A systematic, predictive and robust study requires a multi-faceted first-principles approach that combines the microscopic physics of the light-semiconductor medium interaction with the proper resolution of the spatial and temporal degrees of freedom of the running laser. In this lecture, I will stress features that highlight the need to treat the behavior of the laser at this more sophisticated level beyond the simple and well-used phenomenological single or coupled mode models. The picture that will emerge is that, although in many cases the laser may be running nominally as a single longitudinal mode device, the other suppressed modes are still active and can grow under the influence of weak external perturbation. Wide aperture lasers, designed for high power, high brightness applications, are intrinsically dynamically unstable and display multi-longitudinally-moded chaotic filamentation instabilities. After briefly reviewing various analytic approaches to studying spatio-temporal dynamics in semiconductor lasers, I will discuss specific examples of a low-power external cavity device and a variety of high power wide aperture systems.

INTRODUCTION

A proper description of the semiconductor laser requires a full microscopic approach that accounts for the fact that the system behaves as a multi-component plasma. Coupling the microscopic degrees of freedom to the optical propagation problem is a daunting task and not feasible at present. Some attempts have been made to include the microscopic degrees of freedom at a computationally manageable level and are typically restricted to the so-called "free-carrier" model. Inclusion of many-body Coulomb effects at the level of a 2 or 3 band model, which makes a relaxation rate assumption, are equally unlikely to be of predictive value. The idea put forth in this lecture is that it is more practical to de-couple the microscopic physics from optical propagation and use a quasi-equilibrium assumption to generate gain tables that feed into the latter. In this way, various many-body effects can be accounted for while making the approach feasible. Important dynamical influences such as strong shifts of gain peak and bandwidth with carrier density, plasma and lattice temperature then can be included. The microscopically computed gain/index spectra can be independently validated from experimental measurements. Our main emphasis will be on going beyond conventional approaches, to exploit recent rapid advances in computer processor speed and high performance graphics. The goal will be to build powerful graphics-based interactive simulation and visualization tools for the study of semiconductor lasers in isolation or as part of a complex integrated opto-electronic network.

It is well established that external perturbations such as injection or optical feedback that are often exploited to stabilize low power semiconductor lasers and reduce the laser linewidth, can also induce the system to behave in an erratic chaotic fashion. This chaotic behavior has now been documented in a variety of semiconductor laser devices including edge-emitting bulk, Quantum Well (QW) and DFB devices. Broad area edge-emitting devices introduce additional lateral degrees of freedom and exhibit complex spatio-temporal dynamics in the absence of external perturbation. The latter instability manifests itself most commonly as a strong broadening of the far-field emission angle well beyond the expected diffraction-limited spot size. Instabilities in these lasers originate from the destabilization of traveling wave solutions, which were first identified in 1991-1992 for two-level lasers[1,2]. These off-axis emission solutions were later referred to as "tilted waves" by a number of authors. The key difference between a semiconductor laser and the two-level laser is the fact that the refractive index of the semiconductor medium is nonzero at peak gain and moreover shows a strong variation with carrier density and plasma temperature. A phenomenological measure of this dependence is the so-called Linewidth Enhancement or α-factor that can be determined experimentally by various means. The microscopic approach mentioned above allows for an explicit and quantitative determination of this quantity, although at significant computational complexity. This factor provides a measure of the strength of coupling between the amplitude and phase of the optical field and is responsible for strong frequency chirping of optical pulses generated with semiconductor amplifiers and modulated lasers and for the lateral filamentation instability observed in high-power wide aperture devices.

In the next section, I will review the features of the microscopic physics that have a profound influence on the actual gain, refractive index and α-factor spectra of a particular semiconductor material. We will see that carrier-carrier and carrier-phonon scattering events between various sub-bands in the material that occur of femtoseconds timescales, profoundly modify the shape and magnitude of the semiconductor gain. Technical details will be avoided as the subject would form the basis for an entire course rather than a single lecture. Instead, references will be given to the appropriate literature. The issue of carrier transport from confining barrier regions into narrow QWs is of great current interest, especially in telecommunications applications. This topic will be briefly touched upon. With the microscopically computed, gain and index spectra in hand, we then proceed to incorporate these into the optical propagation problem. This will be the topic of Section III. The latter problem will be described by a set of coupled nonlinear partial differential equations

(PDEs) for the counter-propagating optical fields, the total carrier density, lattice temperature etc. By retaining the PDE description, our approach becomes highly flexible and reflects realistic experimental situations. This is critical as we will want to build a robust opto-electronic system simulator of our nonlinear dynamical system, incorporating various optical components such as HR-, AR-facet coatings, gratings (DBR, DFB, straight, curved, first- and second-order), external optical injection and feedback, coupled lasers and so on. Moreover, the strong dynamical variation in carrier density within operating semiconductor lasers and the associated shifts in gain peak and gain dispersion, etc cannot be adequately captured within a coupled mode reduction. Some illustrations will be provided of the role of external optical feedback in promoting multi-longitudinal modes dynamics and power-dropouts in a low-power external-cavity semiconductor laser. Section IV will focus on high-power wide aperture lasers and the added complexity of filamentation instabilities. Here intensity hot-spots generated within the device greatly increase the far-field emission angle beyond the ideal diffraction limit and can lead to facet or internal damage of the device. These lasers are best viewed as weakly optically turbulent systems. One approach to improved system performance involves introducing flared semiconductor amplifier sections with the view to filtering the lateral higher order modes. Another approach would be to accept that this optically turbulent behavior is inevitable and try to implement control schemes, which take advantage of our knowledge that the underlying spatio-temporal dynamics involves unstable traveling wave solutions [3].

SEMICONDUCTOR GAIN AND INDEX SPECTRA

The simple textbook picture of an inverted semiconductor laser medium is the two parabolic band model with electrons occupying the higher conduction band and holes the lower valence band. The next level of complexity beyond this is to recognize that the lowest lying conduction band, in say GaAs, has s-type orbital symmetry whereas the highest occupied valence band has p-type orbital symmetry. Ignoring electron spin, we conclude that there should be a single conduction band and a triply-degenerate valence band at the so-called Γ-point ($\mathbf{k} = 0$, where $\mathbf{k}$ is the electron momentum) at the center of the crystal Brillouin zone. Spin-orbit angular momentum coupling splits the triply degenerate valence band into a light-hole, heavy-hole and split-off band. The split-off band is so strongly detuned from the light and heavy hole bands that its contribution to optical transitions can be safely ignored. In a QW structure, quantization effects along the material growth direction further mix the light-hole (lh) and heavy-hole (hh) valence bands at the zero crossing points shown in Fig. 1. The final three-band picture of an inverted semiconductor medium is as shown in this figure where the hh and lh bands are actual admixtures. An excellent discussion of this topic can be found in the textbook by Chow, Koch and Sargent [4].

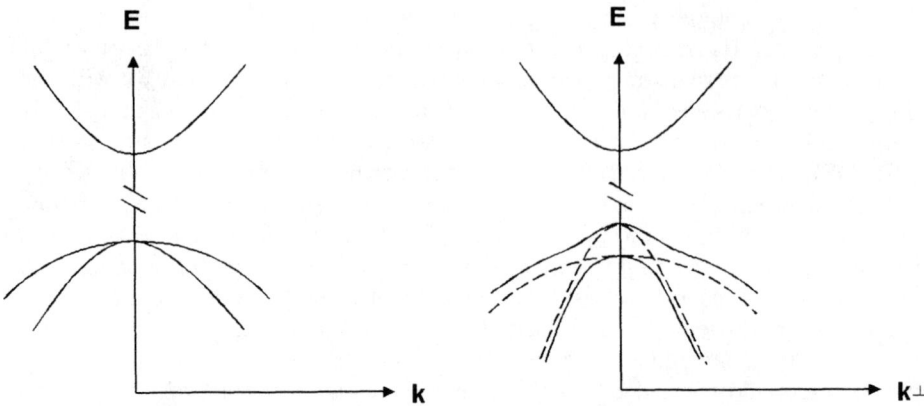

Figure 1 Schematic drawing of a three-band inverted semiconductor medium. The dashed curves depict the light-hole (lh) and heavy-hole (hh) bands of a bulk medium whereas the solid curves correspond to a quantum-well material. The latter valence bands are non-parabolic.

This three-band picture accounts qualitatively for most phenomena observed in edge- and surface-emitting semiconductor lasers. Transverse electric polarized (TE) emission, where the electric field lies in the plane of the laser heterojunction, essentially involves dipole transitions between the conduction and heavy-hole (e-hh) valence band. Transverse magnetic polarized (TM) emission, where the electric field lies along the growth direction, essentially involves dipole transitions between the conduction band and the light-hole (lh) valence band (e-lh). A many-body calculation of the gain and refractive index for this 3-band material would involve solving the Semiconductor-Bloch equations [4] coupled to a quantum kinetic equation (Quantum Boltzmann) that describes the microscopic carrier-carrier and carrier-phonon scattering between individual bands and the lattice vibrations. Replacing the full solution of the quantum kinetic equation by a rate assumption, which is often the case in many-body approaches, introduces non-physical artifacts into the gain spectra. The best known of these is finite absorption below the renormalized bandgap. Such an assumption also yields the incorrect gain lineshape and shows significant deviations in the peak gain value [5]. This subject is highly technical and will not be elaborated on further here. Instead, we will show graphs of experimentally validated gain and index spectra and simply use such spectra in the simulation.

The computation of realistic gain spectra requires that one includes all of the relevant occupied valence and conduction sub-bands within the QW as well as unconfined barrier states in the surrounding barrier regions. As an illustration we will consider an InGaAs QW with AlGaAs barrier regions. Two cases will be considered where we simply change the well width by a factor of two and keep everything else the same. These examples provide a nice illustration of why bandgap engineering at the materials level is particularly vital in designing semiconductor lasers for specific applications. The figure below shows the material gain/absorption and refractive index spectra for a 5 nm $In_{0.2}Ga_{0.8}As$ and a 10 nm $In_{0.2}Ga_{0.8}As$ QW with an Al_xGa_{1-x} As GRINSCH barrier region, where x ranges from 0.1 to 0.6 over a distance of 85 nm [6]. Each curve represents a fixed carrier density and the value ranges from a sheet density of 1×10^{12} cm^{-2} to 8×10^{12} cm^{-2}. The figures show the low density absorption and high density gain with the bumps in the gain spectra for the 10 nm QW showing gain on higher sub-bands at the higher sheet density. These spectra were computed at 300° K with a total of 12 hole and 8 electronic sub-bands for the 10 nm QW and a total of 8 hole and 5 electronic sub-bands for the 5 nm QW. For the 10 nm QW, 5 hole and 3 electronic sub-bands are confined in the well whereas for the 5 nm QW, 2 hole and 1 electronic sub-bands are confined in the well.

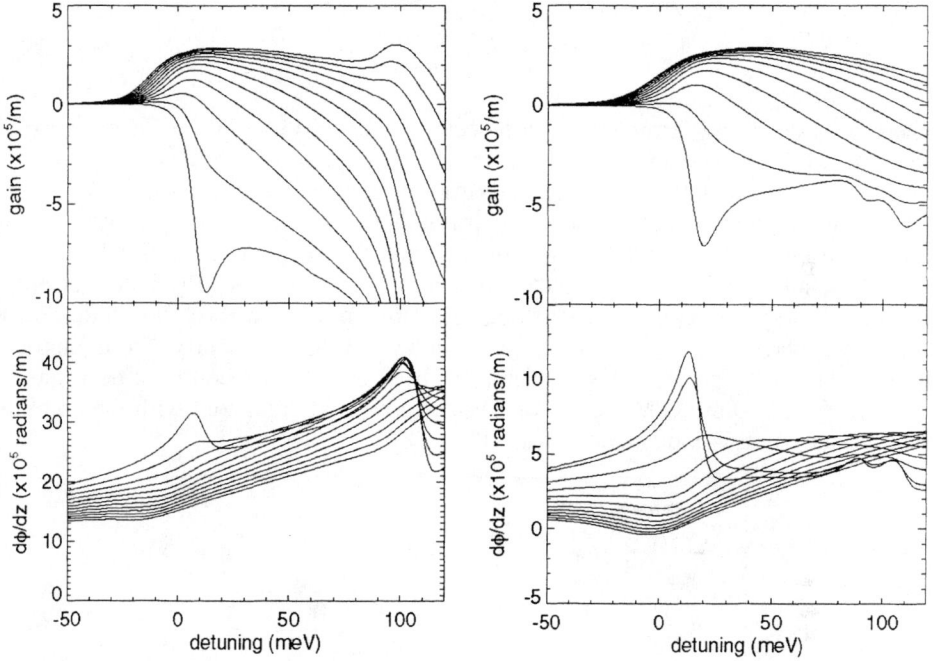

FIG. 2. Computed many-body gain and refractive index spectra for a 10 nm (left) and 5 nm (right) $In_{0.2}Ga_{0.8}As$ QW. The carrier sheet density varies between 0.5×10^{12} – 6.5×10^{12} cm^{-2}. The lowest gain curve at $N = 0.5\times10^{12}$ cm^{-2}, in each top picture shows a strong exciton feature. Reproduced with permission [25].

Note the enormous (> 30 nm or 12 THz) gain bandwidth at the higher densities where the gain flattens out. A 500 μm long semiconductor laser has a longitudinal mode spacing on the order of 70 GHz indicating that about 170 modes lie under the flattened portion of the gain curve. Of course, the very weak gain curvature is sufficient to suppress most of these modes from lasing unless an external perturbation is applied or the laser is extended in the lateral dimension as in a broad area device. Another notable feature is the marked shift in peak gain (approximately 0.5 THz) when the carrier density varies between 1×10^{12} - 6×10^{12} cm^{-2}. Any semiconductor laser dynamic, which involves significant swings in carrier density, will also have large excursions in the peak gain. A unique feature of the semiconductor laser is the nonzero refractive index at peak gain as is obvious in each of the pictures. Unlike the gain, the refractive index decreases with increasing density. A useful qualitative measure of the degree of coupling between the amplitude and phase of the optical field is the so-called Linewidth Enhancement or α-factor. This quantity is defined as:

$$\alpha(N,\omega) = \frac{\left.\dfrac{d\Re e\{\chi\}}{dN}\right|_{N=N_{th}}}{\left.\dfrac{d\Im m\{\chi\}}{dN}\right|_{N=N_{th}}}$$

where, $\chi(N,\omega)$ is the semiconductor dielectric susceptibility. The α-factor is usually assumed to be a constant, independent of density and measured at peak gain. However, semiconductor lasers typically run at frequencies down-shifted from peak gain by one or more longitudinal mode spacings and the carrier density can undergo strong swings in magnitude. Rather than freeze the α-factor to a fixed value, we instead include the full semiconductor dielectric response in our PDE simulations. In this way, the system can adapt the local gain/absorption and refractive index on the fly. Moreover, the dispersion of the gain, index and consequently, the α-factor are automatically incorporated. The α-factor has been measured recently at peak gain for both the 5 and 10 nm QW devices discussed above and compared with the combined bandstructure/many-body calculations. The results are displayed in Fig. 3.

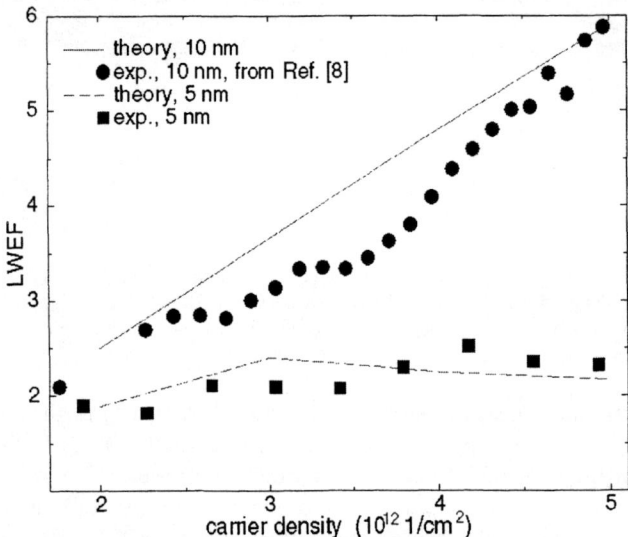

Figure 3. Linewidth enhancement factor for the 10 nm and 5 nm In$_{0.2}$Ga$_{0.8}$As QWs measured at peak gain as a function of carrier sheet density. Reproduced with permission [25].

The LWEF for the 10 nm QW shows a linear increase with carrier density whereas the narrower QW shows a clamping of this quantity around a value of 2.0 – 2.5. Measurement of the α-factor is a particularly stringent constraint on the theory as it allows little or no flexibility in the carrier density. Inspection of the gain plots in Fig. 2 show why there is such a different behavior between the two QW structures. The differential gain is roughly the same for both QWs but the differential refractive index ($\propto \chi'$) is much larger for the wider well (Compare the scalings of the ordinate in both cases). The refractive index is much more sensitive to non-resonant states as its influence falls off as ω^{-1}, whereas the absorption/gain falls off as ω^{-2}. Getting the α-factor right therefore requires the inclusion of many more sub-bands than are needed to compute the gain.

THE OPTICAL PROPAGATION PROBLEM

The semiconductor optical response can be computed from the microscopic many-body theory for single or multiple QW media with different constituent materials, varying material composition or varying well widths. As emphasized above, the confining barriers can play a prominent role in obtaining convergent results. In order to properly describe propagation of the optical fields, carrier density and temperature within the laser itself, we will need an algorithm which can accurately resolve the enormous gain bandwidth of the semiconductor medium itself while, at the same time, being able to follow that large gain peak shifts with density. For many realistic lasers, the low density absorption features in the above spectra, will play an important role in determining the laser dynamics. For example, one may wish to describe a multiple contact device with one section being run as a saturable absorber section. In wide aperture high power lasers, the unpumped region outside the contact itself, plays an important role determining how the carriers are bleached at high current levels. It is important therefore to be able to capture the gain/absorption and refractive index dispersion features over that bandwidth which encompasses the full dynamical range of the device starting at switch-on from noise. This is a major computational challenge requiring that we resolve time steps of tens of femtoseconds while wanting to propagate the relevant fields over hundreds of nanoseconds to microseconds.

A Parallel Partial Differential Equation Solver

An earlier pde approach to capturing the dispersion of the semiconductor material response was to use a "pole method" where a background term with a sufficient number of Lorentzian oscillators was taken to mimic the gain around the peak gain and over a sufficient bandwidth around this peak [7,8]. This approach worked quite well as long as carrier densities were not approached where there was strong gain flattening or the appearance of additional bumps in the gain due to higher sub-bands contributing to lasing. Capturing the low density absorption features was a challenge as these curves cannot be adequately approximated by Lorentzian-like lineshapes. The pole approach allowed the lineshape parameters to depend on carrier density so that the varying gain peak and bandwidth could be adequately described. The challenge lies in designing a robust algorithm that can represent the induced semiconductor polarization for widely different shapes of the complex frequency dielectric response. This can be achieved by designing a "memory function" or digital filter that directly approximates the convolution integral, where, $\chi(N,t-\tau)$ is obtained by inverse Fourier transforming the data in Fig. 2.

$$P = \varepsilon_0 \int_{-\infty}^{+\infty} \chi(N, t-\tau) E(\tau) d\tau$$

AN INTERACTIVE PDE SIMULATOR

The goal of this section is to illustrate how implementation of the above algorithm in an interactive setting can be used to a study a variety of semiconductor laser dynamical problems. We will discuss a (1+1)D (space and time) plane wave implementation where we envisage the semiconductor laser as being a single component within an integrated opto-electronic system. The idea here is that the user can interact in real-time with a simulation as it proceeds. This setting mimics a laser running in the laboratory. The "laser" is the simulation running in the background on one or more dedicated fast processors. The user will be able to look into the laser itself by firing up a remote graphical interface which will contain "pull-down" menus allowing one to change critical system parameters on the fly. An optical system editor (OSGE) will allow the user to assemble the opto-electronic system graphically by combining various modules representing semiconductor lasers, optical fibers, AR-, HR- coatings, beam splitters, etc. This will allow us to study interactively, for the first time, a realistic assembled opto-electronic environment as a collective nonlinear dynamical system. Extending the interactive pde simulator to wide aperture high power devices is nontrivial and requires a parallel implementation on a graphical multi-processor computer. This (2+1)D simulation will be dedicated to resolving fully the spatio-temporal evolution of the optical fields, the carrier density and temperature fields within the 2D (y-z) active layer of the device. Plans are underway to extend this pde solver to (3+1)D by coupling the electrical and thermal degrees of freedom in the transverse growth dimension.

Optical Propagation in the Active Layer

The problem of propagating the optical fields, carrier density and lattice temperature within the active region is described by the following set of coupled nonlinear pdes:

$$\frac{\partial E^\pm}{\partial t} \pm \frac{c}{n_g}\frac{\partial E^\pm}{\partial z} = \frac{ic}{2n_g}\left\{\frac{1}{k}\frac{\partial^2 E^\pm}{\partial y^2} + k\Gamma P^\pm\right\} + \frac{i\omega_c}{n_g}\frac{\partial n_b}{\partial T}(T-T_a)E^\pm$$

$$P^\pm = \varepsilon_0 \int_{-\infty}^{+\infty} \chi(N, t-\tau) E^\pm(\tau) d\tau$$

$$\frac{\partial N}{\partial t} = \frac{\partial}{\partial y}\left(D_N \frac{\partial N}{\partial y}\right) - \gamma_{nr} N + \frac{\eta J}{ew} + \left\{P^{+*}E^+ - P^+E^{+*} + P^{-*}E^- - P^-E^{-*}\right\}$$

$$\frac{\partial T}{\partial t} = D_T \frac{\partial^2 T}{\partial y^2} + \gamma_T (T_a - T) + \frac{\hbar\omega_c}{C}\gamma_{nr} N + \frac{SRI^2}{Cw}$$

subject to the boundary conditions,

$$E^+(y,0,t) = -\sqrt{R_1}\, E^-(y,0,t)$$
$$E^-(y,L,t) = -\sqrt{R_2}\, E^+(y,L,t) + \sqrt{1-R_2}\, E_{inj}(L,t)$$

This system of equations is stripped down to reflect the core aspects of the nonlinear evolution. For example, we could add distributed grating terms as linear self- and cross-coupling terms in the counter-propagating optical field ($E^\pm$) equations. For a semiconductor laser extended in the lateral dimension, these gratings could be straight or curved and first-order or second-order. Second-order gratings are included to couple the output light through the surface rather than the ends as in a conventional edge-emitting high power laser. The counter-propagating optical field equations allow for propagation along the laser axis and for diffraction in the lateral dimension. The confinement factor Γ measures the fraction of the optical mode overlapping with the active layer. For bulk materials $\Gamma \approx 0.2 - 0.5$ whereas for a quantum well, $\Gamma \approx 0.02 - 0.05$ typically. The carrier density decays via non-radiative recombination and carriers are pumped into the active layer from an external current source. In addition, the carriers diffuse in the lateral dimension (diffusion coefficient D_N) and are also removed through stimulated recombination (third term on the right).

The polarization source term represents the coupling to the material semiconductor optical response and is constructed from the microscopically computed gain and index spectra, such as those appearing in Fig. 2. This term is represented by the convolution (second line) and its efficient evaluation is critical to the success of the particular algorithm. Originally this term was approximated by a "pole" approach which essentially involved adding a background term and a sufficient number of Lorentzian oscillators to approximate the gain dispersion over a wide frequency bandwidth as needed. As mentioned above, this approach is rather limited and we now approximate the polarization by a "memory" function or digital filter. In effect, the convolution integral is treated as a digital filter acting on past histories of the optical field at each point in space. Each electric field is advanced one time step Δt as follows:

$$E(r,t+\Delta t) = \frac{ic\Delta t \Gamma}{2n_g k \varepsilon_0 \varepsilon_b} \sum_{m=0}^{M-1} W_m(N(r)) E(r, t - m\Delta t)$$

The details of the implementation of this algorithm are rather involved so instead we simply contrast it with the earlier pole method in Fig. 6. In this figure we take the raw microscopically computed gain spectral data at three different densities spanning absorption through to gain and fit the corresponding numerical dispersions for the pole and memory function algorithms.

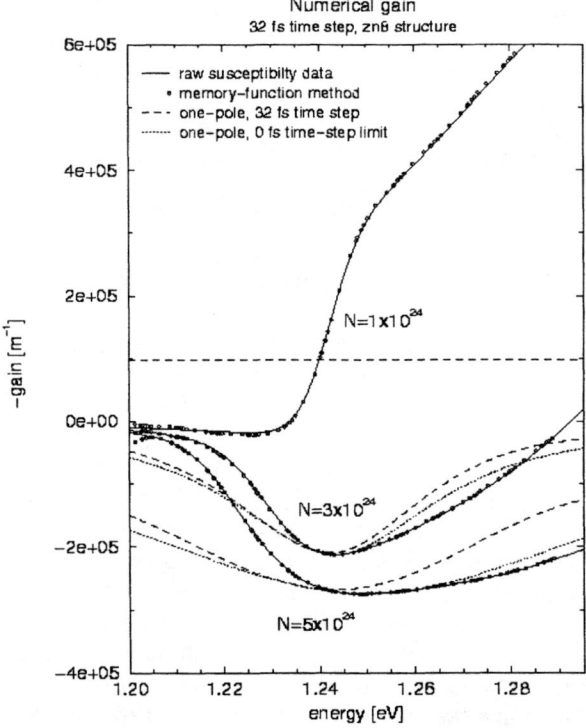

Figure 4. Comparison of the microscopically computed gain dispersion and numerical gain dispersions for the pole and memory function algorithms at sheet densities $N = 1 \times 10^{12}$, 3×10^{12} and 5×10^{12} cm^{-2}. Reproduced with permission [25].

The pole fits, shown for two different time steps, only work for the higher density gain shapes and even then, there is a strong discrepancy between the pole fits (dash – dotted curves) and the actual gain especially at the higher density. In contrast, the memory function algorithm is indistinguishable from the real gain data for both absorption and gain regions. The memory function approach also matches gain spectra with multiple sub-band peaks. Our simulations are typically set up to resolve a 30 nm frequency bandwidth in order to allow for broadband spontaneous emission and strong many-nanometer shifts in gain peak during large swings in the carrier density. In practice, the final lasing state will span a much narrower frequency bandwidth although pulsed high power broad area devices can display spectral emission spanning 6-7 nm. The simulation results presented in the section on high power laser diodes exclusively employ the memory function approach.

An Interactive Opto-Electronic Simulator

A schematic of the layout of the plane wave simulator is presented in Fig. 5. The simulator carries out the dedicated computation and is written in either Fortran, C or C++. The simulator is written so as to be highly portable and can run over a wide mix of platforms. The rate limiting step is the core computation within simulator itself and these may need to be distributed over multiple processors when, for example, two or more semiconductor lasers are coupled together as in a master-slave arrangement or a semiconductor laser is coupled to an optical fiber. The graphical viewer can be initiated on the same machine or remotely, allowing the user to simultaneously view say, the raw or detector-averaged optical power at either laser output facet, the time evolving laser output spectrum, reconstruction of a chaotic attractor on the fly, etc. The optical system graphical editor (OSGE) is the control center, performing the dual function of initial system assembly and intervention as the simulation proceeds. One can imagine a scenario where the semiconductor laser is running chaotically and one wants to perturb the system to test for robustness of the chaotic attractor.

As an illustration, we consider the dynamics of a semiconductor laser in the presence of weak external optical feedback. The Lang-Kobayashi delay differential equation approach captures much of the qualitative dynamical behavior of this laser system, including laser line narrowing at very low optical feedback levels, chaotic dynamics and power drop-out dynamical phenomena under stronger optical feedback. In addition, it accounts for stabilization of the laser under strong external optical feedback. The basic assumption of the L-K model is that the isolated laser is restricted to run in a single longitudinal mode and moreover, any non-uniformities in optical field amplitudes or carrier density are averaged along the laser axis. The former single mode assumption is most likely to be valid in a DFB laser where the imposed grating is designed to select a single lasing wavelength. However, one must keep in mind that even an ideal DFB laser has 2 non-degenerate wavelengths unless one breaks the symmetry of the grating structure by including a phase offset. Other complications with DFB lasers

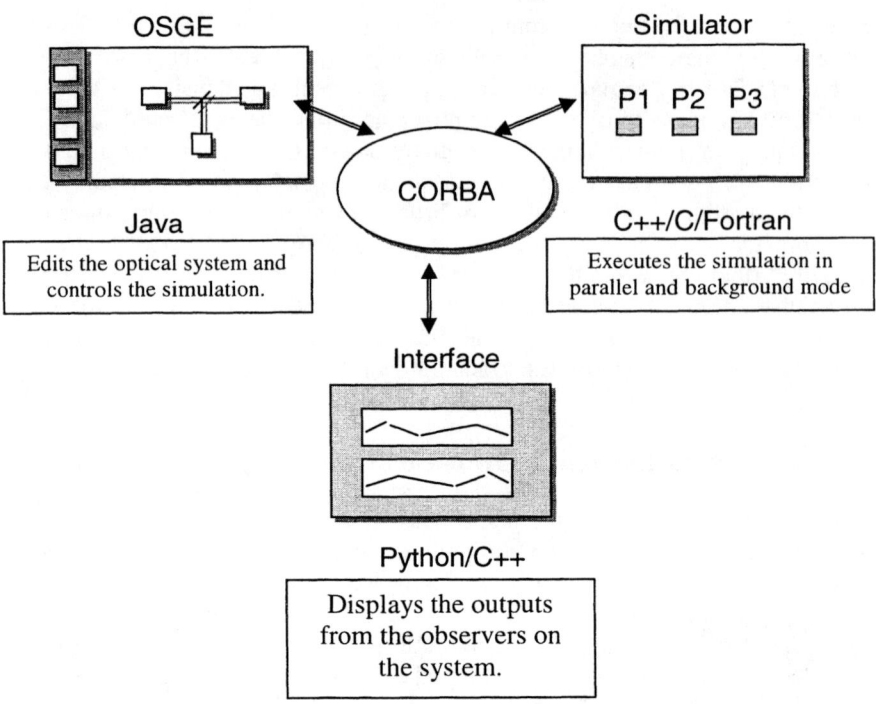

Figure 5. Schematic of the layout of the opto-electronic simulator showing the optical system editor (OSGE), the core simulator and a graphical viewer.

arise due to a mismatch between the underlying laser cavity length and the grating length. In addition, one cannot avoid some residual reflectivity at the laser facets even with the best AR-coatings. Unfortunately, many of the qualitative dynamical features associated with power dropouts do not change significantly whether the laser runs in a single or multiple longitudinal mode. We clearly observe situations where the laser may run in many isolated modes at some parameter settings, but in a single mode under others. The advantage of the pde simulation approach is that one can distinguish between both types of behavior and potentially exploit the multi-mode aspects of the dynamics when these are present. Figure 6 shows a frozen snapshot of an earlier version of the simulator where the cavity length is chosen to be only 10 μm long to ensure single isolated mode operation. The external reflector is placed at a distance of .88 m from the output facet and the external mirror reflectivity is 5%. In order to achieve gain in this system, we increased the mode confinement appropriately, indicating that one would need to grow a multiple QW structure for this device to lase. The basic QW is the 10 nm InGaAS structure for which the gain and index spectra are shown in Fig. 2. This device shows a linear increase in α-factor with increasing carrier density. We remark that the 5 nm InGaAs structure does not display power dropouts due to the strong clamping of the α-factor.

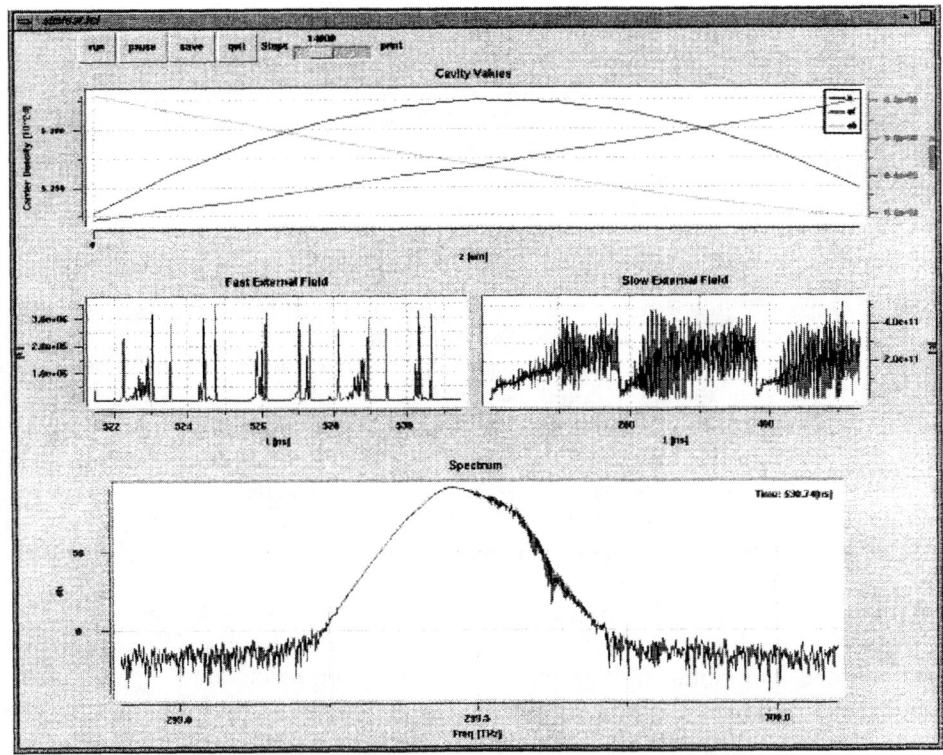

Figure 6. Snapshot of a running simulation. Top Frame: the instantaneous forward, backward optical field intensities and carrier density along the laser axis. Middle Frames: the left frame shows the fine details of the chaotic pulsing while the right one shows a much longer detector averaged (1 ns averaging) output showing power dropouts. Bottom Frame: the evolving output spectrum of the device in an interval between power dropouts.

When the cavity length is increased to 60 μm, the laser tends to run on three isolated longitudinal modes. In the single mode case depicted in Fig. 5, a power dropout is accompanied by the transient appearance of a strong narrow peak at the right edge of the above spectrum. The spectrum shown in this figure is essentially the chaotic spectrum associated with the external cavity modes defined by the external reflector. This narrow spectral spike is the isolated mode of the original laser. In the longer device, we observe similar features across the three isolated modes but then details are different. Prior to a power dropout, one of the modes becomes visibly stronger and the recovery is accompanied by the appearance of an isolated cavity mode at the right edge of one of the other suppressed broadened spectra.

An Interactive High Power Simulator

One of the limitations of conventional narrow aperture semiconductor lasers (3μm – 10 μm contact stripe widths) is that one cannot extract much power from the device due to the occurrence of catastrophic optical damage (COD) which tends to occur primarily at one of the output facets. This damage is initiated by the high intensity local fields generated at the facet. One of the early proposed solutions to this problem was to widen the output stripe contact from say, 10 μm to 100 μm or more. The hope was that the device would emit a wide uniform lateral intensity output thereby avoiding high local intensities at the output facets. In the growth direction, the strong waveguiding enforced by the relatively large index steps across the various layers would guarantee a single confined optical mode in that dimension. It was observed from the outset that the laser output was highly irregular showing strong intensity spiking which could again degrade the output facet. Most simulations of these devices, with few exceptions [9, 22], used the steady state beam propagation method (BPM) which misleadingly suggested that the intensity filaments were stationary. Unfortunately, the BPM is still being used extensively to simulate wide aperture semiconductor lasers despite the fact, as we shall show, that the outputs of broad area devices are nowhere stationary.

The issue of patterns or lateral modes in wide aperture lasers has been discussed extensively in the literature over the past decade. Unlike other laser systems such as gas lasers and some solid state lasers, the issue of pattern formation in semiconductor lasers is quite another issue. In the former, the lateral modes are essentially dictated by the resonator geometry and the active layer occupies a relatively small fraction of the overall mode volume. The active medium essentially weakly couples the transverse modes. In a broad area semiconductor laser, the cleaved end facets are flat. This would correspond to the critical transition between stable and unstable resonator design in conventional systems and any level-headed laser designer would recommend avoiding this situation. However, in semiconductor lasers the active region essentially fills the entire mode volume (the one exception being in the growth direction where the narrow gain stripe overlaps a small fraction of the strongly confined optical mode. In this situation, the lateral modes are dictated by the gain medium rather than some external geometric influence such as a curved reflector. The lateral modes in these devices are very different from the usual TEM transverse modes of optical resonators. In a series of papers in the early to mid nineties [1,2,10,11,12,13,14], we laid the fundamental groundwork for analyzing such spatially extended systems. The first observation was that the solutions to the laser equations compactly supported traveling rather than standing waves. These traveling waves appear as an off-axis emission of light relative to the laser axis. Some authors later referred to such traveling waves as 'tilted waves'. In laterally extended two-level or Raman lasers, the traveling waves were shown analytically to be stable within a finite range of off-axis emission angles. By restricting the wide cavity to a short longitudinal dimension (single or two longitudinal modes), we were able to derive complex order parameter equation descriptions of the wide laser. Moreover, we were able to show self-consistently, that the solutions to various order parameter equation descriptions (Complex Ginzburg-Landau, Complex

Swift-Hohenberg, Complex Newell-Whitehead-Segel) in different limiting cases were consistent with the solutions of the original physical pdes. When that particular analysis is extended to broad area semiconductor lasers, we find that there are nowhere stable traveling wave solutions for values of the magnitude of the α-factor lying within the experimentally measured range [3]. The picture of a broad area semiconductor laser that emerges from this analysis is that of a weakly turbulent spatially extended system once the laser turns on. The aforementioned body of work is very important in interpreting what happens in real high-power semiconductor lasers.

Experimental Evidence for Optical Turbulence

A series of elegant experiments in the early nineties, carried out by DeFreez et al. [15, 23, 24] using a streak camera which could capture a 5 ns time window while simultaneously resolving the near-field output, showed unambiguous evidence for the existence of randomly occurring intensity "hot-spots" in the output of various wide aperture semiconductor lasers. These intensity spikes were of very short duration (tens of picoseconds). In more recent experiments on a 500μm long by 100μm wide broad area device, Bossert et al. [16] were able to measure the spectrally resolved far-fields of the device at various pump current levels ranging from below to above threshold. Most of these results remain unpublished but are presented here with the permission of the author. Figure 7 is a schematic of the experimental arrangement used to measure the spectrally resolved far-field.

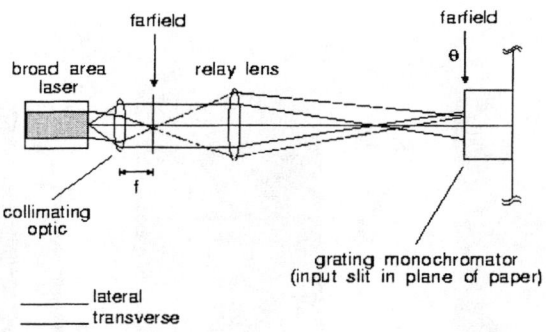

Figure 7. Schematic of experimental arrangement to measure the spectrally resolved far-field of a broad area semiconductor laser.

Actual far-field measurements are shown in Fig. 8. The laser was operated in pulsed mode with current pulses lasting about 500 ns and a repetition rate of 100 kHz to avoid thermal influences. The pictures from left to right show the spectrally resolved far-field just below, just above and at six times threshold. The laser runs simultaneously on 30-40 longitudinal modes and only about 5 are shown here. Associated with each longitudinal mode is a parabolic shaped dispersion curve, which reflects the lateral filamentation. Just below threshold, we are observing the amplified spontaneous emission spectrum (ASE) of the device. This picture in Fig. 8 (a) shows the incredibly rich dynamically accessible mode longitudinal and lateral mode content of the device. Just above threshold, the angular divergence of the spectrally resolved far-field narrows significantly, indicating that the laser is running in a spatially extended mode, which sloshes back and forth chaotically. The double lobes are a reflection of this behavior. As the pump is increased well beyond threshold, the far-field divergence increases strongly and the parabolas associated with each longitudinal mode begin to extend out and eventually overlap the next highest order longitudinal mode. This signifies that the average chaotically spiking filament size is decreasing. The filament size is controlled by the combination of diffraction and diffusion and typically is of the order of 5 – 10 μm. The ideal diffraction limited spot size should be of the order of λ/D, where λ is the lasing wavelength and D is the characteristic lateral dimension (here 100 μm) of the laser. Assuming that λ is around 1 μm, the angular divergence would be about 0.6°. In effect, the output beam is chaotically steering from left to right typically filling out an angle between 8° - 15° depending on pump level above threshold. The infinite dimensional attractor of the system can be viewed as excursions across a continuum of unstable compactly supported traveling waves with chaotic hopping amongst longitudinal modes. Combined external feedback/spatial filtering has been proposed as a means of controlling the laser output [3, 17]. This control scheme has been shown to work for a short, wide laser confined to oscillating on a single longitudinal mode. This would correspond to controlling the traveling wave on a single parabolic curve such as one of those shown in Fig. 8.

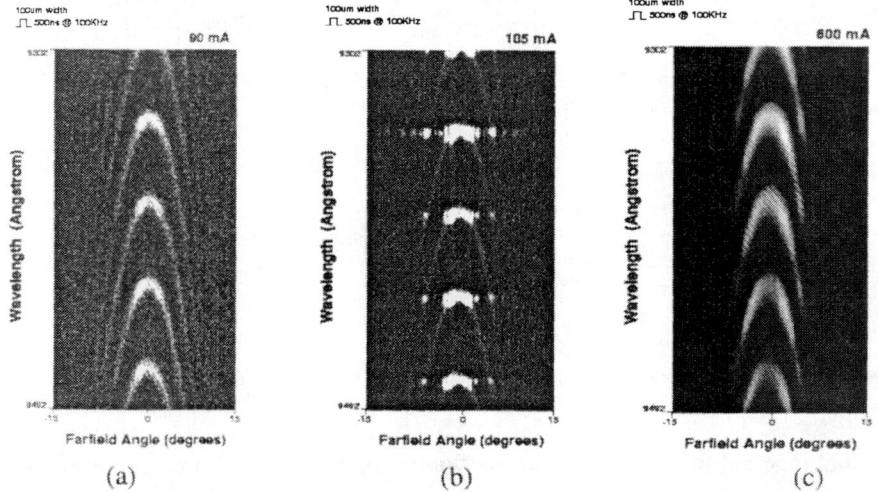

Figure 8. Experimentally measured spectrally resolved far-fields of a broad area semiconductor laser at different pump levels.

Simulation of High Brightness Semiconductor Lasers

Simulation of these broad area semiconductor lasers is extremely challenging, requiring the resolution of extremely short time steps simultaneously with the longitudinal and lateral spatial dimensions. In order to simulate spontaneous emission, we introduce a noise seed as a random number at each point on the space grid and update this at each time step. The noise, which is introduced in the polarization, is assumed to be delta-correlated and as the noise builds up, it is colored by the gain dispersion of the semiconductor material. As we want to simulate the build-up of the ASE as lasing threshold is approached, we will choose our grid resolution so as to span a 30 nm frequency bandwidth. Our goal is to see if our simulation can capture the details shown in the experimental spectrally resolved far-fields shown in Fig. 8. In order to achieve interactive performance, the simulation is run in parallel on 8 processors of a Silicon Graphics ONYX2 graphical supercomputer. The simulator allows one to view the time space evolution of the relevant optical, carrier and temperature fields within the device as well as various outputs such as time evolving lateral near-fields, far-fields, emission spectra and total output power.

Figure 9 shows an instantaneous snapshot of the simulator running just below threshold. There are 10 viewing panels, which show different aspects of the simulation of the broad area device as the simulation proceeds. The top two left-hand panels, display the forward (top) and backward (below) propagating optical field intensities within the device. The facet reflectivity on the left is 90% and on the right 5%. The light intensity is therefore brightest near the low reflectivity facet. The carrier density distribution within the device is shown in the third panel from the top left. This is flat in the pumped region as the laser has not turned on yet. A cross-hair (white vertical line on the left of the top three panels) allows the user to select a lateral cross-section of each of the top three left-side panels and view the relevant details of the fields at that location. The second panel from the bottom, on the left, shows the lateral intensity distribution of the forward and backward optical fields, whereas the bottom panel on the left, shows a lateral cross-section of the carrier density (and lattice temperature) at the same location. The top frame on the right displays the near-field output intensities at the right and left facets. The second from top frame on the right displays time evolving the far-field angular distribution of the emission from the right facet. The middle frame on the right shows the time evolution of the total output power from both facets and the second from bottom frame on the right shows the evolution of the lasing emission spectrum. Finally, the bottom frame on the right shows the spectrally resolved far-field emission and should be compared with Fig. 8(a) above. This displays the ASE just below threshold.

Figure 9 displays the initial noisy features of the broad area laser just before it switches on. If we increase the pump current so that we are slightly above threshold, the details within each panel change considerably.

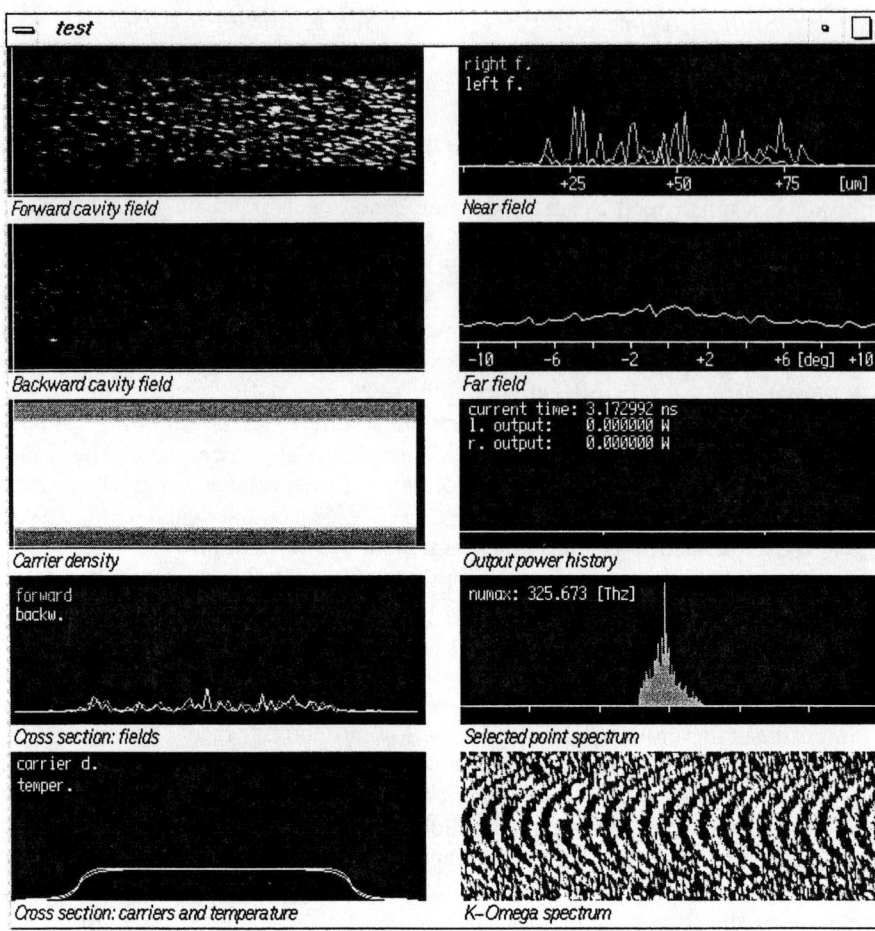

Figure 9. Instantaneous snapshot of the simulator running just below threshold.

Figure 10 shows an instantaneous snapshot of the simulator panels when the broad area laser is operating just above threshold. The laser is now on and the intensity filaments are clearly evident (bright spots) within the structure (top left panel and top right panel). These filaments are rather fat and the far-field (second from top right) is relatively narrow. The angular divergence of the spectrally resolved far-field in the bottom right panel has now reduced and the double-lobed feature shown in the experimental data in Fig. 8(b) is now clearly evident.

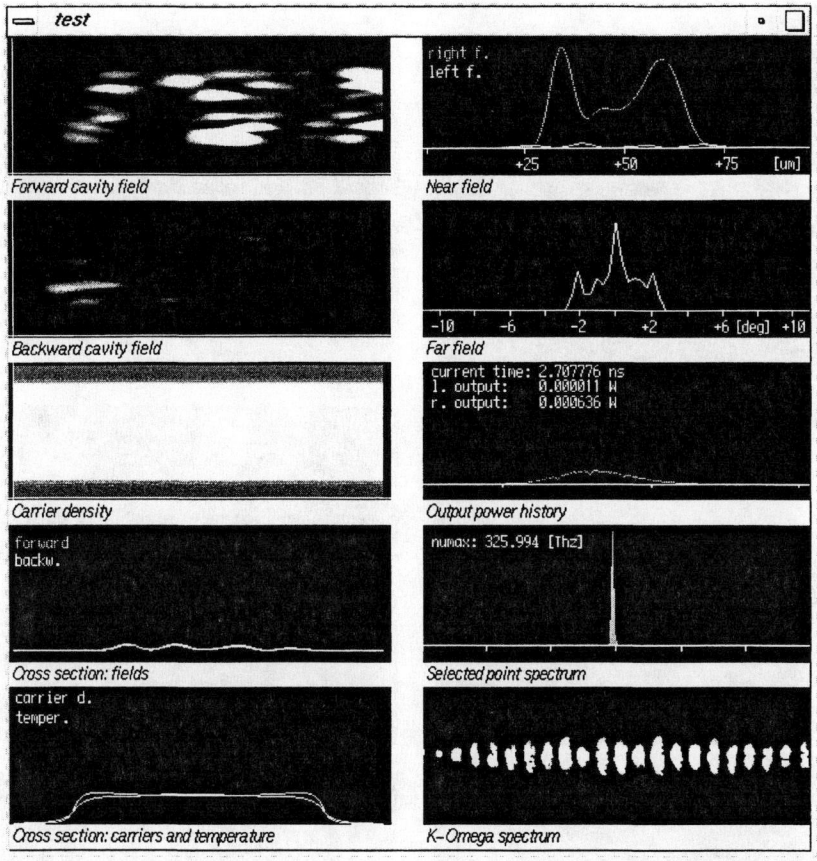

Figure 10. Instantaneous snapshot of the simulator running just above threshold.

Finally Fig. 11 shows a snapshot of the simulator panels when the laser is running around six times threshold.

The new features that emerge at strong pumping levels are narrower intensity spikes (top left and right panels), broadened far-field emission (second from top left and bottom left). The bottom right panel should be compared to the experimental data in Fig. 8(c). The cross-section of the carrier density begins to show spikes at the lateral extremities of the contact. These spikes are more pronounced at the right lossy facet and are indications that unsaturated carriers build up at the edges of the pumped region. These high levels of unsaturated carriers at the edge of the pumped region tend to promote lateral filamentation instabilities, even in specially flared amplifier sections. The latter are introduced in an effort to spatially filter higher order lateral modes but are only partially successful in doing so.

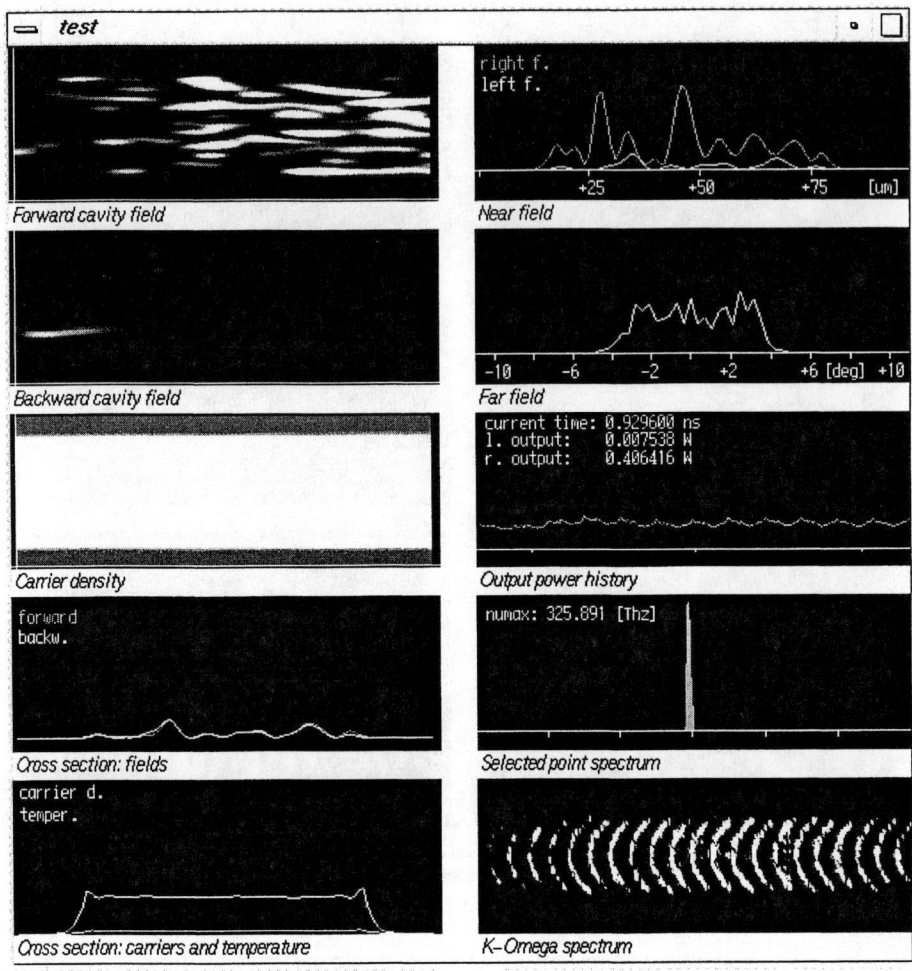

Figure 11. Instantaneous snapshot of the simulator running at six times threshold.

The obvious advantage of the full pde formulation of the problem is that there are no particular restrictions to the geometry of the device that we consider. The filamentation instabilities associated with broad area lasers makes them unsuitable as high brightness sources where high power needs to be delivered in near diffraction limited spots. Multi-section monolithically integrated devices, such as the Master Oscillator – Power Amplifier (MOPA), were introduced in an effort to suppress lateral intensity filamentation. The idea behind MOPA geometries is that the master oscillator provides a single longitudinal-lateral mode low power output, which is then swept through an expanding flared amplifier. The flared amplifier is typically 1 – 2 mm in length and expands from about a 4-5 μm aperture near the master oscillator end to a 100 – 200 μm aperture at the wide end of the flare. Many variations on this geometry have been tried but with only limited success. Figure 12 provides a nice illustration of the problems that arise with flared devices and in addition, highlights the need to have accurate low density absorption data as well as gain data when modeling realistic semiconductor lasers.

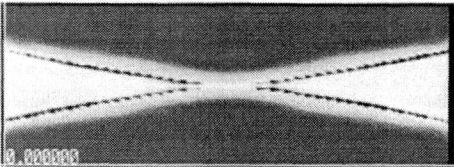
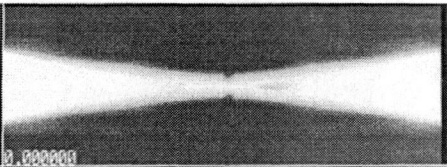

Figure 12. Instantaneous snapshot of the carrier density distribution within a bow-tie laser. (a) with no current profiling and cavity spoiler. (b) with lateral current profiling and a cavity spoiler.

These pictures display the internal carrier density distributions within a bow-tie laser, which consists of two expanding flares, coupled to a narrow DBR pumped grating section. There are three separate contacts on this device and each can be controlled individually. The output facet reflectivities are low (1%) and then DBR grating is meant to act as a wavelength filter for longitudinal mode control. In the left picture, the applied current is flat in the lateral dimension. This device is susceptible to filamentation instabilities as the pump current is increased. The dark ridges at the edge of each flare are high densities of unsaturated carriers. These "bat-ears" [18] tend to grow with current pumping and provide strong gain for photons reflected backwards from the wide flared facets. This in fact is the crux of the problem. In reality, it is impossible to make the reflectivity of a facet identically zero. The best Ar-coated facets have reflectivities around 0.05%. Such weak reflectivities are sufficient to cause MOPA or flared structures such as those shown in Fig. 12 to go unstable at high pump levels. The second problem is bleaching of the carriers outside of the pumped region. The lightly shaded regions outside of the dark ridges represent unpumped regions, which are transparent. With stronger pumping these extend out further and eventually the device resembles a classical Fabry-Perot cavity which can support many higher order lateral modes. Recent efforts to improve the stability of flared devices have involved modifying the electrical contacts so as to introduce a smoothed lateral current profile in the active layer [20, 21] and the introduction of a cavity spoiler to block the light from propagating in the bleached transparent regions. The cavity spoilers are created by etching through the active layer. The picture on the right shows the carrier density distribution in a bow-tie laser where the current has been profiled so as to remove the unsaturated carriers at the edge of the flares and a cavity spoiler has been introduced in the middle of the narrow DBR section to block the light in the bleached regions. The position of the cavity spoiler corresponds to the narrow symmetric indentation at the middle of the narrow section. This device shows a considerably improved performance maintaining a single lateral mode output to high pump levels.

As a final illustration of the combined use of the microscopic gain calculations and full scale simulation, we contrast the behavior of a broad area laser under strong pumping conditions. The two devices for which the gain and refractive index spectra are shown in Fig. 2 are used for the case study. The broad area laser has low facet reflectivities (1%) for maximal output coupling. Figure 13 shows the time-averaged far-fields for the 5 nm and 10 nm InGaAS QW device.

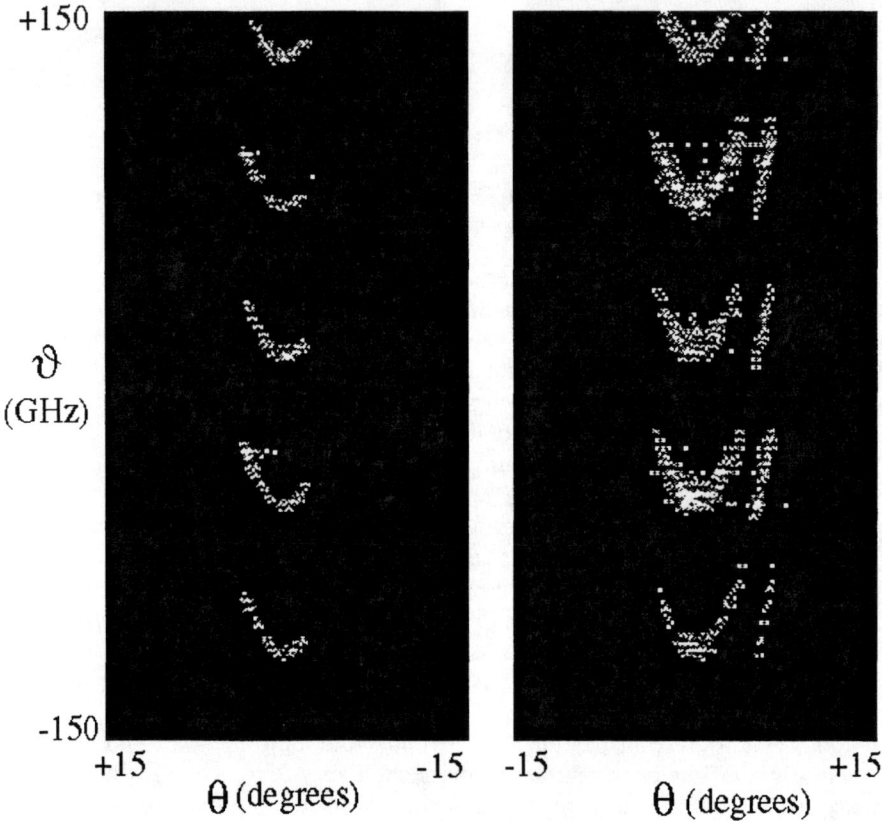

Figure 13. Spectrally resolved far-fields for the α-clamped 5 nm InGaAs device (left frame) and for the linearly increasing α 10 nm InGaAs QW device (right frame). Reproduced with permission [26].

One can estimate the average far-field divergence of the laser output by looking at the extension of the parabolas in the angular dimension. The 10 nm QW device has a time-averaged far-field divergence, which is about twice as large as the 5 nm QW device. Reducing the average far-field divergence, even though the output still remains weakly turbulent, is crucial for improving the imaging quality of the laser.

This is an illustration of where the predictive property of the combined microscopic physics and simulation approach should be of great value in choosing the optimal design of the device before proceeding to growth and fabrication.

SUMMARY AND CONCLUSION

The illustrations presented in this article of semiconductor device simulation should convince the reader that this whole area of research is an extremely fertile area for the study of nonlinear dynamical behavior in complex spatially extended nonlinear systems. As mentioned in the introduction, we have amassed a considerable body of analytical techniques for studying the main qualitative features that emerge in the study of semiconductor lasers. Beyond providing qualitative insight into what we can expect, these techniques are of limited utility. Certainly, knowing that the underlying lateral nonlinear cavity modes are compactly supported traveling waves and not, standing waves or traditional TEM-like resonator modes, is of the utmost importance. The fact that these are intrinsically unstable in broad area semiconductor lasers is also of great importance. However, if we are ever to be able to provide a first-principles predictive tool that will have a direct impact in limiting wasteful growth and fabrication of real lasers, we must invest in gaining a quantitative understanding of all the important physics that is operative in a real laser system. Otherwise our theories are necessarily built on a post-mortem analysis of experimental measurements and inference of the physics from the experimental observations. The approach put forth in this article goes a significant way in achieving this end.

What we have shown is that semiconductor lasers are marvelously rich nonlinear dynamical systems with enormous potential for controlling their dynamical behavior through bandgap engineering at the fundamental materials level. The plane wave simulator was illustrated by an application to power dropouts in an external cavity semiconductor laser. An area where both understanding and control of growth at the materials level is becoming essential is in the exploding telecommunications market. Building robust and reliable interactive simulation tools is a natural progression that opens up a vast landscape of fundamental and applications areas. The role of thermal influences (plasma and lattice heating) has not been discussed here but there is significant literature on this subject.

ACKNOWLEDGMENTS

The author acknowledges financial support for the work presented in this article from the U.S. Air Force Office of Scientific Research under grants AFOSR F49620-00-1-0002 and AFOSR DURIP F49620-00-1-0190, and partial support from the National Science Foundation, under grant DMS 9811466. Many colleagues and graduate students were involved in the research reported on here. I would like to acknowledge the contributions of Robert Indik, Miroslav Kolesik, Marcelo Matus, D. Bossert, J. Hader, and S.W. Koch.

REFERENCES

1. Jakobsen, P.K., Indik, R.A., Newell, A.C.,and Moloney, J.V., "Transverse Instabilities in Laser Diode Arrays Under Free-Running and Injection Locking Conditions." *OSA Proceedings on Nonlinear Dynamics in Optical Systems 1990*, Oklahoma,7, pp. 132-136 (1991).
2. Jakobsen, P.K., Moloney, J.V., Newell, A.C., and Indik, R.A., *Phys. Review A*, **45**(11), pp. 8129-8137 (1992).
3. Hochheiser, D., Moloney, J.V., and Lega, J., *Phys. Review A,* **55**(6), pp. R4011-R4014, (1997).
4. Chow, W.W., Koch, S.W., and Sargent, III, M., "Semiconductor Laser Physics," *Springer-Verlag* (1994).
5. Hughes, S., Knorr, A., Koch, S.W., Binder, R., Indik, R.A., and Moloney, J.V., *Solid State Communications*, **100**, pp. 555-559 (1996).
6. Hader, J., Bossert, D., Stohs, J., Chow, W.W., Koch, S.W., and Moloney, J.V., *Appl. Phys. Lett.*, **74**(16), pp. 2277-2279 (1999).
7. Ning, C.Z., Moloney, J.V., Egan, A., Indik, R.A., *Quantum Semiclass Opt.*, **9**, pp. 681-691, (1997).
8. Moloney, J.V., Indik, R.A., and Ning, C.Z., *IEEE, Photonics Technology Letters*, 9(6), pp. 731-733, (1997).
9. Adachihara, H., Indik, R.A., Moloney, J.V., and Hess, O., *J. Opt. Soc. Am. B.*, **10**, pp. 496-506 (1993)
10. Jakobsen, P.K., Lega, J., Feng, Q., Staley, M., Moloney, J.V., and Newell, A.C., *Phys. Review A*, **49**(5), pp. 41189-4200 (1994).
11. Lega, J., Jakobsen, P.K., Moloney, J.V., and Newell, A.C., *Phys. Review A*, **49**(5), pp. 4201-4212, (1994).
12. Feng, Q., Moloney, J.V., and Newell, A.C., *Phys. Review A*, **50**(5), pp. R1-R4 (1994).
13. Lega, J., Moloney, J.V., and Newell, A.C., *Phys. Review Letters*, **73**(22), pp. 2978-2981 (1994).
14. Lega, J., Moloney, J.V., and Newell, A.C., *Physica D*, **83**, pp. 478-498 (1995).
15. DeFreez, R.K., Bossert, D.J., Yu, N., Hartnett, K., Elliott, R.A., and Winful, H.G., *Appl. Phys. Lett*, **53** (24), pp. 2380-2382 (1988).
16. Bossert, D.J., and Gallant, D., *IEEE Photonics Technology Letters*, **8**, pp. 322-324 (1996).
17. Bleich, M.E., Hochheiser, D., Moloney, J.V., and Socolar, J.E.S., *Phys. Review E*, **55**(3), pp. 2119-2126 (1997).
18. White, J.K., McInerney, J.G., and Moloney, J.V., *Electorn. Lett.*, **31**(1), pp. 38-39 (1995).
19. Ning, C.Z., Indik, R.A., and Moloney, J.V., *IEEE-JQE*, **33**(9), pp. 1543-1550 (1997).
20. O'Brien, P., Skovgaard, P.M., McInerney, J.G., and Roberts, J.S., *Electron. Lett.*, **34**, pp. 1943 (1998).
21. Skovgaard, P.M., O'Brien, P., and McInerney, J.G., *Electron. Lett.*, **34**, pp. 1950 (1998).
22. Adachihara, H., Hess, O., Abraham, E., Ru, P., and Moloney, J.V., *J. Opt. Soc. Am. B.* **10**(4), pp. 658-665 (1993)
23. DeFreez, R.K., Bossert, D.J., Yu, N., Hunt, J.M., Ximen, H., Elliott, R.A., Carlson, N.W., Lurie, M., Evans, G.A., Hammer, J.M., Bour, D.P., Palfrey, S.L., Amantea, R., Winful, H.G., and Wang, S.S., *IEEE Photonics Tech Letters*, **1**(8), pp. 209-211 (1989).
24. Yu, N., DeFreez, R.K., Bossert, D.J., Wilson, G.A., Elliott, R.A., Wang, S.S., and Winful, H.G., *Applied Optics*, **30**(18), pp. 2503-2513 (1991)
25. Moloney, J.V., Kolesik, M., Hader, J., and Koch, S.W., *Modeling high-power semiconductor lasers: from microscopic physics to device applications, Advanced High-Power Lasers*, SPIE, **3889** (2000).
26. Moloney, J.V., Indik, R.A., Hader, J., Koch, S.W., *J. Opt. Soc. Am. B*, **16** (11), pp. 2027 (1999).

High Brightness Semiconductor Lasers with Reduced Filamentation

John G McInerney[*], Peter O'Brien[**], Peter Skovgaard[+]
Mark Mullane, John Houlihan, Eamonn O'Neill

*Department of Physics/Institute for Nonlinear Science
National University of Ireland, University College, Cork, Ireland*

Jerome V. Moloney[*], Robert A. Indik

Arizona Center for Mathematical Sciences, University of Arizona, Tucson, Arizona 85721, USA

[*]: also with the Optical Sciences Center, University of Arizona, Tucson, Arizona 85721, USA
[**]: present address: National Microelectronics Research Centre, University College, Cork, Ireland
[+]: present address: Research Center COM, Technical University of Denmark, 2800 Lyngby, Denmark

Abstract. We outline physical models and simulations for suppression of self-focusing and filamentation in large aperture semiconductor lasers. The principal technical objective is to generate multi-watt CW or quasi-CW outputs with nearly diffraction limited beams, suitable for long distance free space transmission, focusing to small spots or coupling to single mode optical fibres. The principal strategies are (a) optimisation of facet damage thresholds, (b) reduction of the linewidth enhancement factor which acts as the principal nonlinear optical coefficient, and (c) design of laterally profiled propagation structures in lasers or amplifiers which suppress lateral reflections.

INTRODUCTION

High power, nearly diffraction limited semiconductor lasers are attractive for many scientific and engineering objectives. Recent dramatic advances in long haul optical fibre communication systems, particularly those based on dense wavelength division multiplexing (DWDM), have created the need for high power diode lasers with sufficient brightness to couple efficiently to single mode fibre amplifiers [1]. Other pressing applications for high brightness lasers include free space communication and energy transfer, cutting and welding, surgical and medical applications. Due to the onset of catastrophic optical damage (COD) above some threshold irradiance at the emitting facets, and of slower degradation phenomena at lower irradiances, CW and quasi-CW powers in excess of 1 W require emitting apertures of ~100 μm or more. Unfortunately these lasers exhibit strongly turbulent spatio-temporal dynamics, resulting in highly irregular behaviour and thus emit beams which may diverge at one or two orders of magnitude more than the diffraction limit [2]. It is essential to understand and control the spatio-temporal dynamics of these large aperture semiconductor lasers. In this paper we describe theoretical models and approaches to suppressing these phenomena which are based primarily on filamentation, a nonlinear optical process where carrier-induced self-focusing is opposed by diffraction and diffusion.

In Section 2 we frame the principal strategies for suppressing filamentation and associated spatio-temporal dynamics. Section 3 describes a many-body microscopic model for computing the linewidth enhancement factor, which is the major nonlinear coefficient in the fast self-focusing process. Suggestions for minimising this factor are made. Section 4 contains a spatial perturbation analysis indicating how to minimise the modulational instabilities leading to filamentation. In Section 5 we describe modelling of light propagation in large aperture lasers. We compare the performance of broad area lasers with lateral mode selective structures such as curved facet or flared devices. Section 6 provides a summary and some conclusions.

STRATEGIES FOR SUPPRESSING FILAMENTATION IN SEMICONDUCTOR LASERS

Self-focusing and Filamentation

Filamentation occurs in an optical medium when self-focusing is counterbalanced by diffraction. A beam self-focuses when it creates a positively focusing index distribution, as for example in the presence of a Kerr effect for which the index n is given as a function of the local intensity I by [3]

$$n(I) = n_0 + n_2 I$$

where n_2 is a positive constant directly proportional to the Kerr coefficient or third-order susceptibility. If n_2 is negative, the effect is called self-defocusing and tends to spread the beam as it propagates. The more usual situation is self-focusing, with a positive n_2, in which the beam creates a progressively stronger lens as it propagates and focuses, until the spreading effect of diffraction exactly compensates the self-focusing and the beam becomes a filament, maintaining a constant profile. Filamentation is thus a nonlinear optical phenomenon which occurs when the beam power exceeds a critical threshold proportional to the nonlinear coefficient n_2 [3]. A broad beam with power well above the filamentation threshold tends to break up into multiple filaments each carrying approximately the threshold power. It is the spatial analog of a temporal soliton or solitary wave pulse formed by the counterbalancing of temporal dispersion and self-phase modulation [4]; both systems are governed by the nonlinear Schrödinger equation.

In a semiconductor laser, coherent Kerr effects last only for a very short time due to dephasing by carrier-carrier scattering. However, a beam propagating in a distribution of excited carriers causes local spatial and spectral hole burning which results in self-focusing [5]. This process is not instantaneous (as a coherent Kerr effect would be) but is delayed by carrier recombination which occurs on a nanosecond time scale – we will refer to this as fast self-focusing in the semiconductor laser. A different process, in which the injected carriers and propagating beam produce a thermal profile which causes lensing, occurs on a much longer time scale (hundreds of nanoseconds) and is referred to as slow self-focusing. In this paper we shall concentrate on fast self-focusing which is the cause of the rapid irregular spatio-temporal dynamics observed experimentally and theoretically in broad area lasers. Slow self-focusing is the cause of a different set of problems including current-dependent beam steering, virtual source movement and lateral mode wavelength tuning.

The background index n_r in the presence of gain guiding becomes
$$n = n_r + i(g/2k_0)$$

with carrier- and temperature-induced changes
$$\Delta n_r(r) = \frac{\partial n_r}{\partial N}\Delta N(r) + \frac{\partial n_r}{\partial T}\Delta T(r)$$

with [6] $\quad \frac{\partial n_r}{\partial N} = -e^2/2n_0\varepsilon_0\omega^2 m^*$ and $\frac{\partial n_r}{\partial T} = 4 \times 10^{-4}\,\text{K}^{-1}$

The importance of self-focusing and filamentation can be gauged by determining the nominal Fresnel number F of the laser, a measure of the number of transverse modes supported, and also determines the complexity of the spatio-temporal dynamics in a similar manner to the Reynolds number in fluid flow past an obstacle [7]. F is given by

$$F = a^2/\lambda L$$

where the resonator has characteristic lateral dimension a, length L and transmits light of wavelength λ. This is a crude approximation for the highly modulated and aberrated beams produced by semiconductor lasers, but it does give an idea of the conditions under which one can expect to encounter serious filamentation. We note that the viewpoint of studying filamentation of lateral modes of the empty cavity (or, equivalently, transverse modes produced by the laser at low power) is equivalent to considering lateral modes of the nonlinear resonator as some authors have done [2]. To achieve high brightness – defined here as intensity per unit solid angle, expressed in watts per square meter per steradian – it is necessary to select a single lateral mode, preferably the fundamental mode, and then to suppress filamentation. We propose suppression of filamentation based on three complementary strategies: (a) minimise the Fresnel number (ie maximise the optical damage threshold and thus minimise the aperture size for a given power), (b) minimise the nonlinearity (linewidth enhancement factor), and (c) develop propagation structures which minimise the modulational instability causing filamentation.

Minimising the Fresnel Number: Maximising the Facet Damage Threshold

Minimising the Fresnel number involves using long resonators of minimal lateral extent consistent with the optical power which the laser is required to generate, and with the desired service lifetime. This strategy involves the use of surface passivation and coating techniques to maximise the threshold intensity for catastrophic optical damage (COD) at the facets. Such techniques as non-absorbing mirrors (NAM) are part of this strategy, given that COD involves rapid and runaway local facet heating resulting in mechanical ejection of material from the facet, destroying its optical integrity. The use of laser materials free of aluminum (*eg* InGaAsP/InGaAs instead of AlGaAs) has resulted in significant improvements in routine damage thresholds – although AlGaAs lasers can in principle have damage thresholds of ~10 MW cm^{-2} they very seldom do so due to the reactivity of Al: numbers closer to 1 MW cm^{-2} are more common. There are several passivation techniques for stabilising Al-containing facets, including sulfur dipping [8,9] and electrochemical treatments [10], but each method has drawbacks. Most recipes for surface passivation are proprietary and this makes a broad scale scientific assault on the problem exceedingly difficult. We will not treat this issue in detail here, except to say that the higher the COD threshold, the smaller the aperture which can be used to generate a given CW power level. For pulsed power generation at low duty cycles, we note that the pulsed COD threshold generally equals the CW value for pulse widths down to the thermal time constant of the facet heating, ~0.1-1 μs. For shorter pulses, the COD threshold increases by an amount approximately equal to the square root of the ratio between the thermal time constant and the pulse width. For example, if the thermal time constant is 1 μs and we wish to generate sparse trains of 100 ps pulses, the effective COD threshold is 100 times the CW value. For dense trains of pulses, the effects of time-averaged heating have to be included.

MINIMISING THE NONLINEAR COEFFICIENT: THE LINEWIDTH ENHANCEMENT FACTOR

Linewidth enhancement factor

The linewidth enhancement factor, or antiguiding parameter, is a simple representation of how the index of refraction seen by a beam propagating in a semiconductor varies with the carrier density [11]. It is a measure of the asymmetry of the gain lineshape, and is defined by

$$\alpha(N,\omega) = \frac{\partial \chi'/\partial N}{\partial \chi''/\partial N} = -\frac{2\omega}{c} \frac{\partial(\delta n)/\partial N}{\partial g/\partial N}$$

where χ' and χ'' represent respectively the real and imaginary parts of the complex electric susceptibility χ. The physics of α derive from a combination of the Kramers-

Kronig relations and the dispersion of the carrier plasma in the semiconductor. α is notoriously difficult to measure, as it varies significantly with operating wavelength, carrier density and other factors. Measured values for semiconductor lasers typically lie between 1 and 7, although higher and lower values have been observed under special conditions [12]. α has a profound influence on the static and dynamic properties of semiconductor lasers, including static spectral linewidth, modulation-induced frequency chirping, sensitivity to external reflections and self-focusing. In the latter case, while it is not straightforward to relate a generally to the Kerr coefficient n_2 used in "normal" self-focusing, this may be done in specific lasers by solving the laterally resolved rate equations with a particular propagating power or intensity, then determining the change in local carrier density profile and computing the susceptibility change $\Delta\chi$ which gives the index change using $1 + \chi = n^2$, from which $\Delta\chi = 2n\Delta n$. Reducing Δn (ie α) is one of the key pathways to suppressing self-focusing, filamentation and spatio-temporal instabilities in lasers.

Reducing α can be accomplished in two ways: reducing the differential index change $\partial(\delta n)/\partial N$, and increasing the differential gain $\partial g/\partial N$. Since the propagating beam samples and spatially averages the indices of all waveguide layers in the transverse (y) direction, the carrier induced index change is generally a relatively slowly varying function of N. Minimising it can be thought of as making the gain spectrum more atomic-like ie symmetric about a fixed point. This may be accomplished in principle by using quantum dots in the active region, but only if the laser can operate on the gain generated by the dots themselves in the first order sub-band – operating in higher order sub-bands or using states in the wetting layer will result in values of α similar to those of bulk lasers. In this paper we concentrate on the more straightforward method of increasing the differential gain by bandgap engineering.

For practical reasons we will concern ourselves primarily with the effects of empirically variable parameters (quantum well width, strain and detuning) on the semiconductor band structure. We use a microscopic many-body model for the electron-hole plasma which includes Coulomb effects. These lead to bandgap renormalisation and excitonic (Coulomb) enhancement of the gain as described below.

Microscopic Physical Theory of the Semiconductor Laser

Microscopic many-body semiconductor theory [13] gives the electric susceptibility in the form

$$\chi(N,\omega) = \frac{-i}{\epsilon V}\sum_{\mathbf{k}} |\mu(\mathbf{k})|^2 \frac{[f_e(\mathbf{k}) + f_h(\mathbf{k}) - 1]}{1 - q(\mathbf{k})} \mathcal{L}(\Delta_{\mathbf{k}}, \tau)$$

where V is the active region volume and we sum over the crystal wave vectors $\mathbf{k}$ of the carriers. ε is the background permittivity, $f_{e,h}$ the quasi-Fermi distributions and ζ a Lorentzian lineshape function with detuning Δ_k (transition energy minus photon energy) and dipole dephasing time τ These are calculated using Luttinger-Kohn theory, together with the Pikus-Bir strain Hamiltonian [14] to include effects of quantum confinement and strain, and giving the dipole matrix elements $\mu(\mathbf{k})$. $1-q(\mathbf{k})$ is the Coulomb enhancement term as explained below.

The bandgap renormalization ΔE_g is a many-body correction due to two distinct contributions. The first arises when we calculate the kinetic energies of carriers in a screened potential, resulting in a fixed correction called the Coulomb-hole (CH) self-energy or Debye shift. In our theory it is given by [13]

$$\Delta E_{g,CH}(\mathbf{k}) = \Sigma[V_s(\mathbf{k}) - V(\mathbf{k})]$$

where the summation is taken over all non-zero $\mathbf{k}$ states; $V(\mathbf{k})$ is the unscreened Coulomb potential and $V_s(\mathbf{k})$ the screened Coulomb potential evaluated using the single plasmon pole approximation [15]. The latter assigns a single plasma frequency to the electrons and holes, and is valid for moderate to high carrier densities where a continuum of states can be assumed.

The second contribution to bandgap renormalization is a Fermionic exchange effect called the screened exchange (SX) energy, given by [13]

$$\Delta E_{g,SX}(\mathbf{k}) = -\Sigma\, V_s(|\mathbf{k'}-\mathbf{k}|)[f_e(\mathbf{k'}) + f_h(\mathbf{k'})]$$

Where the summation is taken over all $\mathbf{k'}$ not equal to $\mathbf{k}$. These two contributions comprise a repulsion between carriers in the same band, reducing their collective energy and hence shrinking the effective band gap. The other significant many-body effect is Coulomb (or excitonic) enhancement, in which attraction between electrons and holes in different bands increases their probability of recombination. (At low carrier densities, the same attraction tends to create excitons, or bound electron-hole pairs.). $q(\mathbf{k})$ is expressed as [13]

$$q(\mathbf{k}) = \Sigma\, V_s(|\mathbf{k'}-\mathbf{k}|)[f_e(\mathbf{k}) + f_h(\mathbf{k}) - 1]\zeta(\Delta_{k,t})$$

where the summation is again taken over all $\mathbf{k'}$ not equal to $\mathbf{k}$ This result can be thought of as an increase in the Rabi frequency. Both the Coulomb enhancement and screened exchange terms involve summation over pairs of $\mathbf{k}$ states, effectively turning the Bloch equations into a very large set of coupled PDEs. The calculation of realistic gain and index change spectra therefore requires considerable computational speed and complexity.

Calculation of the Linewidth Enhancement Factor α

We have calculated α for strained InGaAs quantum wells of various widths, clad by 10 nm InGaAsP barriers, lattice matched to an InP substrate. The value of α at the gain peak is plotted as a function of peak gain (determined directly from the carrier density) for well widths of 3, 5 and 7 nm. The results for compressive strain (Ga mole fraction x < 0.47) are shown in Figure 1(a)-(c). Similar results for tensile strain (x > 0.47) are given in Figure 2 (a)-(c). The strain values used are 0, 0.8% and 1.5%. Compressive strain generates strongest gain for the TE optical polarization, while tensile strain favours TM polarization.

We see that for TE gain in compressively strained InGaAs quantum wells, increasing strain tends to decrease α and increases the gain range where α remains low. Narrow quantum wells should maintain lower α to higher gain values. However, abrupt changes in α can occur due to shifts in the gain peak. A values in the range 1-1.5 should be achievable, which should lead to significant improvements in laser dynamics

For the TM polarization in InGaAs quantum wells with tensile strain, increasing the strain above a critical value tends to decrease α and again increases the gain range for low α operation. In this case, there is little dependence on well width at relatively large strain, and there are no abrupt changes in α. Further reduction in α is possible by detuning from the gain peak towards the differential gain peak, for example by incorporating a grating or other highly dispersive cavity. VCSELs with DBR cavities have achieved values of $\alpha < 0.5$ [16].

SIGNIFICANCE OF SPATIAL PERTURBATIONS OF THE GAIN AND INDEX

Self-focusing and filamentation should not occur in a plane wave propagating in an infinite medium, no matter how nonlinear. Even given that any beam of finite spatial extent must have some finite curvature, a real Gaussian beam filling a broad waveguide should have relatively slow self-focusing, even when well above the critical power threshold for filamentation. However, the system is unstable relative to small perturbations of the lateral gain and index profiles – thus we interpret filamentation is a modulational instability in which small perturbations seed the subsequent filament growth. Suppression of filamentation thus involves eliminating the lateral spatial modulation as far as possible and reducing the growth rate of perturbations which occur.

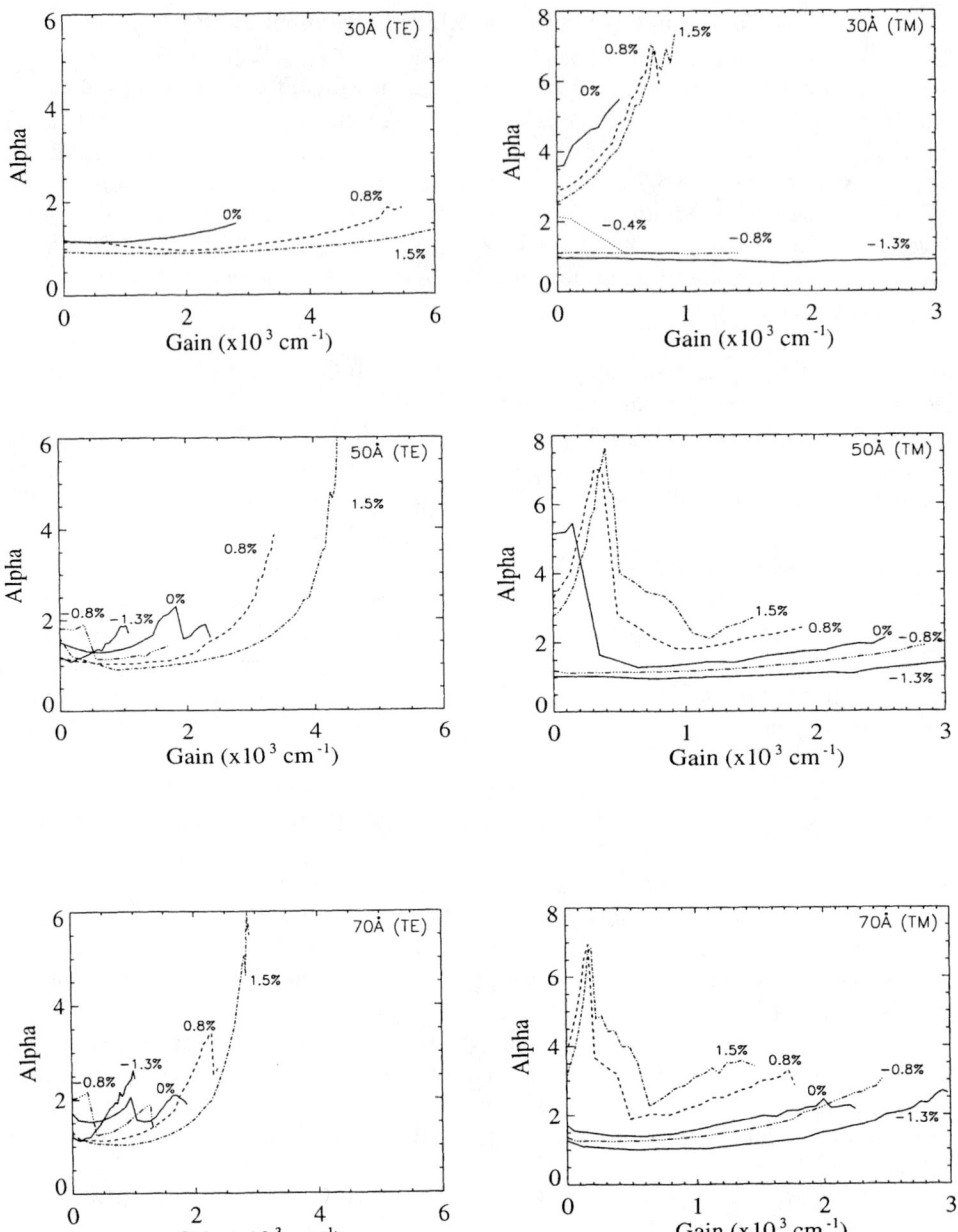

Figure 1: a factor at peak vs. peak gain for various quantum well widths and compressive strains.

Figure 2: similar calculations for tensile strain which favours TM polarised light

To analyse this situation mathematically, we begin from travelling wave rate equations

$$\frac{n}{c}\frac{\delta E}{\delta t} + \frac{\delta E}{\delta z} = \frac{i}{2k_0}\frac{\delta^2 E}{\delta x^2} + \Gamma[g(N,\omega_0) + ik_0\delta n(N,\omega_0)]E - \alpha E$$

$$\frac{\delta N}{\delta t} = D_f\frac{\delta^2 N}{\delta x^2} + \frac{J(x,z)}{ed} - \gamma_{sp}N - \frac{\varepsilon_0 c n g(N,\omega_0)}{2\hbar\upsilon}|E|^2$$

in which the symbols have their usual meanings [17], and the gain and index change spectra can be determined from the Semiconductor Bloch equations [13]

$$\left[i\frac{\delta}{\delta t} - (e_{e,k} + e_{h,k})\right]P_{\vec{k}} = (f_{e,\vec{k}} + f_{h,\vec{k}} - 1)\left[d_{cv,\vec{k}}\frac{E_0(t)}{2\hbar} + \sum_{q\neq k}\frac{V_{|\vec{k}-\vec{q}|}}{\hbar}P_{\vec{q}}\right]$$

We now restrict our attention to lateral perturbations which are slowly varying in time, so that the time derivatives in the rate equations become negligible:

$$\frac{\delta E}{\delta z} = \frac{i}{2k_0}\frac{\delta^2 E}{\delta x^2} + \Gamma[g(N,\omega_0) + ik_0\delta n(N,\omega_0)]E - \alpha E$$

$$0 = D_f\frac{\delta^2 N}{\delta x^2} + \frac{J(x,z)}{ed} - \gamma_{sp}N - \frac{\varepsilon_0 c n g(N,\omega_0)}{2\hbar\upsilon}|E|^2$$

We also consider a saturated gain regime in which the carriers follow the current density

$$\frac{\delta E}{\delta z} = \frac{i}{2k_0}\frac{\delta^2 E}{\delta x^2} + \Gamma[g(x) + ik_0\delta n(x)]E - \alpha E$$

which we reduce to the generic form

$$\frac{\delta E}{\delta z} = \frac{i}{2k_0}\frac{\delta^2 E}{\delta x^2} + f(x)E$$

where $f(x)$ represents the lateral gain and index profile. In a broad area semiconductor laser, the first (diffraction) term on the right side is 2-3 orders of magnitude less than the second (guiding) term prior to the onset of filamentation [18]. We consider small perturbations to the lateral profile of the form

$$E = E^{(0)}\left[1 + \frac{i}{2k_0}E^{(1)} + ...\right]$$

and extract the various orders of perturbation by substitution. The results for the zero-order terms are

$$\frac{\delta E^{(0)}}{\delta z} = f(x)E^{(0)}$$

$$E^{(0)} = E_0\exp(zf(x))$$

as expected, while to first order,

$$E^{(0)}\frac{\delta E^{(1)}}{\delta z} = \frac{\delta^2 E^{(0)}}{\delta x^2} = [z^2 f'(x)^2 + zf''(x)]E^{(0)}$$

$$E^{(1)} = \frac{z^3}{3}[f'(x)]^2 + \frac{z^2}{2}f''(x)$$

which implies that the perturbation grows strongly with propagation length z, and that the rate of growth depends critically on the sharpness of the profile $f(x)$. If $f(x)$ has sharp features, leading to locally high first and second derivatives f' and f'', these will "seed" the filamentation process to produce strong spatial modulation of the field downstream, even in well saturated amplifying media.

This analysis suggests firstly that the laser length should be minimised, consistent with other requirements such as heat sinking. Secondly, the derivatives of $f(x)$ itself are critical in determining filament growth. The gain, built-in index profile and index changes due to carriers and heat should vary slowly in the lateral direction to minimise filamentation. This requirement is in addition to the obvious need to select a single (preferably fundamental) lateral mode for highest brightness. Hence in addition to mode selective structures such as flared or unstable resonators, we also need to produce gradual lateral profiles of the gain and/or refractive index. To illustrate this point, we have calculated the growth of edge perturbations ("bat ears") formed in broad area lasers with traditional top-hat and diffusion-smoothed lateral gain profiles [17]. The presence of moderate to strong diffusion suppresses the edge effects which would otherwise grow to become filaments.

MINIMISING TRANSVERSE SPATIAL MODULATION

Here we examine the effects of profiling the lateral gain distribution. We compare the spatio-temporal dynamics of broad area lasers having traditional top-hat gain distributions with those having various lateral gain profiles. We represent the gain and index change spectra as in the previous section, computing these functions using the Semiconductor Bloch equations as before. In addition, we use laterally resolved travelling wave rate equations to study propagation effects. These are [17]

$$\pm\frac{\partial E}{\partial z} - \frac{i}{2K}\frac{\partial^2 E^\pm}{\partial x^2} + \frac{n_g}{c}\frac{\partial E^\pm}{\partial t} = \frac{iK\Gamma}{2\varepsilon_0\varepsilon_b}\cdot P_{tot}^\pm - \frac{\alpha_{int}}{2}E^\pm$$

$$\frac{dN}{dt} - D_N\frac{\partial^2 N}{\partial x^2} = -\gamma_1 N + \frac{\eta J}{e\omega} + \frac{i}{4\hbar}\left[P_{tot}^{+*}E^+ - P_{tot}^+E^{+*} + P_{tot}^{-*}E^- - P_{tot}^{+-}E^-\right]$$

where the + represents forward and the − reverse propagating fields, P_{tot} the total propagating power, D_N the lateral carrier diffusion coefficient and $\gamma_1 = 1/\tau_{sp}$ the spontaneous recombination rate constant. The polarization dynamics and electric susceptibility are determined using

$$\frac{dP_j}{dt} = \left[-\Gamma_j(N) + i(\delta_j(N))\right]\cdot P_j(t) - i\varepsilon_0\varepsilon_b A_j(N)E(t)$$

$$\chi(n,\omega) \approx \chi_0(N) + \sum_{j=1}^{M}\frac{A_j(N)}{i\Gamma_j(N) + (\delta_0 + \omega + \delta_j(N))} \equiv \chi_0(N) + \sum_{j=1}^{M}\chi_j(N,\omega)$$

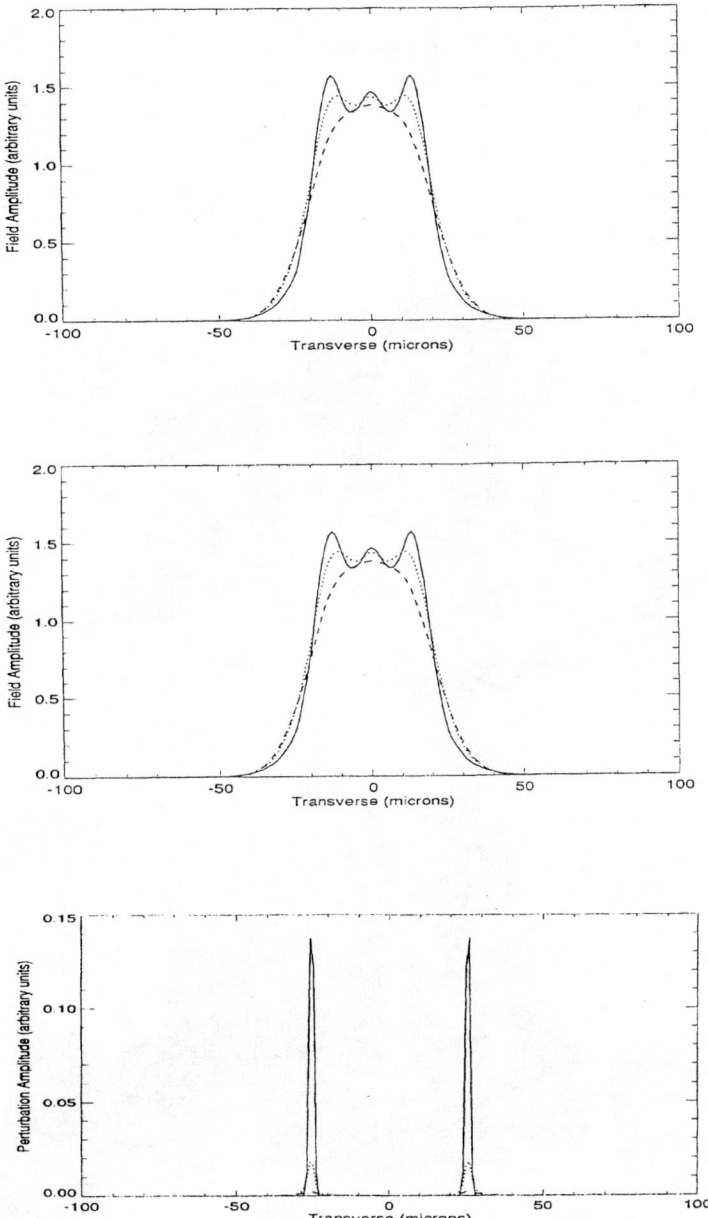

Figure 3: (a) current profiles, (b) electric field amplitudes and (c) perturbation amplitudes produced as the smoothing is increased. Solid lines: top-hat (no diffusion). Dotted lines: moderate diffusion. Dashed lines: strong diffusion

Using this model, we simulate the spatial and temporal dynamics of the photons and carriers in broad area lasers with various lateral gain profiles. Figure 4 shows the current distributions used in the simulations, and Figure 5 the corresponding 2-D carrier density maps $N(x,z)$. The time-resolved near fields are compared in Figure 6. We see

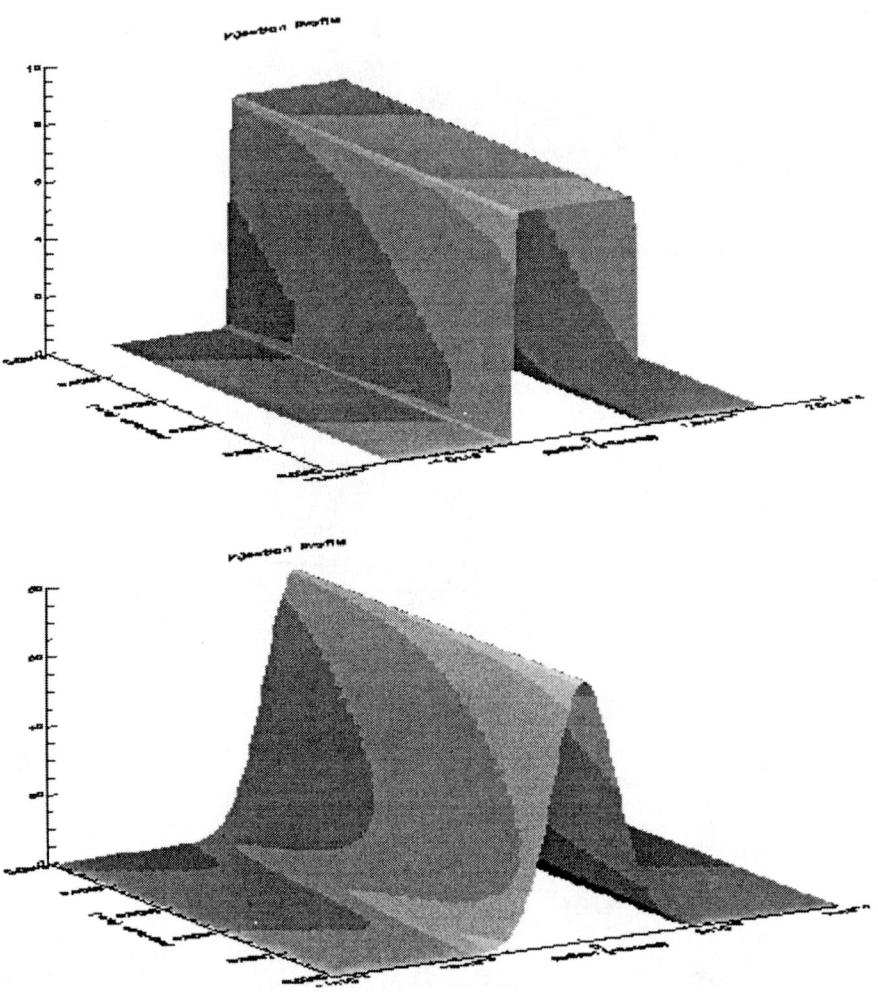

Figure 4: "Top hat" and Gaussian lateral gain distributions used in spatio-temporal dynamical simulations

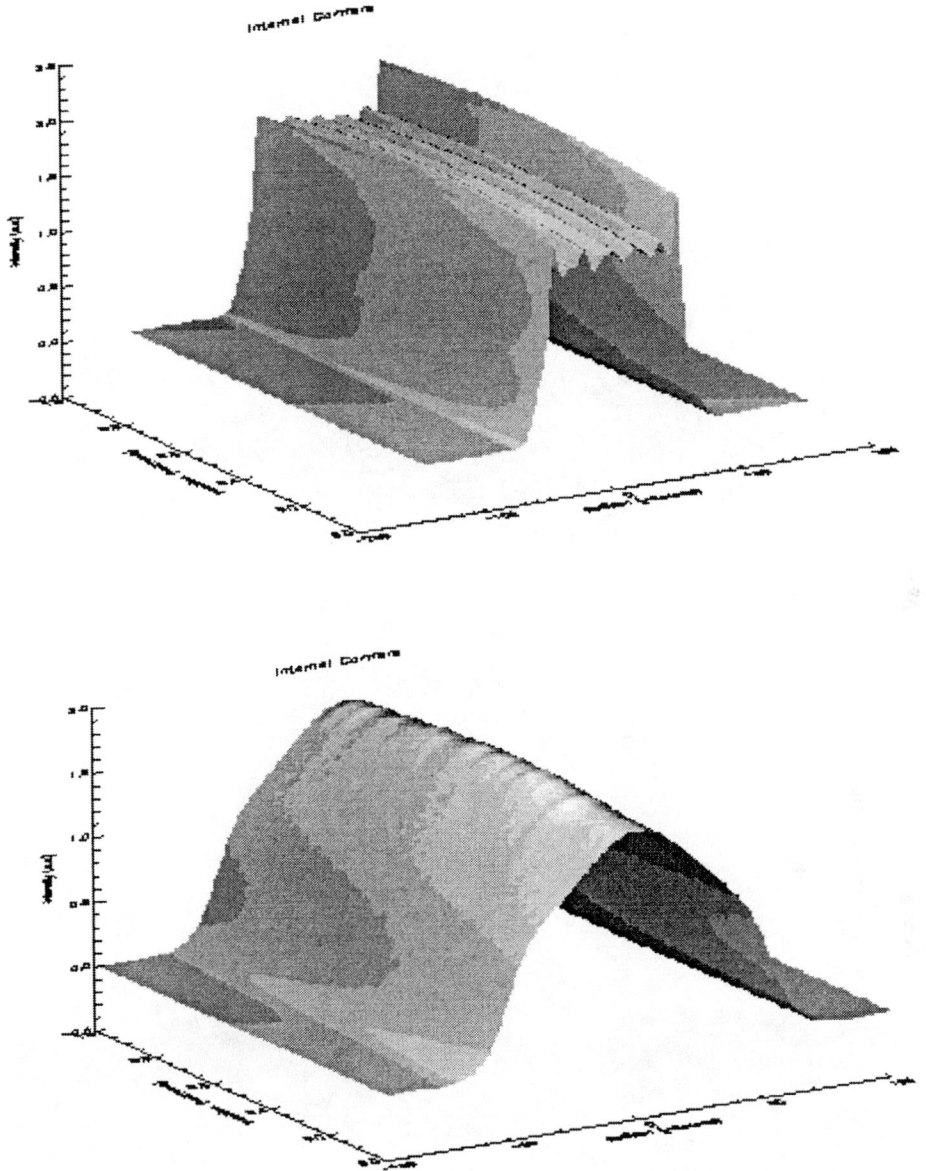

Figure 5: simulated 2-D carrier density maps for 100 μm top hat and Gaussian lateral profiles of Fig 4

that the Gaussian lateral profile has much more stable dynamics. Time-averaged near and far fields are given in Figures 7 and 8: for the Gaussian profile the beam can be nearly diffraction limited even at high power.

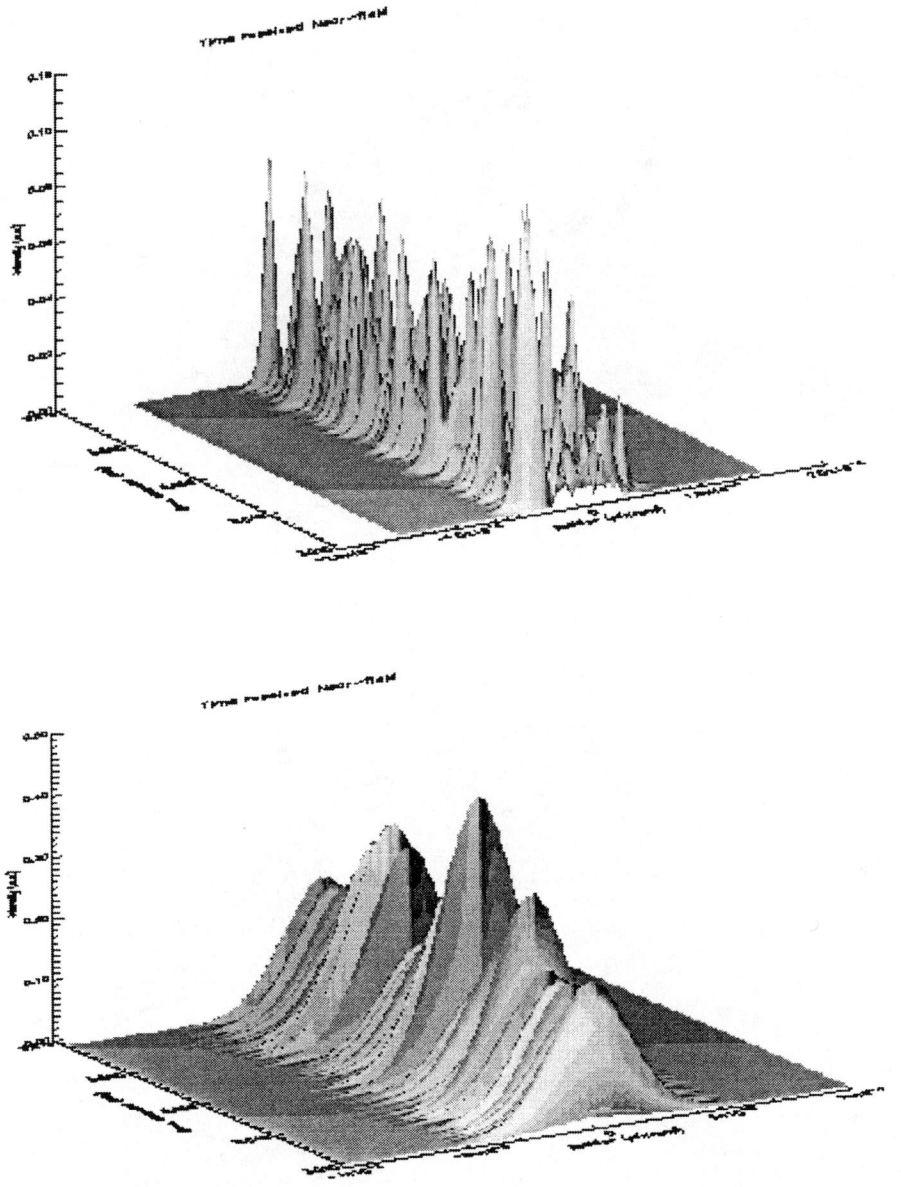

Figure 6: Lateral near fields vs. time over ~1 ns for lasers with top hat and Gaussian lateral profiles driven at 2 I_{th}. Regular temporal modulation in the latter case is due to beating between longitudinal modes which have not been filtered.

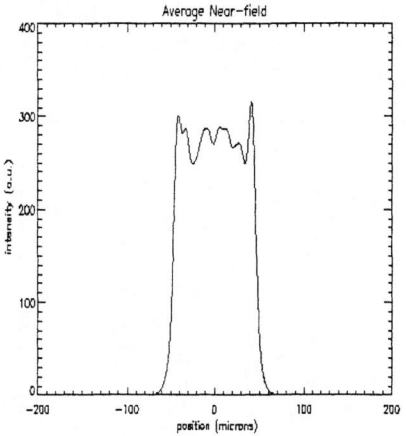
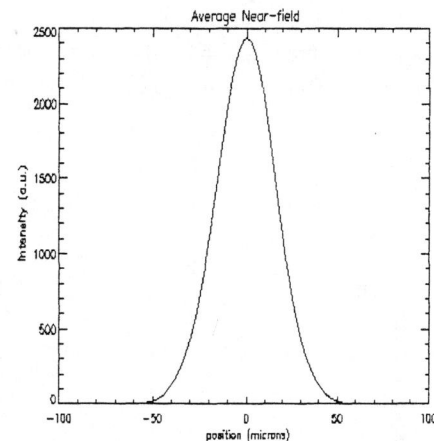

Figure 7: Time-averaged near fields for lasers with top hat and Gaussian lateral profiles under conditions as in Figure 6. Note that the vertical scale for the Gaussian profile is compressed by a factor > 6: this laser has much higher brightness

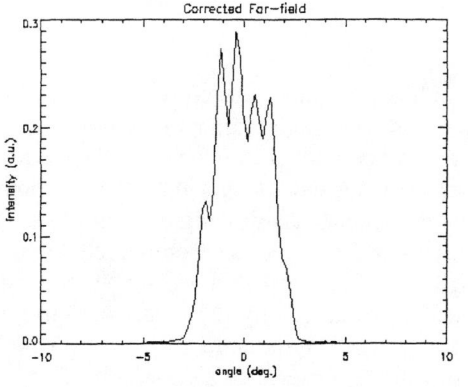
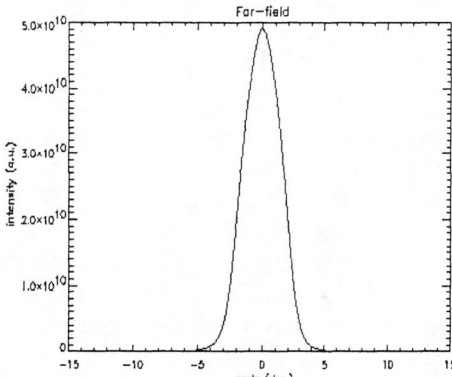

Figure 8: Time-averaged corrected far fields for top hat and Gaussian profiles under conditions as above.

Smoothing the lateral gain profile in large aperture lasers clearly results in dramatically improved spatial coherence. Similar results have been obtained for other lateral profiles, and for lateral index and temperature profiles. In practice, gain profiling is the easier technique to implement, involving modifying the lateral distribution of the injected carriers. We have demonstrated two principal methods for profiling the gain in large aperture semiconductor lasers: (1) enhanced current spreading (ECS), and (2) segmented or digitated contacts. In the first case, the injected carriers are spread transversely in a thick, heavily p-doped layer to produce a Lorentzian-like profile [19]. The extent of profiling is determined by the thickness and doping level of the current spreading layer. Since this technique is implemented epitaxially, it is relatively inflexible, but it has generated very satisfactory initial experimental results. Figure 9 shows experimental far fields from a normal (unprofiled) broad area laser and from an ECS laser with a 10 μm thick p+-doped spreading layer [20]

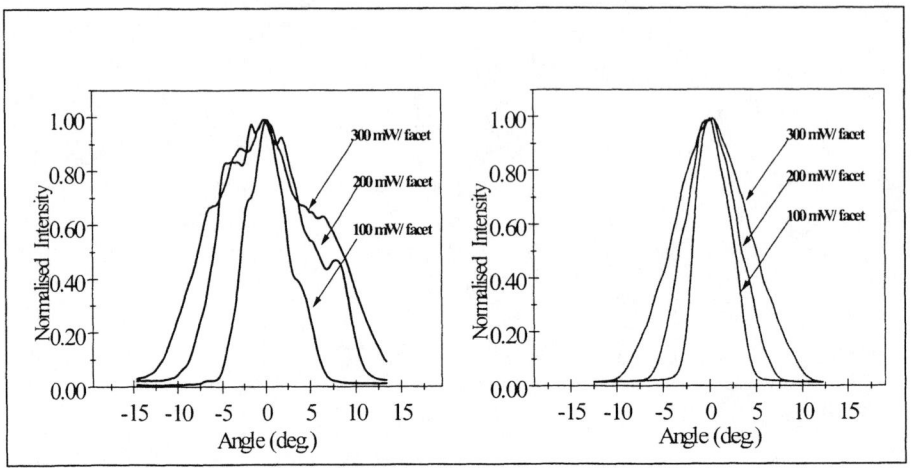

Figure 9: Experimental time-averaged far fields for normal and ECS lasers at various output powers (from [20], © 1998 IEEE).

The second technique is implemented lithographically and is therefore more flexible and less expensive to implement. In place of the normal rectangular metal contact for injecting the carriers, a segmented or patterned contact is used to distribute the carriers laterally and possible longitudinally [21]. Such a patterned contact may take various forms, for example a half-tone pixellated image, a central solid region with dots or holes whose densities vary laterally, or a central rectangular section from which tapered fingers extend laterally. Since the carriers diffuse laterally on their way to the active region, the spatially averaged carrier density will be modified by the local area ratio of current injection (*ie* metal) to unpumped material (*ie* bare semiconductor), a ratio which can readily be varied arbitrarily on a scale of the lateral carrier diffusion length [22].

SUMMARY AND CONCLUSIONS

We have presented and discussed three complementary theoretical design approaches to minimising filamentation in broad area semiconductor lasers. Maximising the facet damage threshold (and minimising other intensity related degradation mechanisms) allows the use of higher intensities, that is smaller apertures to generate a given CW power. This reduces the tendency to lase in multiple lateral modes and also minimises complex spatio-temporal dynamics which cause beam pointing instabilities and flickering on sub-nanosecond time scales. Minimising the effective Kerr coefficient, that is the linewidth enhancement factor, has a similar beneficial effect on the laser spatial coherence. This is most simply accomplished by choosing the quantum well width and strain to yield minimum index change and maximum differential gain at the desired operating wavelength. Sophisticated theoretical treatments including many-body effects such as bandgap renormalisation and excitonic enhancement will be essential design tools for this purpose. Using a spatial perturbation analysis we have shown that filaments may be "seeded" by sharp edges in the lateral gain or index profiles. Detailed spatio-temporal dynamical simulations based on microscopic many-body representations of the gain and index, as well as spatially resolved travelling wave rate equations, have confirmed the value of lateral gain profiling, which has been demonstrated experimentally by including current spreading layers as well as segmented or digitated contacts. Further experimental validations of these models are necessary and are currently underway.

ACKNOWLEDGEMENTS

The authors wish to thank Messrs R Gillen, J Sheehan and J O'Riordan for technical assistance. Important contributions to the theory of spatial perturbation growth were made by Dr K White (now at Nortel Networks, Ottawa). We acknowledge financial assistance from the US Air Force Office of Scientific Research, Forbairt (the Irish Science and Technology Agency), Optronics Ireland and the Commission of the European Union.

REFERENCES

1. Walpole, J.N., *Opt. Quantum Electron.* **28**, 623 (1996)
2. Lang, R., Hardy, A., Parke, R., Mehuys, D., O'Brien, S., Major, J., Welch, D., *IEEE J. Quantum Electron.* **29**, 2044 (1993)
3. Boyd, R.W., "Nonlinear Optics", Academic Press, 1992.
4. Agrawal, G.P., "Nonlinear Fiber Optics", Academic Press, 1989.
5. Kirkby, P.A., Goodwin, A.R., Thompson, G.H.B. and Selway, P.R., *IEEE J. Quantum Electron.* **QE-13**, 705 (1977)
6. Cook, D.D. and Nash, F.R., *J. Appl. Phys.* **46**, 1660 (1975)
7. Weiss, C.O. and Vilaseca, R., "Transverse laser dynamics", VCH, 1996.
8. Sandroff, C.J., Hegde, M.S. and Chang, C.C., J. Vac. Sci. Technol. B7, 841 (1989)
9. Lee, H.H., Racicot, R.J. and Lee, S.H., Appl. Phys. Lett. 54, 724 (1989)
10. El-Bahnasawy, R.F., PhD thesis, National University of Ireland, Cork, 1999.
11. Henry, C.H., *IEEE J. Quantum Electron.* **QE-18**, 259 (1982)
12. Osinski, M. and Buus, J., *IEEE J. Quantum Electron.* **23**, 9 (1987)
13. Chow, W.W., Koch, S.W. and Sargent, M., "Semiconductor-Laser Physics", Springer-Verlag, 1994
14. Chuang, S.L., "Physics of Optoelectronic Devices", Wiley, 1995.
15. Haug, H. and Koch, S.W., *Phys. Rev.* **A 39**, 1887 (1989)
16. Li, H., McInerney, J.G. and Malloy, K.J., CLEO 94, post-deadline paper
17. Skovgaard, P.M.W., McInerney, J.G., Moloney, J.V., Indik, R.A., Ning, C.Z., *Photonics Technol. Lett.* **9**, 1220 (1997)

18 White, J.K., McInerney, J.G. and Moloney, J.V., *Opt. Lett.* **20**, 593 (1995)
19 O'Brien, P.A., PhD thesis, National University of Ireland, Cork, 1999
20 O'Brien, P.A., Skovgaard, P.M.W. and McInerney, J.G., *Electron. Lett.* **34**, 1943 (1998)
21 Skovgaard, P.M.W., O'Brien, P.A. and McInerney, J.G., *Electron. Lett.* **34**, 1950 (1998)
22 Skovgaard, P.M.W., PhD thesis, National University of Ireland, Cork, 1998

Nonlinear Dynamics of Semiconductor Lasers: Theory and Experiments

A. Gavrielides

Air Force Research Laboratory Nonlinear Optics Center 3550 Aberdeen Ave. SE, Kirtland AFB, NM 87117, U.S.A.

Abstract. The dynamics of semiconductor lasers under external influences are examined. In particular, the focus is on two possible configurations: the semiconductor laser under optical injection, and under delayed optical feedback. The injection model for the case of weak injection and large detuning is analyzed and experimental results are used to first determine the parameters that appear in the dynamical equations. A second injection is then used to probe the injection locked system and investigate theoretically and experimentally the response of this system.

The stability boundaries of both configurations are analyzed and limit points and Hopf bifurcations points are determined. For the delayed feedback model a large number of external cavity modes can become available to the system for strong feedback and long external cavities, and a series of cascades close to threshold appears to be the route into the Low Frequency Fluctuations. A state that shows similarities with the Low Frequency Fluctuations but appears to be more regular is characterized experimentally and numerically.

INTRODUCTION

The traditional approach to dynamical system investigations relies on time series analysis of measurements of one or more of the system's dynamical coordinates. In this manner bifurcations can be observed as a function of a bifurcation parameter by the changes in the amplitudes of the signals, and chaos can be demonstrated by computing the Lyapunov exponents of the time series by well known methods. This approach, although used extensively for CO_2, Nd:YAG and other lasers, fails in the case of semiconductor (or diode) lasers because the intrinsic time-scales are very fast, of the order of tens of picoseconds or smaller. It is very uncommon to have measurements of representative time series of the dynamical state of the diode laser, except in the case when a streak camera is used, but then only a fixed number of points is available over the time interval. Thus either there is very good resolution and a very short time segment or poor resolution and a long time segment; see also the chapter by *Fischer et al.*

Therefore, it is not unusual that the preferred method of bifurcation examination in the case of semiconductor lasers relies mainly on spectral analysis. The most commonly used spectra are the optical spectrum of the field obtained by a Fabry-Perot and the intensity noise spectrum (RIN) obtained by a fast detector through a spectrum analyzer. Naturally, such spectra are time averaged quantities and must be interpreted carefully. Bifurcations can be identified and routes to chaos established in this way, but the identification of chaos must be inferred in general.

Optical and power spectra can be combined in a number of ways to produce the spectral response or impulse response of the laser system. One such particular method, for example, is to look at the regenerative peak of an optical signal injected into a semiconductor laser. As we will see later, such analysis can produce a wealth of information about the system's resonances and their location. From this we can then expect to observe such effects as resonance enhancement, entrainment, bistability, period doubling bifurcations etc. and clarify the regions in the parameter space in which such behavior is possible.

A drawback of this approach is the identification of possible coexisting attractors with the same spectral content or attractors in which the system does not stay for sufficient time, so that the spectral signatures are averaged. Sometimes numerical calculations of the equations that approximate the system's dynamics must be used to actually interpret the spectra. In such cases we must use equations that have been tested before and have good predictive power and also be very certain of the specific values of the parameters that appear in these equations. One method that yields good estimates to the system parameters and, in particular, to the linewidth enhancement factor α will be described here.

The material in this chapter is organized as follows. The first main section is focused on the solitary laser problem, the equations of the laser system and the estimation of the values of the minimum number of parameters necessary to describe the dynamics of this system. The next section deals with the semiconductor laser subject to injection and an analysis is presented of the response of the system to a weak external probe. Then we describe the semiconductor laser with external feedback, outline the stability boundaries and discuss some interesting, complex and unusual behavior that can be found in this system. Finally, the last section is a short summary of this chapter.

SOLITARY LASER MODEL

The main objective of this section is to discuss the model appropriate for semiconductor lasers, the parameters appearing in this model, their values, and to present a robust experimental method of measuring these parameters. Fortuitously, this method relies on using a weak optical probe injected into the laser under consideration and, therefore, at this point it is necessary to outline in detail the injection model. The model we will use to describe the diode laser under injection by a master laser is a class B laser model that takes into account the effect of the Fabry-Perot resonant cavity [1]. In this model the field amplitude and the carrier density are described by the equations

$$\frac{dA}{dt} = -\frac{\gamma_c}{2}A + i(\omega_0 - \omega_c)A + \frac{\Gamma}{2}(1 + i\alpha)GA + fA_i e^{-i\Omega t} \tag{1}$$

$$\frac{dN}{dt} = \frac{J}{ed} - \gamma_s N - \frac{2\epsilon_0 n_0^2}{\hbar \omega_0} G|A|^2. \tag{2}$$

The gain function G is given by

$$G = \gamma_N(N - N_t) - \gamma_p(|A|^2 - A_0^2), \tag{3}$$

where A is the complex electric field, γ_c is the cavity decay rate, ω_c is the longitudinal mode frequency of the cold cavity, Γ is the confinement factor and α is the linewidth enhancement factor. The frequency of the injected light relative to the slave laser is Ω and the carrier density is given by N, while J is the injection current density. The injected field is denoted by A_i and f is the coupling efficiency of the injected light into the laser cavity. The remaining parameters are the electronic charge e, the active layer thickness d, the spontaneous carrier lifetime $\gamma_s = 1/\tau_s$, the permittivity of free space ϵ_0, and the refractive index n_0 of the semiconductor laser for the lasing mode. In the gain function the constant γ_N denotes the differential carrier relaxation rate and γ_p the nonlinear carrier relaxation rate.

The constant N_t is the inversion at transparency and A_0^2 is the steady state solution of Eqs. (1)–(3) and is given in Appendix A. Performing a rather lengthy normalization, Eqs. (1)–(3) can be recast into a simpler form with a reduction of the necessary parameters to describe the problem to four. From Appendix A the reduced problem becomes

$$\frac{dE}{ds} = (1 + i\alpha)[n - \varepsilon P(|E|^2 - 1)]E + \eta e^{-ivs} \tag{4}$$

$$T\frac{dn}{ds} = P - n - P[1 + 2n - 2\varepsilon P(|E|^2 - 1)]|E|^2. \tag{5}$$

The parameters that appear in these equations are defined explicitly in Appendix A. They are P, which is proportional to pumping above threshold, the ratio of the cavity lifetime to the carriers lifetime T, the linewidth enhancement factor α, and the gain saturation parameter ε.

In order that this model be useful in computations all the parameters appearing in these equations need to be determined to good accuracy. To this end we will rely on a number of experiments that were performed expressly for this purpose [1]. The experiment takes advantage of the injection configuration and an extremely small amount of light is injected into the slave laser at large detuning. For a given detuning the optical spectrum consists of three main peaks as shown in Fig. 1. The frequency of the injected light is at approximately 2.5GHz relative to the slave laser

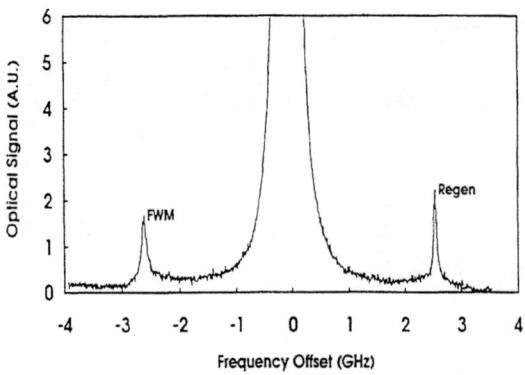

FIGURE 1. Typical four-wave mixing spectra of the output of the slave laser. The injection signal is on the high frequency side with respect to the frequency of the slave laser. From [1] © 1994 IEEE.

frequency. This light is amplified in the laser medium and its presence is indicated by the regenerative peak, and it corresponds to the radiation of the injected signal from the master laser. There is a large peak at the slaver laser frequency corresponding to the unlocked slave light, part of which has been truncated to bring out the rest of the spectrum. A four-wave mixing signal also appears at about -2.5GHz as a result of the beat frequency between the slave and master laser signals in the inversion. As a result of this interaction there are small frequency shifts, i.e. frequency pulling or pushing that the center line and the four-wave mixing signals experience. However, they are of $O(\eta^2)$, very small at this injection strength, and will be neglected in the subsequent calculations.

A useful method of exhibiting a large collection of spectra, is to form a compound spectrum of the peak of the regenerative signal or of the four-wave mixing signal as a function of detuning. To obtain a truthful picture and make an absolute comparison of the four-wave mixing and the regenerative signals, however, the linewidth difference between them should be taken into account. Therefore the measured peak values have to be scaled with their respective linewidths [2]. Fig. 2 shows such spectra of the measured regenerative signal and the four-wave mixing signal as a function of detuning. The asymmetry of the regenerative spectrum at the relaxation frequency of the laser is evident, whereas the four-wave mixing signal is symmetric. This asymmetry is due to the non-zero value of the linewidth enhancement factor, and it will allow us its final determination. In essence these spectra can be regarded as the impulse response of the linear solitary laser system, that is, the response of the solitary laser to a weak oscillatory perturbation. The spectra reveal all the resonances of the solitary laser, i.e. at the relaxation frequency and

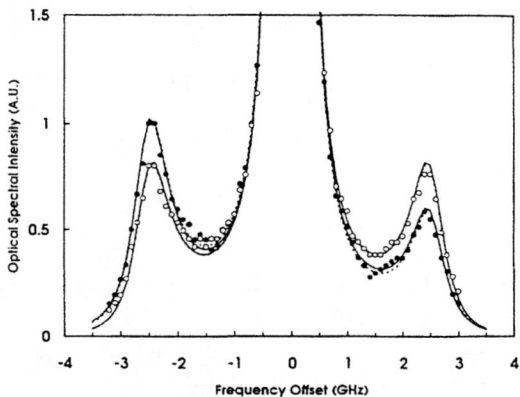

FIGURE 2. Calculated and measured regenerative and four-wave mixing signals plotted as a function of detuning $-v$. The full and open circles are the experimentally measured data for the regenerative and the four-wave mixing signals. The locking range is less than 100MHz. From [1] © 1994 IEEE.

also the large increase of the amplitude as the detuning approaches zero, indicating the Fabry-Perot resonance close to the locking regime.

We can use Eqs. (4)–(5) to calculate the four-wave mixing and regenerative response of the laser with injection. For this we set

$$E = 1 + \eta u + .., \qquad n = \eta N + .. \qquad (6)$$

and obtain the leading order problem in the injection rate η:

$$\frac{du}{ds} = (1 + i\alpha)[N - \varepsilon P(u + u^*)] + e^{-ivs} \qquad (7)$$

$$\frac{dN}{ds} = -\frac{(1 + 2P)}{T}N - \frac{P}{T}(1 - 2\varepsilon P)(u + u^*). \qquad (8)$$

The new parameters that we will find useful can be defined as

$$\zeta = \frac{1 + 2P}{T}, \qquad \omega_r = \sqrt{2P(1 + \varepsilon)/T} \qquad (9)$$

and they denote the laser damping and the relaxation frequency of the solitary laser. There is a small correction to the relaxation frequency that is proportional to the gain saturation strength. However the main correction will be in the damping and it will come from the field equation. Similar calculations were carried out in

Ref. [3], in which some of the nonlinear terms appearing in Eqs. (4)–(5) were also retained.

We now look for harmonic solutions of the form

$$u = R(v)e^{-ivs} + F(v)e^{ivs} \tag{10}$$

$$N = N_1 e^{-ivs} + c.c \tag{11}$$

where R and F refer to the regenerative and four-wave mixing amplitudes. The solutions are obtained in terms of

$$\sigma = R + F^* = \frac{(iv - \zeta)}{[v^2 + iv(2\varepsilon P + \zeta) - \omega_r^2]} \tag{12}$$

and

$$\delta = i(R - F^*) = -\frac{(v^2 - \omega_r^2 + 2\varepsilon Pav) + i[a\omega_r^2 + v(\zeta + 2\varepsilon P)]}{v[v^2 + iv(\zeta + 2\varepsilon P) - \omega_r^2]}. \tag{13}$$

The absolute value of Eq. (12) is also proportional to the spectrum of the intensity and that of Eq. (13) to the phase modulation. Notice that the modulation spectra have a pole at zero indicating that the amplitude of the modulation becomes very large as the detuning approaches zero and rendering this approximation invalid. A different approximation is therefore required as the locking regime is approached for values close to zero detuning. In Appendix B we give the explicit expressions for the spectra in terms of the original quantities appearing in Eq. (1) through Eq. (3). Notice also that there are poles at $v \approx -(i\xi/2) \pm \omega_r$ with a damping given by $\xi = (\zeta + 2\varepsilon P)$. For large values of T the laser damping is dominated by the gain saturation term $2P\varepsilon \gg \zeta$. These correspond to the free running laser relaxation frequency and damping.

Using the spectral data indicated by circles in Fig. 2 the basic parameters that characterize the laser can be easily deduced. First fitting the regenerative spectrum, the values of the parameters ω_r, γ_r, and α are determined. The value of γ_p can be assumed to be zero or any other reasonable guess, as this will not affect the determination of the rest of the parameters. Indeed, if we look at the asymmetry of the spectrum at the location of the relaxation frequency resonance we find that for the regenerative spectrum we have

$$\frac{|R(\omega = -v = -\omega_r)|}{|R(\omega = -v = \omega_r)|} \approx 1 + \frac{4\xi\alpha}{1 + \alpha^2} \tag{14}$$

where the damping has been normalized to the relaxation frequency as $\xi = \zeta/\omega_r$. The four-wave mixing spectrum, however has no such asymmetry. Once the values of ω_r, γ_r, and α have been obtained the value of γ_p is obtained by fitting the power spectrum, given by the square of the amplitude of Eq. (12). Of course, by taking a series of spectra at various levels of pumping, the power dependences of these quantities can be obtained. From this power dependence the value of γ_c, and γ_s can be determined. That is, from the intercept of γ_r at $P = 0$ (threshold) the value of γ_s is determined since both γ_N, and γ_p have a linear dependence on P. The value of γ_c is obtained from Eq. (63). This is an advantage over other techniques, which require a separate measurement of γ_c before the value of γ_p can be determined [4]. The values shown in Table 1 were determined for a Spectra Diode Labs Laser SDL-5301-G1 single longitudinal and transverse mode GaAs/AlGaAs quantum well laser operating at 827.6 nm, and were reported in [1].

TABLE 1. Laser parameters; the laser power P in the power dependent parameters is in mW.

Parameters	Symbol	Value
Linewidth enhancement factor	α	4
Relaxation resonace frequency	f_r	$\sqrt{0.954P}\ GHz$
Spontaneous carrier relaxation rate	γ_s	$1.458 \times 10^9\ \sec^{-1}$
Differential carrier relaxation rate	γ_N	$0.155P \times 10^9\ \sec^{-1}$
Nonlinear carrier relaxation rate	γ_p	$0.28P \times 10^9\ \sec^{-1}$
Cavity decay rate	γ_c	$2.4 \times 10^{11}\ \sec^{-1}$

Notice that from a large number of parameters in the original equations Eqs. (1)–(3) only the ones shown in the Table 1 are required with the exception of the relaxation frequency, which is a combination of the last four rates. Thus the five quantities are the minimum set of numbers required to specify the dynamical behavior of the semiconductor laser. In terms of normalized quantities we need only four, P, α, T, and ε, since the time is normalized to the cavity decay rate. These are shown in Table 2.

TABLE 2. Normalized laser parameters; the pumping level is for an optical power of 9 mW.

Parameters	Symbol	Value
Pumping	P	0.48
Linewidth enhancement factor	α	4
Ratio of carrier lifetime to cavity decay	T	165
Gain Saturation	ε	0.011

From the dynamic response of the laser view point, the most critical parameters appear to be the linewidth enhancement factor α and the gain saturation parameter ε. In the next few sections we will have occasion to comment on this.

INJECTION MODEL

In an analogous way to the discussion of the previous section, we can effectively use weak four-wave mixing to probe the properties and dynamical behavior of the system of semiconductor laser under injection. In particular, we now require three lasers, a master laser a slave laser and a tunable probe laser to act as a secondary injection. The slave-master system is adjusted to a given dynamical behavior by selecting the injection strength rate and detuning, and then the system is probed and the regenerative and four-wave mixing spectra are obtained and analyzed.

The field equation Eq. (4) is now modified to take account of the secondary injection and becomes

$$\frac{dE}{ds} = (1+i\alpha)[n - \varepsilon P(|E|^2 - 1)]E + \eta e^{-ivs} + \eta_1 e^{-iv_1 s} \quad (15)$$

where η_1 denotes the probe injection rate and v_1 the probe detuning. The carrier equation, of course, is not altered by the addition of a secondary probe and Eq. (5) is still used in combination with Eq. (15). Now the computation of the four-wave and regenerative spectra becomes more involved, but it remains straightforward. For purposes of illustration we reproduce here the analytical spectra with the gain saturation parameter set to zero. We use an expansion in terms of the weak probe η_1 and expand the field and the inversion about the injection problem:

$$E = E_0 + \eta_1 u + .., \qquad n = N_0 + \eta_1 N_1 + .. \quad (16)$$

and obtain for the $O(1)$ problem

$$\frac{dE_0}{ds} = (1+i\alpha)N_0 E_0 + \eta e^{-ivs} \quad (17)$$

$$T\frac{dN_0}{ds} = P - N_0 - P(1+2N_0)|E_0|^2. \quad (18)$$

For the $O(\eta_1)$ problem we obtain:

$$\frac{du}{ds} = (1+i\alpha)(N_0 u + E_0 N_1) + e^{-iv_1 s} \quad (19)$$

$$T\frac{dN_1}{ds} = -\zeta N_1 - \frac{\omega_r}{2}(E_0^* u + E_0 u^*) \quad (20)$$

These equations are similar to the linearized version of Eqs. (7)–(8) except that the linearization is about the injection problem and the relaxation frequency and damping have been redefined to

$$\zeta = \frac{1 + 2P|E_0|^2}{T}, \qquad \omega_r = \sqrt{2P(1+2N_0)|E_0|^2/T}. \quad (21)$$

We will compute the regenerative and four-wave mixing spectra of the probe signal when the laser subject to injection is at the locked steady state. The steady state solutions of Eqs. (17)–(18) are given by the implicit equations [8] for $-1/2 < N_0 \leq P$

$$\eta^2 = \sqrt{\frac{P - N_0}{P(1 + 2N_0)}}[N_0^2 + (v + \alpha N_0)^2] \tag{22}$$

and

$$|E_0|^2 = \frac{P - N_0}{P(1 + 2N_0)}, \tag{23}$$

while the periodic solutions for u and N_1 are described by

$$u = R(v_1)e^{-iv_1 s} + F(v_1)e^{iv_1 s} \tag{24}$$

$$N_1 = n_1(v_1)e^{-iv_1 s} + c.c. \tag{25}$$

The regenerative and four-wave mixing spectra of the probe radiation are given by

$$R(v_1) = -\frac{(iv_1 - \zeta)[iv_1 + (1 - i\alpha)N_0] + (1 - i\alpha)\omega_r^2/2}{D(v_1)} \tag{26}$$

$$F^*(v_1) = (1 - i\alpha)\frac{E_0^*}{2E_0}\frac{\omega_r^2}{D(v_1)} \tag{27}$$

where the denominator is given by

$$D(v_1) = (iv_1 - \zeta)[-v_1^2 + 2iv_1 N_0 + (1 + \alpha^2)N_0^2] + \omega_r^2[iv_1 + (1 + \alpha^2)N_0]. \tag{28}$$

The amplitude and phase modulation spectra are given in terms of the regenerative and four-wave mixing spectra by

$$\sigma = \frac{E_0^* R + E_0 F^*}{|E_0|^2} \tag{29}$$

$$\delta = i\frac{E_0^* R - E_0 F^*}{|E_0|^2}.$$

These equations reduce to Eqs. (12)–(13) in the limit of a solitary laser, i.e. $N_0 = 0$. Experiments were carried out with a slave laser of the same type as the one that was characterized before [5]. It has very similar dynamical properties

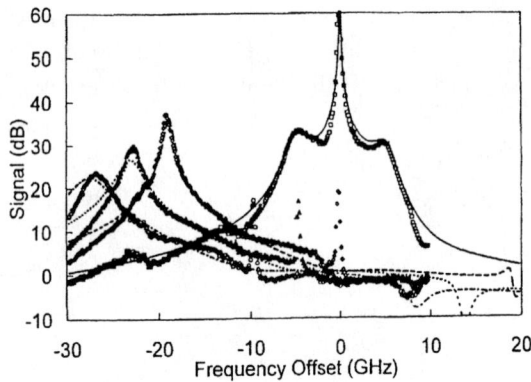

FIGURE 3. Regenerative spectra of a weak probe of the injection locked system as a function of probe detuning $-v_1$. Symbols are experimental data and curves are calculated spectra for four different operating conditions. The solid curves and squares are for free running laser, the dashed curve and diamonds are for injection locking at zero detuning, the dotted curve and triangles are for injection locking at -4.8GHz, and the dot-dashed curve and circles are for injection locking at -9.7GHz. From [5] © 1996 IEEE.

and we will not repeat the parameters, except to mention that the linewidth enhancement factor for this laser was determined to be 5. The probe laser was a New Focus external cavity semiconductor laser whose frequency was scanned over approximately 40GHz. Figure 3 shows the regenerative spectra of the injection locked system when the injection rate is fixed at $\eta = 0.4$ for three operating points. Detunings $(-v)$ of the master laser relative to the slave laser of 0, -4.8 and -9.7 GHz are shown. In addition, the familiar regenerative spectrum of the free running laser, which was used to obtain the laser parameters, is also included. With the amplitude of the signal adjusted to fit the free-running spectrum there are no free parameters left for fitting the detuned injection spectra. Notice the strongly asymmetric shape of the spectrum, reflecting the shift of the Fabry-Perot resonance corresponding to the pole at $v_1 = 0$ in the free running laser spectrum. For large values of injection and detuning $N_0 \sim O(1)$ while the rest of the parameters appearing in Eqs. (26)–(27) remain of order $O(1/T)$. The regenerative spectrum reduces to

$$R(v_1) \approx \frac{1}{i(v_1 + \alpha N_0) + N_0} \qquad (30)$$

a radically asymmetric spectrum, the magnitude of which shows a maximum at $v_1 = -aN_0$. This is a good approximation to the location of the resonance. In addition, a resonance feature can be discerned at the injection frequency in the

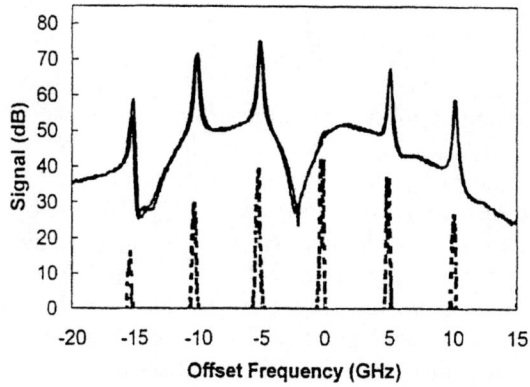

FIGURE 4. Spectrum of the slave laser output (lower dashed line) and the regenerative amplification of a weak optical probe (upper solid line) when the master-slave system is on a limit cycle.

experimental data but this corresponds to the injection peak, and it is not part of the modulation spectrum [6].

We complete this section by including in Fig. 4 the regenerative spectrum of the probe laser when the injection locked system is on a limit cycle [7]. Both the optical spectrum and the regenerative spectrum are plotted in Fig. 4 for comparison. The optical spectrum shows sidebands close to the relaxation frequency of the free running laser and higher harmonics. The regenerative spectrum shows a dramatic increase in the response at these frequencies indicating the possibility of locking of the limit cycle frequency to the external (probe) frequency. This problem is more difficult to analyze, and numerical or analytical results will not be presented here.

Stability Boundaries

We will now address the primary bifurcations of the injection model and the determination of the stability boundaries as obtained from equations Eqs. (4)–(5) with $\varepsilon = 0$. The system of equations admits steady state solutions given by

$$E = A_s e^{i(\Phi - vs)} \qquad n = N_s. \tag{31}$$

The carrier density $N_s = N_s(\eta^2)$ is given in implicit form by

$$\eta^2 = \frac{P - N_s}{P(1 + 2N_s)}[N_s^2 + (v + \alpha N_s)^2] \tag{32}$$

and we find for the amplitude of the laser field

$$A_s = \sqrt{\frac{P - N_s}{P(1 + 2N_s)}}, \qquad (33)$$

provided that the quantity under the square root is positive. These are the identical equations Eqs. (22) and (23) used in the previous section.

A stable steady state may loose stability through a limit point bifurcation (also called saddle-node or fold bifurcation) or a Hopf bifurcation. In the former case one of the real eigenvalues of the characteristic equation passes through zero from negative to positive values., and in the latter case a pair of complex eigenvalues cross the real axis and acquire a positive real part; see also the chapter by *Krauskopf*. The limit point boundary that delineates the locking of the phase Φ and the emergence of the steady state satisfies the quadratic equation for $(v + aN_s)$

$$\frac{(1 + 2P)}{(1 + 2N_s)}[(v + \alpha N_s)^2 + N_s^2] - 2(P - N_s)[N_s + \alpha(v + \alpha N_s)] = 0. \qquad (34)$$

Solving this equation for $v = v(N_s)$ and determining the real roots gives us along with Eq. (32) a parametric representation of the limit point curve in the (η, v) parameter space. The Hopf bifurcation boundary corresponds to the emergence of time periodic solutions, and satisfies

$$(v + \alpha N_s)^2 N_s + \epsilon b(v + \alpha N_s)(P - N_s) - \epsilon[\epsilon(P - N_s) + 2N_s^2]\frac{1 + 2P}{1 + 2N_s}$$

$$+ N_s^3 + \epsilon N_s(P - N_s) + \epsilon^2 N_s \frac{(1 + 2P)^2}{(1 + 2N_s)^2} = 0, \qquad (35)$$

where $\epsilon = 1/T$ is a small parameter. Solving this equation for $v(N_s)$ produces none, one or two real roots. For the real roots we use Eq. (32) to obtain a parametric representation of the Hopf bifurcation curve in the (η, v)-space. Using Eq. (32) and the solutions of Eq. (34) and Eq. (35) we draw the two stability boundaries as shown in Fig. 5 for representative laser parameters $P = 0.25$, $T = 1000$, and $\alpha = 5$. The possible range of the inversion is $-1/2 < N_s \leq P$ as it can be inferred from Eq. (33). However, Eq. (35) suggests that the interesting range of $|N_s|$ is perhaps either $O(\epsilon)$ or $O(\epsilon^{1/2})$ if $|P| = O(1)$. In Fig. 5(a) the curve marked (LP) is the limit point curve that denotes the emergence of the locked steady state. Below this curve the solutions are pulsating four-wave mixing solutions. The curve marked SHB (stable Hopf bifurcation boundary) marks the Hopf bifurcation boundary. Notice that it folds back and allows the possibility of two consecutive Hopf bifurcations as a function of injection rate for some values of fixed detuning. The curve marked UHB corresponds to a unstable Hopf bifurcation boundary and leads to unstable limit cycles. Both UHB and SHB meet the LP curve at a degenerate bifurcation point, called a saddle-node Hopf bifurcation point, located at $v = v_c \simeq -(1 + a^2)\epsilon(1 + 2P)/(2a)$ and denoted by an open circle in the magnified plot of Fig. 5(b). This point has interesting consequences in the dynamics of the semiconductor laser subject to injection and they are discussed at some length in Ref. [9].

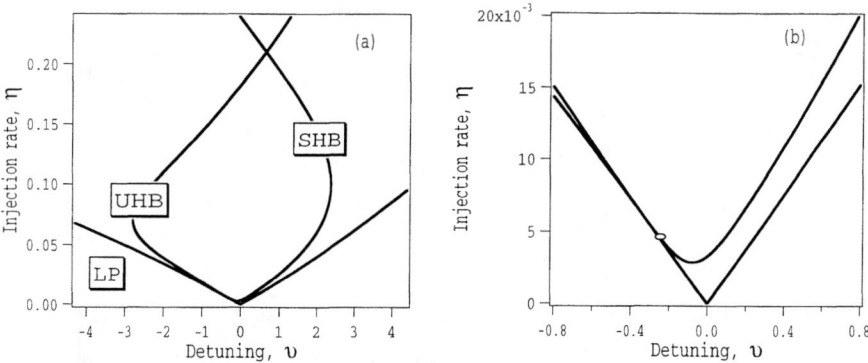

FIGURE 5. Limit point and Hopf bifurcation curves. The values of the parameters are $T = 165$, $\alpha = 4$, and $P = 0.48$.

THE LANG-KOBAYASHI MODEL

One of the particular models of semiconductor laser under external influences, which has attracted great interest and has generated a vast amount of literature in the past twenty years, is the semiconductor laser subject to delayed feedback. Because the dynamics of the semiconductor laser are very fast, typical timescales are of the order of tens of picoseconds, feedback of the emerging radiation from a reflector situated even a few millimeters form the laser facet can have appreciable delay relative to the cavity dynamics. Therefore, the light introduced back into the cavity will show a time lag equal to the external roundtrip time. Such a model was introduced by Lang and Kobayashi to explain unusual dynamical behavior observed experimentally [10].

$$\frac{dE}{ds} = (1 + i\alpha)[n - \varepsilon P(|E|^2 - 1)]E + \eta E(s - \tau)e^{-i\omega_0 \tau} \qquad (36)$$

$$T\frac{dn}{ds} = P - n - P[1 + 2n - 2\varepsilon P(|E|^2 - 1)]|E|^2 . \qquad (37)$$

The last term in Eq. (36) represents the delayed electric field reinjected into the cavity after a delay of τ while the equation for the carriers is identical to the equation used previously. In addition, a phase $\omega_0\tau \in [-\pi, \pi]$ acquired by the feedback field after one round trip in the external cavity is explicitly included. For simplicity in the following analysis we will only consider the case of $\varepsilon = 0$. These equations are already scaled versions of the original solitary laser equations with the addition of the time delayed field; see also the chapter by *Lenstra & Yousefi*.

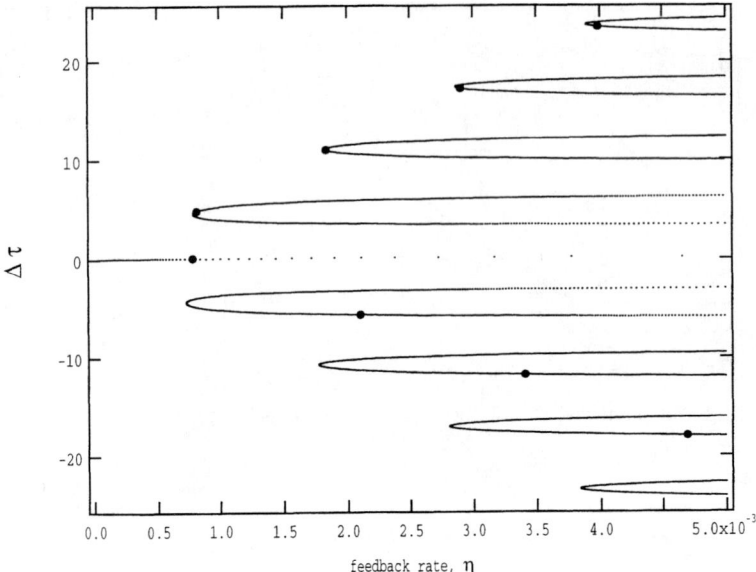

FIGURE 6. Steady state solutions of the Lang-Kobayashi system of Eqs. (36)–(37). The frequency of the external cavity modes and antimodes are plotted as a function of feedback strength. The values of the parameters used are $T = 1000$, $P = 1.391$, $\alpha = 6$, $\tau = 1000$, and $\omega_0 \tau = -1.57$.

Stability Boundaries

The systems admits steady state solutions [11] given by

$$n = N_s \qquad E_s = A_s e^{i(\Delta - \omega_0)s} \qquad (38)$$

with solutions obtained from a transcendental equation for Δ

$$\Delta \tau - \omega_0 \tau = -\eta \tau \sqrt{1 + \alpha^2} \sin[\Delta \tau + \tan^{-1}(\alpha)] \qquad (39)$$

and with

$$N_s = -\eta \cos(\Delta \tau) \qquad (40)$$

$$A_s^2 = \frac{P - N_s}{P(1 + 2N_s)} \geq 0. \qquad (41)$$

At weak feedback levels Eq. (39) has only one solution, which is close to the solitary laser frequency. As η increases additional solutions emerge in pairs from saddle-node bifurcations as shown in Fig. 6. The feedback level at which a new pair of fixed-point solutions is created can be determined by approximating the extrema of Eq. (39). Thus we obtain

$$\eta \approx \frac{\beta + \sqrt{\beta^2 - 2}}{2\tau\sqrt{1+\alpha^2}}, \qquad (42)$$

where $\beta = \omega_0 \tau + \tan^{-1}(\alpha) + (2n - 1/2)\pi$, $n = \pm 1, \pm 2 \ldots$ The approximation improves as η increases.

The external modes lie on an ellipse in the frequency Δ and inversion n plane as determined from

$$\eta^2 = N_s^2 + (\alpha N_s - \Delta)^2. \qquad (43)$$

A stability analysis shows that one of the points is inherently stable and is identified with an external cavity mode (ECM) of the system, while the other is a saddle, often called an antimode, and unstable. The stability condition is given by

$$1 + \eta\tau\sqrt{1+\alpha^2}\sin(\Delta\tau + \tan^{-1}(\alpha)) > 0. \qquad (44)$$

Note that the inequality in Eq. (41) implies that some of the solutions are not allowed and gives the condition

$$P > -\eta\cos(\Delta\tau) \qquad (45)$$

if η is small. Therefore negative pumping, i.e. pumping below the solitary laser threshold is possible if $\cos(\Delta\tau) > 0$. This will have important consequences when small values of P are considered.

From the linearized equations about the steady states we find that the condition for a nontrivial solution leads to the characteristic equation for a growth rate λ

$$2\frac{P - N_s}{T}[(N_s F + \lambda) + \eta\alpha F \sin(\Delta\tau)]$$

$$+[\frac{1 + 2P}{T(1 + 2N_s)} + \lambda][\eta^2 F^2 - 2\lambda\eta\cos(\Delta\tau)F + \lambda^2] = 0 \qquad (46)$$

where

$$F = e^{-\lambda\tau} - 1. \qquad (47)$$

These equations determine possible Hopf bifurcations emerging from the steady state; see also the chapter by *Verduyn Lunel & Krauskopf*. Assuming that there is a Hopf bifurcation located at $\eta = \eta_H$ with frequency $\omega = \omega_H$, from Eq. (46) by substituting $\lambda = i\omega$ and separating real and imaginary parts we obtain two equations for (η_H, ω_H). These equations can be solved numerically, but the process is rather tedious. A rather simple expression is obtained by noting that $T = O(10^3)$ is a large number and seek an approximation of Eq. (46) in powers of $\epsilon = 1/T$ [11].

From Eq. (46) with $\eta = 0$ and $N_s = 0$ (the case of solitary laser) it is found that $\omega_H = O(\epsilon^{1/2})$, the relaxation frequency of the solitary laser. To keep F finite as $\epsilon \to 0$, it is assumed that $\tau = O(\epsilon^{-1/2})$. Therefore, a solution of the form

$$\eta_H = \epsilon\eta_1 + \epsilon^{3/2}\eta_2 + ... \qquad \omega_H = \epsilon^{1/2}\omega_1 + \epsilon\omega_2 + ... \qquad (48)$$

is appropriate.

Inserting this equation into the real and imaginary parts of Eq. (46) and equating to zero the coefficients of each power of ϵ leads to consecutive problems for the coefficients of Eq. (48). This leads to the following expressions for the Hopf bifurcation points:

$$\eta_H = -\frac{\xi}{2B\sin^2(\frac{\omega_r \tau}{2})} \qquad (49)$$

$$\omega_H = \omega_r + \frac{\xi}{2}\cot(\frac{\omega_r \tau}{2}) \qquad (50)$$

where the $\omega_r = \sqrt{2P/T}$ is the relaxation frequency of the solitary laser, $\xi = (1+2P)\omega_r/(2P)$ the laser damping, and $B = \alpha\sin(\Delta\tau) + \cos(\Delta\tau)$.

In Fig. 6 these Hopf bifurcation points are shown by filled circles indicating that the stable modes are destabilized by this bifurcation. The approximation is very good for the primary mode that emerges from the solitary laser mode, but it becomes less and less accurate for larger values of the feedback rate and for higher order modes.

In addition to this simple picture, more complicated Hopf bifurcation scenarios become possible if we consider the case of $\varepsilon \neq 0$. For specific values of the gain saturation parameter more than one Hopf bifurcation from the same external cavity mode is possible [12]. In particular, it was found numerically and determined experimentally that after a Hopf bifurcation to an oscillating solution a second Hopf bifurcation, with a frequency that differs by the external cavity round trip frequency, can occur. This bifurcation is at first unstable since it emerges from an unstable steady state destabilized by the first Hopf bifurcation. However, it stabilizes and coexists with the first limit cycle.

Low Frequency Fluctuations

Feedback induced delay systems can be characterized by very complex and high dimensional behavior; see also the chapter by *Fischer et al.* This is attributed to the delay terms appearing in these equations and the necessity to specify initial conditions over the complete time interval of the delay; see also the chapter by *Verduyn Lunel & Krauskopf*. Specifically reducing the delay differential problem to a set of ordinary differential equations requires an infinite set of differential equations for closure.

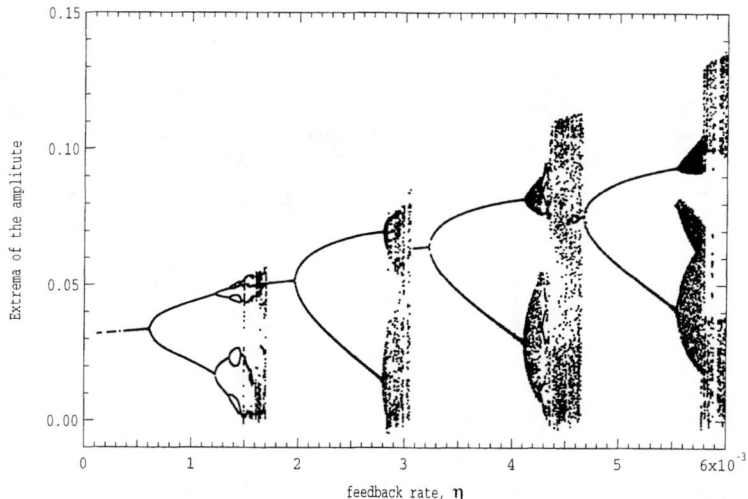

FIGURE 7. Numerically computed bifurcation cascade pumped close to threshold. The values of the parameters used are $T = 1000$, $P = 0.001$, $\alpha = 4$, $\tau = 1000$, and $\omega_0 \tau = -1.0$.

In the earlier sections and, in particular in Fig. 6, a rather simple and regular behavior is evident in which the laser operates first on a steady state which can be destabilized through a Hopf bifurcation. Further increasing the feedback strength, the laser can undergo either a period doubling route to chaos or a quasiperiodic route. The possibility that after this the laser can move into another steady state, i.e. the next mode generated through a saddle-node bifurcation, has been recently recognized. It can be seen in Fig. 7 in which a numerically computed bifurcation cascade is shown for a semiconductor laser pumped close to threshold. As the feedback strength is increased from zero the first ECM is stable, becomes unstable through a Hopf bifurcation and then chaotic via a sequence of period-doubling bifurcations. The next higher ECM (mode 2) becomes stable and coexists with the chaotic attractor of the unstable mode 1. Then mode 2 becomes unstable and follows a quasiperiodic route to chaos. The stable mode 3 is created and becomes accessible to the system and coexists with the attractor of mode 2. Then mode 3 becomes unstable and the bifurcation cascade continues. At some value of the feedback rate all the attractors from the previous modes merge into a single large attractor giving birth to what in the literature is referred to as the Low Frequency Fluctuation attractor (LFF). For example, for $\eta \sim 0.0043$ a sudden increase in the size of the chaotic attractor can be clearly seen. This region of the attractor, even though the LFF attractor is composed of only three ECM, is characterized

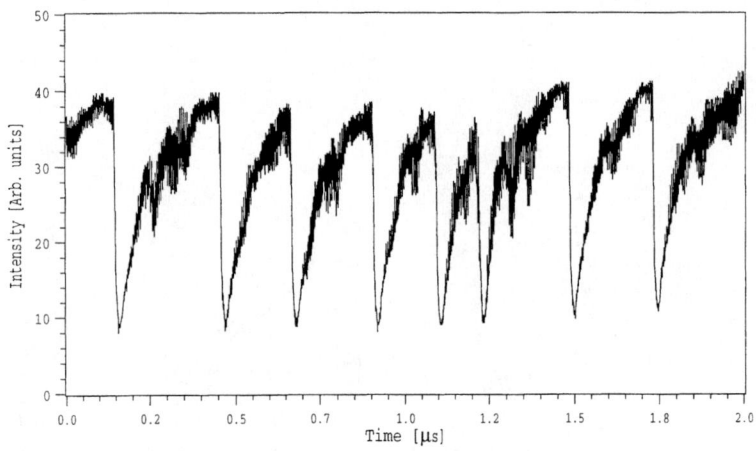

FIGURE 8. Time trace of LFF taken with a low bandwidth instrumentation of 400MHz.

by irregularly occurring intensity dropouts which were first observed by Risch and Voumard [13]. Such a time trace is shown in Fig. 8 taken with low bandwidth resolution. The trace shows the characteristic slow increase of the intensity followed by a sudden dropout event in which the intensity of the laser drops almost to zero. Fischer *et al* [14] performed high bandwidth streak camera experiments and discovered that the intensity displays irregular pulsations with a duration of a few picoseconds which had been predicted by Tartwijk *et al* [15] by detailed numerical simulations; see also the chapter by *Fischer et al.* Sano [16] discovered that the dropout events are initiated as a result of a crisis, a collision of the trajectory that wanders through the LFF attractor with an unstable saddle. This event could be either deterministic or noise influenced as the trajectory passes close to the saddle [17]. Once the collision takes place the trajectory follows the unstable manifold of the saddle towards low intensities and the chaotic ruins that emerged from the solitary laser mode.

The particular cascade computed numerically in Fig. 7 has been verified experimentally in Ref. [18], which is shown in Fig. 9. As the feedback strength is increased, as measured by the amount of threshold reduction, the laser undergoes a cascade of bifurcations as one ECM becomes unstable and is replaced by the next ECM. Traces (b), (d), (f), (h), (j), and (l) in Fig. 9 are the optical spectra of the laser in a stable state composed of a narrow line with no sidebands. In traces (c), (e), (g), (i) and (k) in Fig. 9 the laser is undergoing LFF and has a broadened line and sidebands. In particular the coexistence between the LFF and a stable ECM is shown in an experimental time trace in Fig. 10. This time trace corresponds to a threshold reduction of 9.5% and is in the region between traces (k) and (l) of Fig. 9. It shows

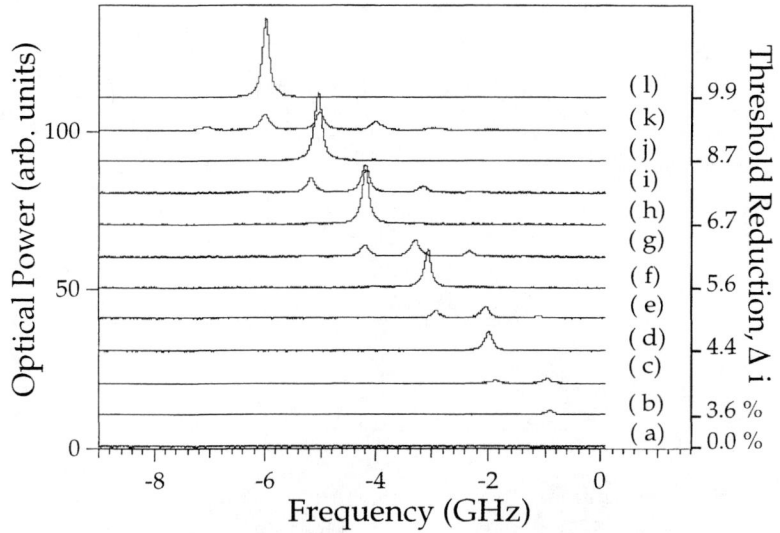

FIGURE 9. Experimentally observed bifurcation cascades.

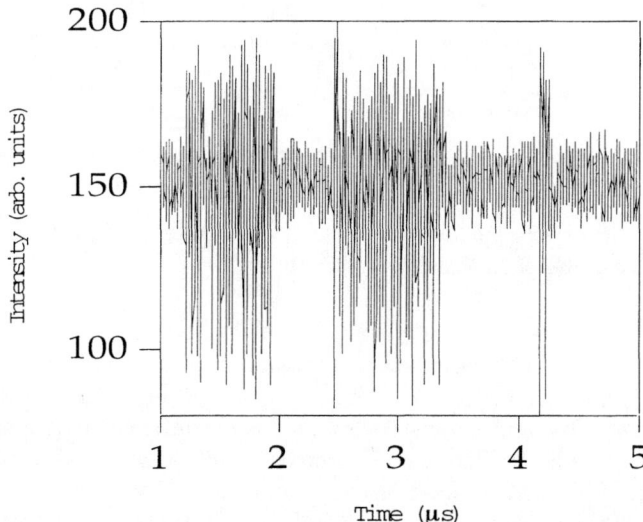

FIGURE 10. Experimental time traces showing the coexistence of LFF and steady state.

that the intensity exhibits regions of large fluctuations, corresponding to the laser being on the LFF attractor, and regions of stable operation. In particular, the

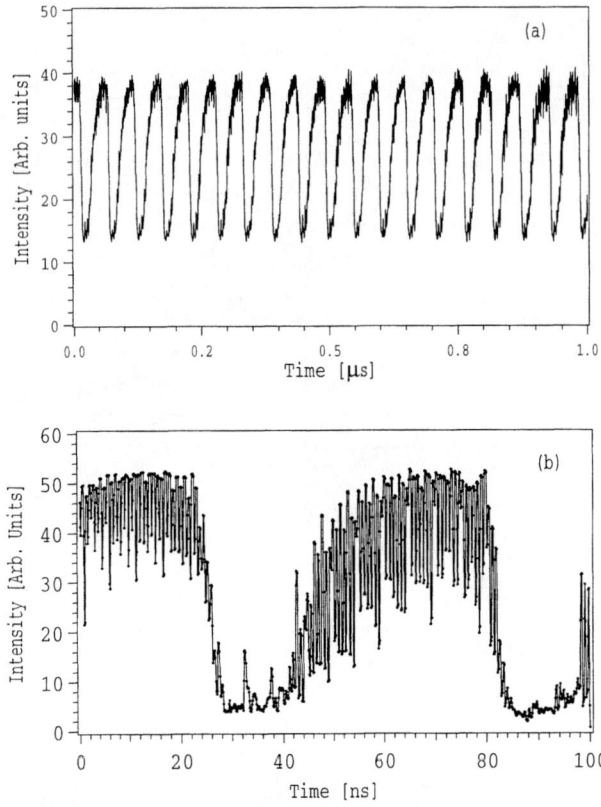

FIGURE 11. Time trace obtained with a 400MHz detection system (a), and time trace taken with a transient digitizer of 4.5 GHz bandwidth (b).

laser appears to switch between stable emission and LFF by a process that is noise dominated. Similar behavior has been seen by Fischer *et al* [19] and the regimes of stable lasing, LFF and coexistence have been studied experimentally and their boundaries outlined; see also the chapter by *Fischer et al*. The deterministic LK equations Eqs. (36)–(37) indeed provide solutions which show the experimentally observed coexistence, as is shown in Fig. 10. By analyzing the regime of coexistence of stable emission and low-frequency fluctuations [21] one finds window-like structures in which sustained LFF regions alternate with transient LFF regions as a function of the feedback parameter. In the transient LFF regions the asymptotic dynamical state is stable emission from the maximum gain mode. The particular structure of these windows depends on the system's parameters and very strongly on the value of the linewidth enhancement factor.

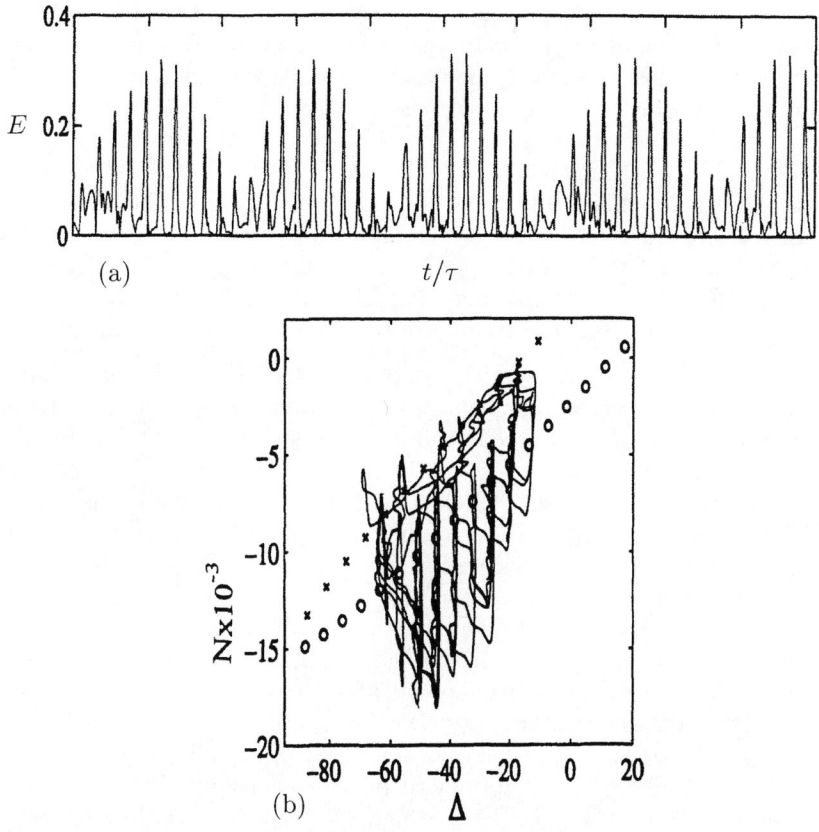

FIGURE 12. Periodic time locked state obtained for $\alpha = 6$, $P = 0.001$, $T = 1000$, $\tau = 1000$, $\omega_0\tau = -1$, and $\eta = 0.015$ (a). The trajectory plotted in the inversion n and phase difference $\Phi(t) - \Phi(t-\tau)$ plane (b). Open circles denote modes and x's denote antimodes.

Experimental observation of LFF below or near the solitary laser threshold has revealed an intriguing very regular structure that appears as an almost sinusoidal oscillation if examined with low resolution instrumentation [20]. Such a time trace is shown in Fig. 11(a). The laser used in the experiment was a Toshiba TOLD9140 index guided InGaAlP diode laser lasing at 693nm. The regime of current pumping and feedback level at which this regular LFF was observed is very small. Even though the dropout events seem to occur very periodically, nevertheless the shape of the oscillations shows small period-to-period differences. When this oscillation was observed with a transient digitizer of sufficient bandwidth (4.5GHz) to resolve the underlying structure it was discovered, as can be seen in Fig. 11(b), that it was composed of regular short pulsations. The envelope of these pulsations appears very regular because the drop off events now occur at very regular times. The actual drop in the intensity occurs in only a few nanoseconds and is followed by a period in which the laser intensity appears to be weak, but not zero. The pulsations appear in bunches of eight pulses and are separated by about 5.1 ns which matches the round trip of the external cavity. This is a feature that is seen also in numerical calculation using the LK equations. This behavior has been called the Synchronous Sisyphous effect as an allusion to the traditional LFF in which the random dropout events have been replaced by a very regular structure. Indeed, if one looks at the time statistics or of the low-frequency fluctuation peak in the rf spectrum one discovers that while the laser is undergoing LFF the peak is very broad. However, there is a dramatic reduction in its linewidth when regular LFF appear.

Computation with the LK equations Eqs. (36)–(37) verify the experimental results and reveal that the regular behavior is characterized by a sequence of time locked pulses that can be observed for moderate levels of feedback, for pumping currents below threshold and for a relatively large linewidth enhancement factor [22]. There is strong numerical evidence that when these conditions are satisfied the underlying structure of the low-frequency fluctuations is regular periodic or quasiperiodic dynamics associated with the interaction among a reduced set of the external cavity modes available to the laser at this feedback level. Thus, under these conditions the commonly observed random LFF are a mixture of these regular dynamics and chaotic or stochastic induced mode hoping. In particular, we show in Fig. 12(a) the time evolution of the amplitude of the regular periodic state for the values of the parameters $\alpha = 6$, $P = 0.001$, $T = 1000$, $\omega_0 \tau = -1$, and $\eta = 0.015$. It can be seen that the motion is quite periodic, but there are small period-to-period differences attributed to possible quasiperiodic dynamics. This can be seen effectively in Fig. 12(b) in which the trajectory is plotted in the plane of inversion n and phase difference $\Phi(t) - \Phi(t-\tau)$. Several trajectories are shown along with the location of the modes (open circles) and antimodes (x's). It is very clear that in this depiction significant differences from cycle to cycle can be seen, however, the number of modes visited in each cycle appears to be the same. In addition, the dropout event always seems to happen in a collision with the same saddle which is not located close to the maximum gain mode, which is in distinct contrast to LFF motion.

To determine the nature of this state and the possibility that this behavior is not a spontaneous stabilization of an otherwise unstable quasiperiodic/periodic orbit, a Poincaré section at $\dot{E} = 0$ and $\ddot{E} < 0$, that is, at the maxima of the pulsations of the amplitude of the electric field is constructed. Thus the system's evolution is characterized by the maxima of the field and the instantaneous frequency $\omega_i \tau = \tau \dot{\Phi}$. The plot in Fig. 13(a) reveals a definite pattern traced by the pulses that appears to indicate that some low-dimensional dynamics are taking place. As the pumping is reduced to below threshold the pattern becomes better defined in Fig. 13(b), and at $P = -0.012$ in Fig. 13(c) the system's dynamics are essentially quasiperiodic, as can be seen from the line trace in the configuration space. The

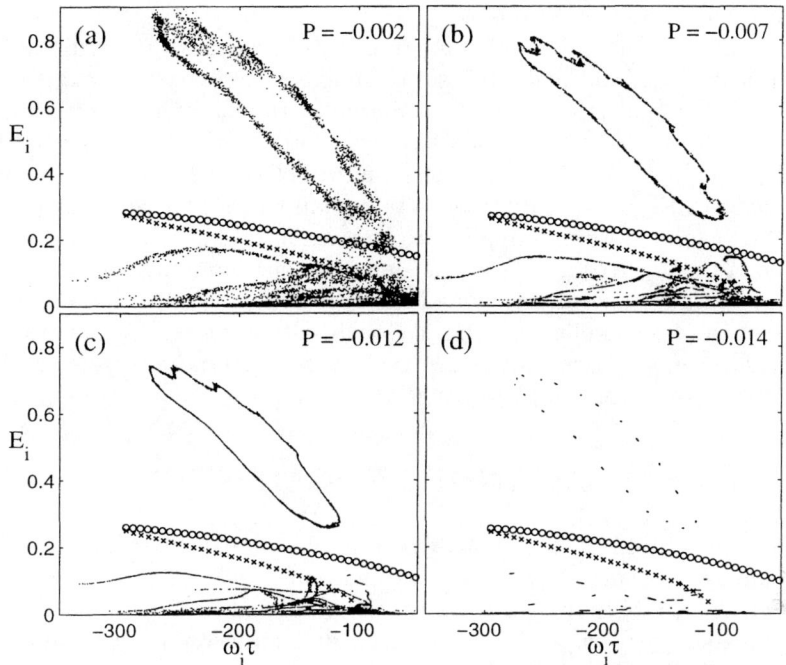

FIGURE 13. Numerical solution of the LK equation for $\alpha = 6$, $T = 300$, $\tau = 700$, $\eta = 0.07$, and $\omega_0 \tau = 0.0$. Low dimensional dynamics for $P = -0.002$ (a), sharply defined pattern for $P = -0.007$ (b), quasiperiodic evolution for $P = -0.012$ (c), and periodic motion for $P = -0.014$ (d).

quasiperiodic motion is extremely long (100-200τ). Finally at $P = -0.014$ the period essentially tends to infinity so that the system's evolution is a periodic motion in Fig. 13(d). As the pumping level is further reduced the motion again becomes quasiperiodic.

SUMMARY

In this chapter the appropriate models for a semiconductor laser under injection and external feedback were discussed. A novel method of optical modulation was used which probes the laser and generates the compound four-wave mixing and regenerative spectra. This allows the determination of the laser parameters with extremely good precision. In general, it was found that the fitting error to the spectra is of the order of a few percent. In addition, optically probing the injection system is of great interest since it can reveal the stability properties and the complex response of the system's dynamics.

The stability boundaries and the primary bifurcations of the diode laser with optical injection and of the diode laser in an external cavity were examined. It was determined that multiple ECM are available to the laser and it was found that a series of cascades close to threshold appears to be the route into the low frequency fluctuation regime. We examined the structure of the LFF and, in particular, the coexistence of LFF and steady state, a regime in which LFF appears as a transient process. A regular state that appears to be a universal backbone of the complicated evolution of the low frequency fluctuations when the pumping level is below threshold, was identified experimentally and described theoretically. The complexity and richness of the laser under external feedback, although extensively studied in the past twenty or more years, still seems to yield new and more surprising phenomena that have not yet been completely understood.

ACKNOWLEDGMENTS

The author would like to acknowledge a number of fruitful collaborations, during which some of the material in this publication was developed, namely with T. Erneux at Université Libre de Bruxelles, Y. C. Lai and R. L. Davidchack at Arizona State University at Tempe, R. B. Harrison at Heriot-Watt University, and T. B. Simpson at Jaycor Corp.

REFERENCES

1. Simpson, T.B., and Liu, J.M., *J. App. Phys.* **73**, 2587 (1993); Liu J.M. and Simpson T.B., *IEEE J. Quant. Electron.* **30**, 957 (1994).
2. Hui, R.and Mecozzi, A., *Appl. Phys. Lett.* **60**, 2454 (1992); Varangis P.M, Gavrielides A., Kovanis V., and Lester L.F., *Phys. Lett.* **A 250**, 117 (1998).
3. Lenstra, D., van Tarwijk, G.H.M., van der Graaf, W.A., and De Jagher, P.C., "Multiwave mixing dynamics in a diode laser", in *Chaos in Optics*, edited by R. Roy, Proceedings of SPIE , San Diego, 1993, pp. 11-22; van Tartwijk, G.H.M., and Lenstra D., "Nonlocking behavior of an optically injected diode laser" in Physics and *Nonlinear Dynamics in Lasers and Optical Systems*, edited by Abraham, N.B., et al., Proceedings of SPIE **2099**, Moscow-Nizhny Novgorod, Russia, 1994, pp.89-99.
4. van Exter, M.E., Hamel, W.A., Woerdman, J.P., and Zeijlmans, B.R.P., *IEEE Journ. Quant. Electron.* **28**, 1470 (1992); Su, C.B., Eom, J., Lange, C.H., Lauer, R.B., Rideaout, W.C., and LaCourse, J.S., *IEEE Journ. Quant, Elector.* **28**, 118 (1992).
5. Simpson, T.B., Liu, J.M., and Gavrielides, A., *IEEE J. Quant. Electron.* **32**, 1456 (1996); Simpson, T.B., and Liu, J.M., *IEEE Photon. Techol. Lett.* **9**, 1325 (1997).
6. Simpson, T.B. , Liu, J.M., Huang, K.F., Tai, K., Clayton, C.M., Gavrielides, A., and Kovanis, V., *Phys. Rev.* **A 52**, R4348 (1995).
7. Gavrielides, A., Kovanis, V., Erneux, T., and Simpson, T.B., In preparation.

8. Gavrielides, A., Kovanis, V., and Erneux, T., *Opt. Comm.* **136**, 253 (1997); Erneux T., Gavrielides V., and Kovanis V., *Quant. Semiclass. Opt.* **9**, 811 (1997).
9. Krauskopf, B., van der Graaf, W.A., and Lenstra, D., *Quant. Semiclass. Opt.* **9**, 797 (1997); Krauskopf, B., Tollenaar, N., van der Graaf, W.A., and Lenstra, D., *Opt. Comm.* **156**, 158 (1998); Wieczorek, S.M., Krauskopf, B., and Lenstra, D., *Opt. Comm.* **172**(1-6), 279–295 (1999).
10. Lang, R., and Kobayashi, K., *IEEE Jour. Quant. Electon.* **QE-16**, 347 (1980); Kobayashi K., *Trans. IECE Japan*, **E 59**, 8 (1976).
11. Tromborg, B., Osmundsen, J.H., and Olsen, H., I*EEE Journ. Quant. Electron.* **QE-20**, 1023 (1984); Erneux, T., Tartwijk, G.H.M., and Lenstra, D., "Determining Lang and Kobayashi Hopf bifurcation points", in *Physics and Simulation of Optoelectronic Devices*, edited by Osinski, M., et al., Proceedings of SPIE **2399**, San Jose, 1995, pp. 170-181 and unpublished results (1996).
12. Mork, J., Tromborg, B., and Mark, J., *IEEE Journ. Quant. Electon.* **28**, 93 (1992).
13. Risch, C., and Voumard, C., *Journ. Appl. Phys.* **48**, 2093 (1977).
14. Fischer, I., van Tartwijk, G.H M., Levine, A.M., Elsäßer, W., Göbel, E., and Lenstra D., *Phys. Rev. Lett.* **76**, 220 (1996).
15. Tartwijk, G.H.M., Levine, A.M., and Lenstra, D., *IEEE Journ. Sel. Top. Quant. Electron.* **1**, 466 (1995).
16. Sano, T., *Phys. Rev.* **A 50**, 2719 (1994).
17. Henry C. H., and Kazarinov R. F., *IEEE Journ. Quant. Electron.* **22**, 294 (1986); Hohl A., van der Linden H. J. C., and Roy R., *Opt. Lett.* **20**, 2396 (1995).
18. Hohl, A., and Gavrielides, A., *Phys. Rev. Lett.*, 82, 1148 (1999).
19. Heil, T., Fischer, I., and Elsäßer, W., *Phys. Rev.* **A 58**, R2672 (1998); Heil T., Fischer I., and Elsäßer W., *Phys. Rev.* **A 60**, 634 (1999).
20. Gavrielides, A., Newel, T.C., Kovanis, V., Harrison, R.G., Swanston, N., Yu, D., and Lu, W., *Phys. Rev.* **A 60**, 1577 (1999).
21. Davidchack, R.L., Lai, Y.C., Gavrielides A., and Kovanis V., *Phys. Lett. A* **267**, 350 (2000).
22. Davidchack, R.L., Lai, Y.C., Gavrielides, A., and Kovanis, V., submitted to *Phys. Rev. E* (2000).

APPENDIX A

Eqs. (1)–(3) can be reduced to a simpler set by normalizing electric field and inversion. In particular, we compute the steady state solutions of the system, $N = N_0$, $A = A_0$ without injection:

$$\frac{J}{ed} - \gamma_s N_0 = \frac{2\epsilon_0 n^2}{\hbar \omega_0} g_N (N_0 - N_t) A_0^2 \tag{51}$$

$$-\gamma_c + \Gamma g_N (N_0 - N_t) = 0 \tag{52}$$

$$\omega_0 - \omega_c = -\frac{\Gamma}{2} \alpha g_N (N_0 - N_t). \tag{53}$$

The first two equations determine the steady state inversion and intensity in terms of the pumping current and the inversion at transparency, and the last one determines the lasing frequency ω_0 in terms of the linewidth enhancement factor α. We proceed by defining

$$Z = \frac{N - N_0}{N_0 - N_t}, \qquad E = \frac{A}{A_0} \tag{54}$$

and normalizing the time in terms of the cavity losses by

$$s = \gamma_c t \tag{55}$$

we find:

$$\frac{dE}{ds} = \frac{(1 + i\alpha)}{2}[XZ - YJ_0(|E|^2 - 1)]E + \eta e^{-ivs} \tag{56}$$

$$T\frac{dZ}{dt} = J_0 - Z - J_0|E|^2 - XJ_0Z|E|^2 + YJ_0^2|E|^2(|E|^2 - 1). \tag{57}$$

The various constants appearing in the equations now can be expressed in terms of the constants that appear in the original equations.

$$\gamma_N = \frac{2\epsilon_0 n_0^2}{\hbar \omega_0} g_N A_0^2, \qquad \gamma_p = \Gamma g_p A_0^2. \tag{58}$$

The pumping above threshold is given by

$$J_0 = \frac{1}{\gamma_s(N_0 - N_t)}\left(\frac{J}{ed} - \gamma_s N_0\right). \tag{59}$$

Finally with

$$X = \frac{\gamma_N}{\gamma_s J_0}, \qquad Y = \frac{\gamma_p}{\gamma_c J_0}, \qquad \text{and} \qquad T = \frac{\gamma_c}{\gamma_s} \tag{60}$$

we obtain Eqs. (56)–(57) where the injection strength is given by

$$\eta = f\frac{A_i}{\gamma_c A_0}. \tag{61}$$

Of course, the frequency of the master laser relative the slave laser is now denoted by v and has been scaled by the cavity losses. The terms in the equations proportional to Y are the well-known gain saturation terms and appear here as an expansion of the usual denominator.

Notice that even though γ_N and γ_p appear to depend on the intensity of the solitary laser and, hence, on the pumping, the parameters appearing in the equations have been scaled in such a way that the pumping level above threshold appears explicitly and the rest of the parameters are pump independent. A further reduction of the Eqs. (56)–(57) is possible by defining

$$n = \frac{XZ}{2}, \qquad P = \frac{XJ_0}{2}, \qquad \varepsilon = \frac{Y}{X}. \tag{62}$$

In this case the equations take the particularly simple form of Eqs. (4)–(5) that we worked with earlier.

APPENDIX B

In terms of the original parameters the relaxation frequency and damping for the solitary laser problem are defined as

$$\omega_r = \sqrt{\gamma_N \gamma_c + \gamma_p \gamma_s} \tag{63}$$

and

$$\gamma_r = \gamma_s + \gamma_N + \gamma_p.$$

By scaling back to the same parameters in Eqs. (12)–(13) we obtain the following expressions for the spectra. The power spectrum is

$$\sigma(\omega) = i \frac{\omega + i(\gamma_r - \gamma_p)}{\omega^2 - \omega_r^2 + i\omega\gamma_r}. \tag{64}$$

The absolute value of this corresponds to the laser power spectrum. The phase modulation spectrum is given by

$$\delta(\omega) = -\frac{(\omega^2 - \omega_r^2 + \alpha\omega\gamma_p) + i(\omega\gamma_r + \alpha\omega_r^2)}{\omega(\omega^2 - \omega_r^2 + i\omega\gamma_r)}. \tag{65}$$

From these two expressions we obtain the regenerative spectrum:

$$R(\omega) = i \frac{(2\omega^2 - \omega_r^2 + \alpha\omega\gamma_p) + i(2\omega\gamma_r - \omega\gamma_p + \alpha\omega_r^2)}{2\omega(\omega^2 - \omega_r^2 + i\omega\gamma_r)}, \tag{66}$$

and the four-wave mixing spectrum

$$F^*(\omega) = i \frac{(\omega_r^2 - \alpha\omega\gamma_p) - i(\alpha\omega_r^2 + \omega\gamma_p)}{2\omega(\omega^2 - \omega_r^2 + i\omega\gamma_r)}. \tag{67}$$

Emission Dynamics of Semiconductor Lasers Subject to Delayed Optical Feedback: An Experimentalist's Perspective

Ingo Fischer, Tilmann Heil and Wolfgang Elsäßer

Institute of Applied Physics, Darmstadt University of Technology
D-64289 Darmstadt, Germany

Abstract. Semiconductor lasers subject to delayed optical feedback from a distant reflector are complex systems exhibiting strong dependence on numerous parameters which can be varied in an experiment. These parameters include operational, detection, device structural, and geometrical aspects. In a phenomenological approach, we present a broad overview of the influence of these different parameters on the actually observed dynamical behavior of the system. In particular, we concentrate on the influence of detection bandwidth, device structure, asymmetric coating, mirror tilting, and system parameters including the injection current, the feedback rate, and the nonlinearity parameter α.

INTRODUCTION

In various experimental configurations and under many circumstances semiconductor lasers (SL) experience delayed optical feedback, e.g. from the facets of an optical fiber, an optical disc, or external cavity configurations. A scheme of this situation is depicted in Figure 1. The SL is electronically pumped with the injection current I_{DC} and has the facet reflectivities R_1 and R_2 and the internal round trip time of T_{LD}. It is subject to optical feedback with a delay T_{delay} from a distant reflector with reflectivity R_3. Due to the small extension of SL, typically lying in the range of 500 μm for edge emitting devices, the internal round trip time of the light within the cavity is significantly smaller than the delay time of the external optical feedback. Already twenty years ago Risch and Voumard reported on dynamical response of the laser on optical feedback from an external cavity [1]. Eight years later, Lenstra et al. demonstrated that the optical linewidth of a SL may increase by a factor of about 1000 as a consequence of the delayed optical feedback [2]. They coined the name *coherence collapse* for this phenomenon. Even though numerous works have investigated the dynamical effects induced by optical feedback identifying a huge variety of phenomena, all these phenomena still are summarized by

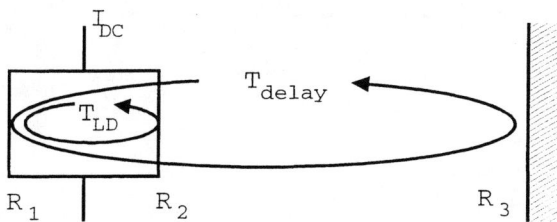

FIGURE 1. Scheme of a semiconductor laser subject to delayed optical feedback.

the name coherence collapse. Tkach and Chraplyvy were the first to provide an overview over different operating regimes of a DFB laser in dependence on the feedback strength. They introduced a numbering of different regimes, labeled regime I to V [3]. In this chapter, we will concentrate on the large regime in which the system shows feedback induced instabilities. This so called regime IV extends over three orders of magnitude of feedback strengths from approximately -45 dB of the output power reflected back into the laser up to -10 dB effective power reflectivity. In order to distinguish different phenomena and scenarios within regime IV, further sub-categories can be introduced that characterize the regimes of weak feedback, moderate feedback and strong optical feedback. Weak feedback denotes a regime typically ranging up to 0.1 percent feedback with respect to power in which a quasi-periodic and a period-doubling route together with frequency locking phenomena have been found [4,5]. The moderate feedback regime typically ranging from 0.1 up to several percent of optical feedback includes the so-called phenomena of low frequency fluctuations (LFF) and fully developed coherence collapse. Finally, the regime of strong optical feedback denotes a range of about 10 to 50 percent of optical feedback in which modifications of the dynamical behavior as compared to the moderate feedback regime may be found as a result of multiple reflections in the external cavity or multi-mode emission of the SL. In what follows, we will concentrate on the moderate feedback regime, since it covers a great variety of fascinating dynamical phenomena. Although similar feedback induced behavior occurs for various laser structures, including also Vertical Cavity Surface Emitting Lasers (VCSEL), as has been shown experimentally [6,7] and numerically [8], we will restrict our discussions to edge emitting devices. Furthermore, we will focus on delay times of the feedback light larger than the laser internal time scale of the relaxation oscillation period.

This chapter is organized as follows. Subsequent to this introduction, we present a detailed discussion of the imposed requirements and eventual shortcomings related to the detection apparatus for the study of feedback induced instabilities. In the first subsection, we describe a typical experimental setup in detail. As an example, we then discuss the important influence of the detection bandwidth on what is actually observed in the experiment. In the third section, we present a de-

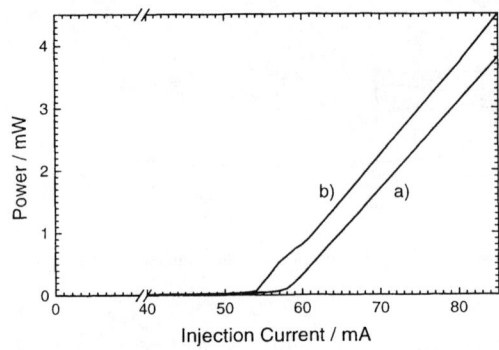

FIGURE 2. Typical P/I characteristics of a SL without delayed optical feedback (a) and with delayed optical feedback (b).

tailed investigation of the dynamical behavior of the system in dependence on the system parameters injection current, feedback rate, and nonlinearity parameter α. In the last section, we discuss the influence of device parameters and geometry. In particular, the role of the laser device structure, asymmetric reflectivities, and misalignment of the external cavity. The chapter ends with some concluding remarks. Let us now continue with a first introduction to some of the striking characteristics of feedback induced instabilities in SL.

First Characteristics of Feedback Induced Instabilities

In order to introduce the phenomena induced by the optical feedback we start by discussing the effect of the feedback on the P/I characteristics of the laser. Figure 2 depicts two P/I characteristics of a semiconductor laser without and with delayed optical feedback. The P/I characteristics exhibit the typical threshold behavior, above which the laser power increases with increasing injection current. Figure 2 demonstrates that the influence of the optical feedback on the P/I characteristics is twofold. First, a reduction of the threshold current can be observed. This is due to the photons being re-injected into the laser, which reduces the total losses. Furthermore, the P/I characteristics of the laser without delayed optical feedback shows a nearly linear increase of the emitted power with increasing injection current, whereas the P/I characteristics of the laser with delayed optical feedback shows a pronounced kink close to the solitary laser threshold.

Usually in the vicinity of this kink, a striking dynamical phenomenon can be found which has been studied by Risch and Voumard for the first time twenty years ago. They described the following phenomenon: induced by the optical feedback, the fluctuations of the light intensity exhibited dominant low frequency contributions. This was surprising since these frequencies were significantly lower than the

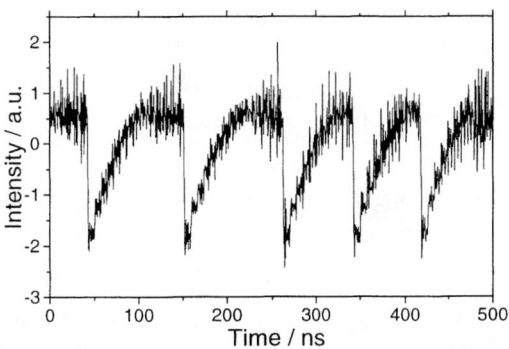

FIGURE 3. Typical LFF dynamics of a SL (HLP 1400) subject to moderate optical feedback as it can be typically observed by a photodiode and 1 GHz scope around the solitary laser threshold, here for 1.05 I_{th}.

other laser characteristic frequencies as e.g. the relaxation oscillation frequency, and the round trip frequency of the light in the external cavity. These low frequency fluctuations, in the following abbreviated as LFF, have been subject to many investigations since then, and still attract enormous interest today. Figure 3 depicts typical LFF dynamics of a semiconductor laser subject to moderate optical feedback measured with a photodiode and a 1 GHz scope. The plot depicts a 500 ns time window of the temporal evolution of the emission intensity. At irregular intervals of about 100 ns fast dropouts of the intensity can be recognized, followed by a significantly slower recovery of the intensity lasting about 10 roundtrips of the light in the external cavity which has been 3.3 ns in this case. The intensity does not reach a stable level, but rather exhibits strong fluctuations before the next dropout occurs. This LFF behavior occurs for the whole moderate feedback regime for injection currents close to the solitary laser threshold. Increasing the injection current leads to more frequent dropouts until they finally cannot be distinguished anymore. The resulting behavior of irregular fluctuations has been called fully developed coherence collapse. For a detailed discussion of the underlying mechanism of LFF and fully developed coherence collapse we refer to the chapter by *Lenstra & Yousefi*, where a rate equation model of delayed feedback instabilities is discussed. Instead, we will report in the following section on the experimental techniques which are necessary to investigate the coherence collapse dynamics in the laboratory.

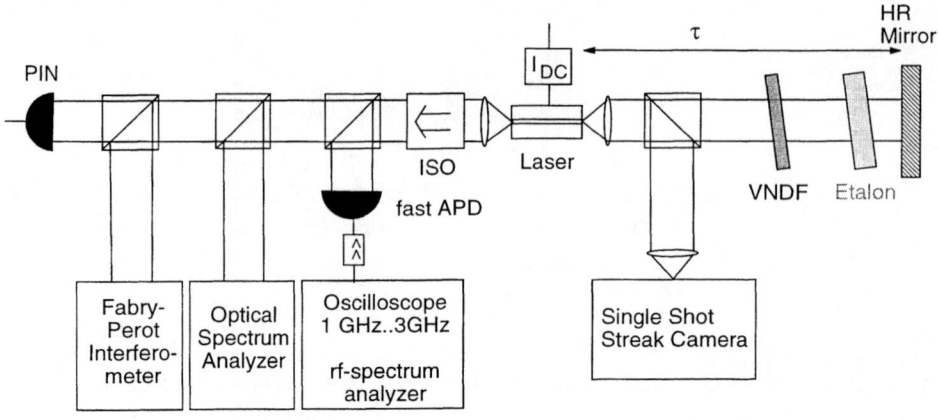

FIGURE 4. Typical experimental setup to study delayed feedback instabilities of a SL.

DETECTION

In the preceeding section, we have already mentioned the different time scales present in the dynamics of semiconductor lasers subject to delayed optical feedback ranging from several GHz for the relaxation oscillations down to some hundreds of kHz for slow LFF. Such wide scales are also encountered in the optical spectra in which both, the diode modes separated by some hundred GHz and the external cavity modes separated by some hundreds of MHz, have to be resolved. These wide ranges of temporal and optical scales impose high demands on the detection setup. In the first part of this section, we will present a typical experimental setup in detail. In the second subsection, where we choose the role of detection bandwidth as an example, we will show that it is necessary for the experimentalist to exercise a certain caution in the interpretation of the experimental results.

Experimental Setup

A typical experimental setup to investigate the dynamical behavior induced by delayed optical feedback is depicted in Figure 4. It consists of the SL in an external cavity configuration and the detection apparatus. The external cavity laser configuration comprises the edge emitting SL which is driven by a dc current source, a high reflecting mirror terminating the external cavity, a lens collimating the laser emission, a variable neutral density filter in order to control the feedback rate, and optionally an etalon by which the emission of the laser can be spectrally tuned. Here, the length of the external cavity typically ranges between 0.15 m and 1.5 m.

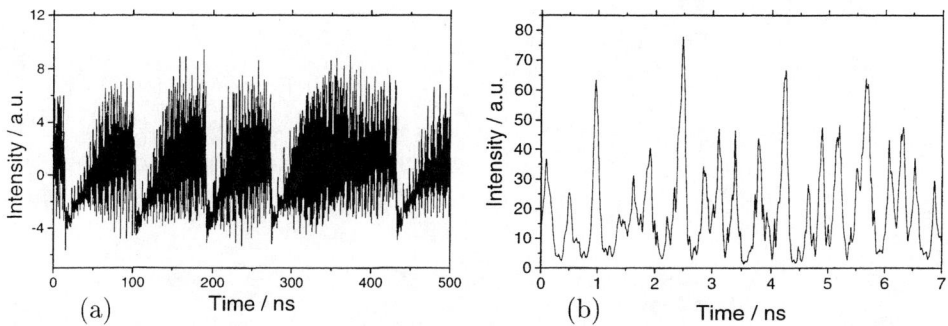

FIGURE 5. Emission dynamics of a HLP 1400 laser in the LFF regime at 1.05 I_{th} measured with a photodiode and a fast digitizer with an analog bandwidth of 3 GHz (a), and measured with a single-shot streak camera with an analog bandwidth of 70 GHz (b).

A striking particularity of a SL subject to delayed optical feedback is the wide range of temporal and spectral scales present in its dynamics. The detection setup has to account for these multiple temporal and spectral scales. The optical spectrum is measured with a grating spectrometer (OSA) resolving the longitudinal modes of the laser diode. Furthermore, a confocal Fabry-Pérot interferometer is used to resolve the modes of the laser with external cavity. A fast photodiode detects the intensity dynamics of the system. Its signal is amplified and analyzed either by a fast digitizing oscilloscope or spectrally by an electrical spectrum analyzer (ESA). Using this method, the temporal resolution is currently limited to 3 GHz. Since even faster time-scales are present in the emission dynamics of SL, a single-shot streak camera providing a resolution bandwidth of up to 100 GHz is important. Finally, a slow integrating p-i-n–photodiode measures the average light intensity, as depicted in the P/I characteristics.

Taking the temporal resolution of the intensity time series as an example, we will show that an explicit consciousness of the limitations of the chosen detection device is indispensable in order to interpret the obtained experimental results correctly.

Role of Detection Bandwidth

Figure 3 shows the LFF behavior recorded with 1 GHz detection bandwidth. The dropout phenomenon can be observed clearly, however, the fast oscillations are not fully resolved. The influence of a higher detection bandwidth is depicted in Fig. 5.

Figure 5(a) shows the LFF dynamics under the same conditions as in Fig. 3, however using a scope with 3 GHz analog bandwidth. The time window is again 500 ns as in Fig. 3. The irregular dropouts of the intensity can still be recognized,

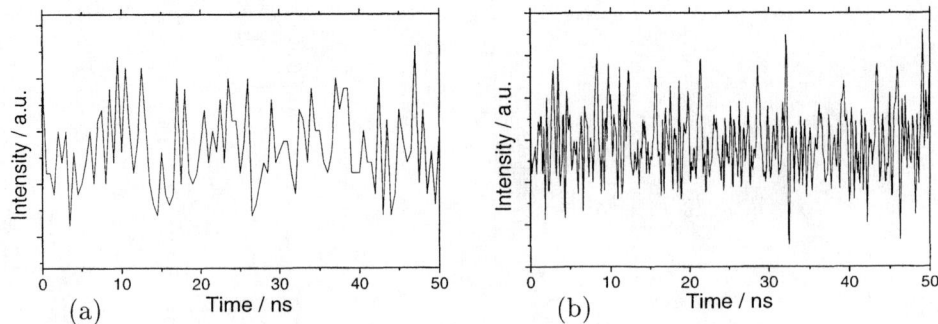

FIGURE 6. Emission dynamics of a HLP 1400 laser in the LFF regime at 1.23 I_{th} measured with a photodiode and a fast digitizer with an analog bandwidth of 1 GHz (a) and of with an analog bandwidth of 3 GHz (b).

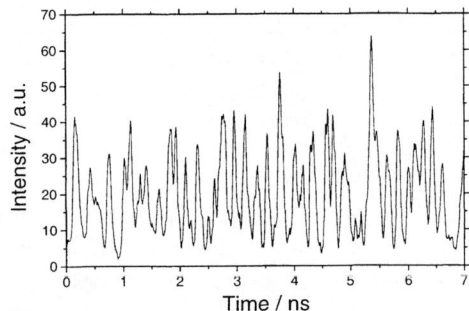

FIGURE 7. Emission dynamics of a HLP 1400 laser in the fully developed coherence collapse at 1.23 I_{th} measured with a single-shot streak camera with an analog bandwidth of 70 GHz.

but now pronounced fast oscillations in between the dropouts are clearly resolved. This indicates that additional fast pulsations on sub-ns timescales are underlying the dropout behavior. Therefore, streak camera measurements are indispensable. Fig. 5(b) shows a plot of the intensity dynamics obtained with a streak camera with 14 ps temporal resolution. Here the time window is much shorter and amounts to 7 ns. In contrast to the oscilloscope measurements the fast dynamics now is fully resolved. Clearly, the emission is composed of irregular pulsations of the intensity on ∼ 50..500 ps timescales. This is in good agreement with the phenomenological modelling by the Lang-Kobayashi equation, as discussed in the chapters by *Gavrielides* and *Lenstra & Yousefi*, which supports the interpretation of the deterministic mechanism underlying this behavior.

The dynamical regime called fully developed coherence collapse is reached by increasing the injection current under otherwise identical conditions. Figure 6 depicts corresponding intensity time series with a temporal resolution of 1 GHZ and 3 GHz, respectively.

On slow time scales the dynamics of the system has drastically changed: the characteristic intensity dropouts present in the LFF regime have disappeared. The dynamics is dominated by oscillations on timescales of the external cavity roundtrip and faster oscillations beyond 1 GHz, as can be recognized from Fig. 6. The streak camera measurement presented in Fig. 7 demonstrates that on a picosecond time scale pronounced qualitative changes of the dynamical behavior of the system do not occur.

Increasing the injection current only leads to some quantitative changes, i.e. the pulse width as well as the average time interval between the pulses decrease, and the modulation depth of the pulses is slightly reduced. We conclude that in order to get a complete picture of the dynamics of SL with optical feedback the whole hierachy of time scales needs to be resolved.

In the preceding paragraph, we have already highlighted one effect of a changing injection current on the dynamical behavior of the system. We continue along this line in the following section.

INFLUENCE OF SYSTEM PARAMETERS

This section intends to give, both, detailed information on and a global overview of the influence of a variation of the system's most relevant parameters on the dynamics. Specifically, we consider the injection current I, the feedback rate γ, and the linewidth enhancement factor α. The information will be summarized in parameter space diagrams defining the parameter regimes of qualitatively similar dynamical behavior. The first subsection concentrates on the influence of the injection current I and the feedback strength γ on the dynamics of the system.

Influence of Injection Current and Feedback Rate

In order to map the dynamical behavior of the system in γ–I–space, we have used the following method. The feedback strength γ is adjusted by using the variable neutral density filter, and calibrated by measuring the effective power reflectivity of the feedback branch. Subsequently, starting below threshold over the whole accessible range, the injection current I is increased. The resulting dynamical behavior is investigated and classified in dependence on I. Repeating this procedure for several different values of γ, we obtain a map of the dynamical behavior of a SL subject to optical feedback in the γ-I-space, which is depicted in Fig. 8.

The intracavity etalon is positioned such that the laser emission is dominated by one longitudinal diode mode at the gain maximum of the laser at 840 nm. Increasing the injection current along the vertical dotted line from "A" to "B" depicted in Fig. 8, corresponding to a feedback strength of $\gamma \approx 25 ns^{-1}$, one reaches every dynamical regime present in the γ-I-space: After passing the feedback reduced

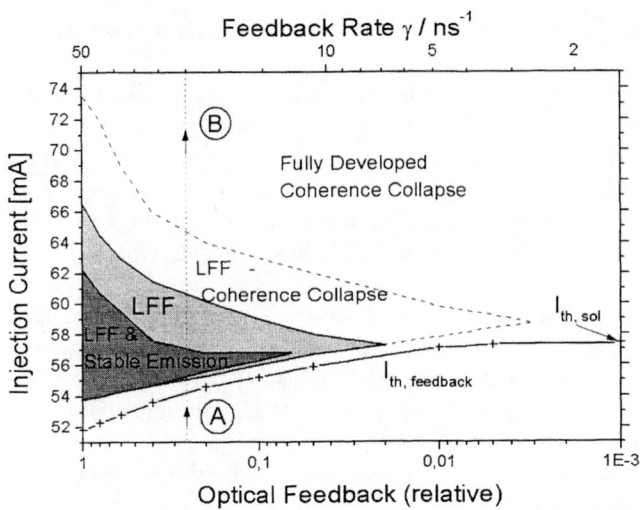

FIGURE 8. The dynamical behavior of a SL subject to optical feedback in feedback(γ) - current(I) - space. The LFF regime is depicted in light-gray. The dark-gray region embedded in the LFF regime corresponds to the region of coexistence of the stable emission state and the LFF state. The unshaded region encompassed by the dashed line corresponds to the continous transition between the LFF regime and the fully developed coherence collapse regime.

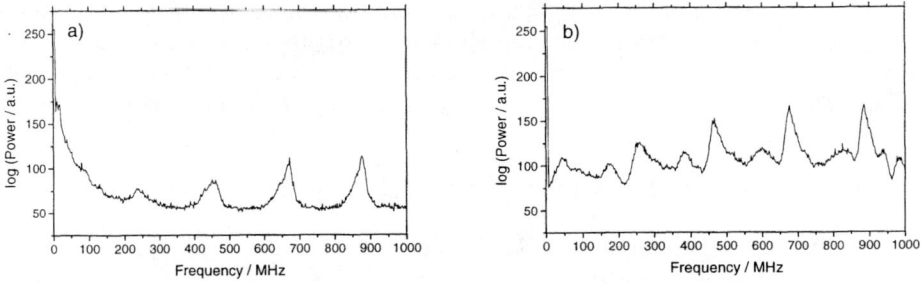

FIGURE 9. Power spectra of the emission dynamics of a HLP1400 laser diode subject to delayed optical feedback showing LFF (a), and fully developed coherence collapse (b). The experimental conditions correspond to the time series depicted in Fig. 5.

laser threshold, the system enters a regime of stable emission on one longitudinal diode mode and several external-cavity modes. The mode beatings of the external-cavity modes appear as sharp peaks in the power spectrum. The LFF regime is reached by increasing the injection current by approximately 1 mA. As depicted in Fig. 9(a), a dominant low frequency contribution builds up in the power spectrum;

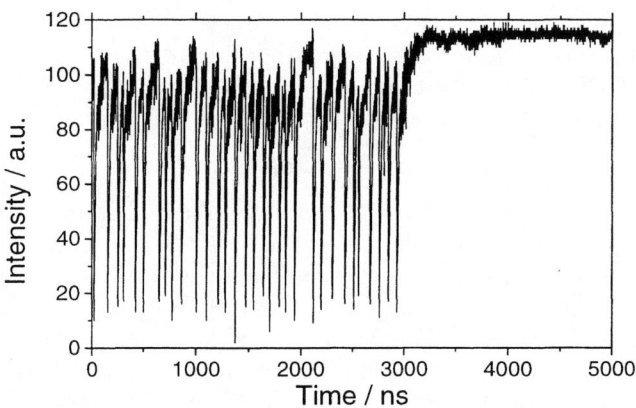

FIGURE 10. Intensity time series for $I = 1.05 I_{th}$ and $\gamma = 45 ns^{-1}$ showing a transition from LFF to stable emission on a high-gain external-cavity mode. Obviously, the intensity stabilizes on a level higher than the intensities reached during the LFF cycles.

the external-cavity mode beatings broaden significantly.

Restricted by the transmission of the etalon, the emission of the laser is still dominated by one longitudinal diode mode. With increasing injection current the time intervals between the drop-outs decrease. Finally, a continuous transition to the fully developed coherence collapse takes place accompanied by a broad power spectrum and completely irregular intensity time series. An example for these broad power spectra is given in Fig. 9(b).

A striking aspect of the phase space diagram depicted in Fig. 8 is the existence of a large region within the LFF regime, where discrete transitions between LFF and stable emission occur. We now focus on this coexistence region embedded in the LFF regime. In a snap-shot-like picture, Fig. 10 shows the intensity time series of a transition from the LFF state to the stable emission state occurring at the time of $3\,\mu s$. The important features accompanied by this transition are, first, that the intensity of the laser stabilizes on a higher level than the LFF. Accordingly, the transition to stable emission is characterized by a sudden increase in the time averaged intensity. Second, the power spectrum is completely flat, no frequency components remain. Third, the stable emission occurs on a single external-cavity mode, resolved by the scanning Fabry-Pérot interferometer. The time intervals of stable emission may exceed one minute, which is more than 10^8 times longer than the typical time intervals between two subsequent LFF events under similar conditions. Thus, a SL subject to optical feedback can operate on a single high-gain external-cavity mode within the LFF regime.

To some extent, the length of the time intervals of stable emission depends on the position in γ-I-space. The time intervals of stable emission are longer for stronger feedback and higher injection current. Only a few tenth of mA above laser threshold the duration of the stable states is comparable to the time interval between two subsequent LFF events. Increasing the injection current, the periods of stable emission become longer by several orders of magnitude, while the time intervals between the LFF drop-outs become shorter. This indicates that the spontaneous emission noise kicks the trajectory off the stable mode in the parameter region right above threshold as it is described in [9]. However, for higher currents, the relative strength of the spontaneous emission noise is reduced and, thus, is not, or only very rarely, sufficient to push the trajectory away from the stable mode. Consequently, the time intervals of stable emission become longer. On the other hand, the time intervals of LFF emission coexisting with the stable emission state become longer with increasing injection current as well. In other words, after being ejected from the stable mode by external perturbations or noise the system undergoes an increasing number of LFF cycles before it returns to the stable emission state. This behavior indicates that the probability to reach the stable mode is reduced with increasing injection current.

In conclusion, the above discussion provides valuable information about the relative importance of spontaneous emission noise and the probability to reach the stable mode. However, we note that the key parameter determining robustness and accessibility of the stable emission state is the linewidth enhancement factor α. We describe the important influence of α on the dynamics of the system in the following subsection.

Influence of the α-Parameter

The linewidth enhancement factor α accounts for the amplitude-phase coupling of the optical field in SL. It is extra-ordinarily strong in SL compared to any other laser and constitutes the main nonlinearity of the system. Thus, the complicated dynamical behavior of SL is closely linked to α.

Physically, the amplitude-phase coupling is caused by the carrier induced variation of real and imaginary parts of the semiconductor material's susceptibility $\chi(n) = \chi_r(n) + i\chi_i(n)$. Thus, α is defined as follows [10,11]:

$$\alpha = -\frac{d(\chi_r(n))/dn}{d(\chi_i(n))/dn} . \tag{1}$$

As one of the fundamental parameters for SL, α causes the enhancement of the laser linewidth [11], is of decisive importance for many dynamical processes like modulation response, frequency chirp, and, in particular, optical feedback effects [12]. Until recently, both, α and the emission wavelength, have been considered as fixed parameters of the investigated semiconductor laser. Indeed, tuning the emission wavelength of the laser by shifting the gain profile of the solitary laser, e.g. by temperature tuning, affects the value of α only very slightly. However, by tuning the emission wavelength of the laser away from its solitary gain maximum

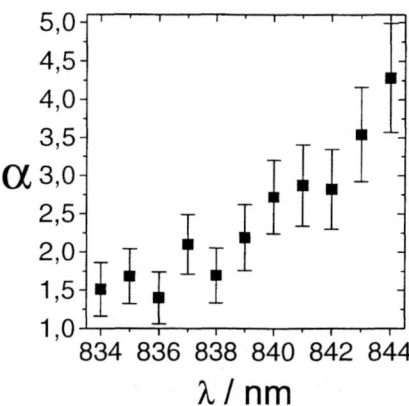

FIGURE 11. Spectral dependence of the linewidth enhancement factor α for a HLP1400 laser diode. The gain maximum of the solitary laser is at 840 nm.

without changing the gain profile of the solitary laser, substantial changes of the value of α can be achieved. Experiments showed that even a transition to negative values of α occurs for wavelengths much shorter than the solitary gain maximum wavelength [13]. We exploit this strong spectral dependence of α [14] in order to provide a detailed overview on the influence of a changing α on the dynamical behavior of the system. Again, we use the setup depicted in Fig. 4. By tilting the intracavity etalon, we tune the emission wavelength of the laser away from its solitary gain maximum, which itself remains unchanged. This allows us to tailor the value of α in a well-controlled way.

Figure 11 gives an overview of the spectral dependence of α of a HLP1400 laser diode. In this case, α has been measured according to the method proposed by Henning and Collins [15]. This method is based on net gain measurements obtained from the modulation depths introduced by Fabry-Pérot resonances into the spontaneous emission spectrum of the laser [16]. Obviously, α decreases continuously with decreasing wavelength. The spectral dependence of α and the absolute values for α presented in Fig. 11 are in good agreement, both, with calculated α-spectra [14] and with previously obtained experimental results for the same type of laser diode [17].

Having selected the desired value of α, we record a γ-I-space diagram using the experimental method described in the previous subsection. Thus, for the selected value of α, we obtain a map displaying the dynamical behavior of the system classified in several distinct dynamical regimes. Comparing these parameter space diagrams recorded for different values of α, we provide both, detailed information and a global survey, about the impact of a changing α on the dynamics of SL subject to delayed optical feedback.

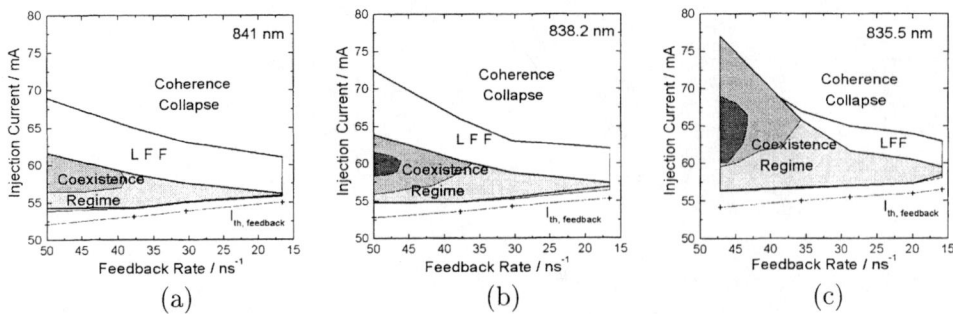

FIGURE 12. The dynamical behavior of SL subject to delayed optical feedback in dependence on I and γ for 841 nm corresponding to $\alpha = 2.8 \pm 5$ (a), for 838.2 nm corresponding to $\alpha = 2.1 \pm 4$. (b), and for 835.5 nm corresponding to $\alpha = 1.6 \pm 3$ (c).

Figure 12 depicts three γ-I-space diagrams for three different emission wavelengths, correspomding to the three different values of $\alpha = 2.8 \pm 5$, $\alpha = 2.1 \pm 4$ and $\alpha = 1.6 \pm 3$. Comparing the panels of Fig. 12 shows that the system exhibits similar dynamical regimes for each of the three different α factors. The dynamical regimes depicted here correspond to those characterized in the previous subsection. However, the extension of these regimes varies significantly depending on α. In particular, we concentrate our attention on the regime of coexistence between stable and unstable emission. As already described in the previous subsection, the relative length of the time intervals of stable emission varies with the position in γ-I-space. In order to account for this parameter dependence, we encode the coexistence regime, as depicted in Fig. 12, by three different shades of grey. Dark grey represents the parameter region in which the duration of the stable emission states is longest, where time intervals of stable emission are typically exceeding one minute, only interrupted by short time intervals of LFF emission. Intermediate grey represents the region in which the system jumps back and forth between LFF and stable emission, typically on a timescale of seconds. Light grey represents the region in which the LFF prevail, and the stable state is only reached for short times, typically milliseconds.

We see that the dynamical behavior of the system is strongly dependent on the value of α. In the following, we compare the three γ-I-space diagrams in order to identify the effects of this α dependence. We find that the following phenomena are associated with a reduction of α:

1. *Enlargement of the Coexistence Regime*
   The regions of coexistence of the LFF state and the state of stable emission on a single external-cavity mode enlargen drastically for decreasing α. Thus, the stable emission state is more easily accessible for decreasing α. This observation is consistent with the notion of an enlarging basin of attraction of the stable state for smaller values of α.

2. *Increased Stability of the Stable State*
   The parameter regime of very long times of stable emission, depicted in dark grey in the γ-I-space diagrams, expands substantially for decreasing α. This means that the stable emission state becomes more robust against external perturbations for decreasing α.

3. *Coexistence of CC and Stable Emission*
   For the lowest investigated value of $\alpha = 1.6 \pm 3$, the coexistence region exceeds the LFF regime. Thus, for stronger feedback and higher injection currents we observe a coexistence of the stable emission state with the fully developed CC. Moreover, the signature of the stable state remains the same while the continuous transition from LFF to the fully developed CC occurs. This experimental result indicates that the dynamical structure of the system is stable although the appearance of the unstable state changes drastically.

4. *Properties of the LFF Regime*
   Within the investigated range of parameters, the extension of the LFF regime does not change as significantly as the coexistence regime with decreasing α. Also, the temporal signature of the LFF and the mean time interval between subsequent intensity dropouts do not exhibit drastic changes. Again, this indicates the structural stability of the system.

A theoretical analysis of the influence of α on the dynamics described above based on the Lang-Kobayashi rate equation model is presented, e.g., in references [17–20]. The experimental results listed above indicate a stabilization of the feedback induced instabilities provided that α can be reduced below a certain threshold level. Indeed, permanent stable emission over wide parameter regimes in SL with optical feedback has been reported on for MQW-DFB lasers [21]. Due to their particular structure, these lasers can exhibit low values of α [22].

INFLUENCE OF DEVICE PARAMETERS AND GEOMETRY

In this section, we review the influence of device parameters like different laser structures and the effect of asymmetric laser facet reflectivities on the dynamical behavior of the system. Their effects have to be kept in mind when investigating the dynamics of SL subject to delayed optical feedback. Furthermore, geometrical aspects of the external cavity configuration, in particular, mirror tilting will be discussed.

Influence of Device Structure

So far, we have discussed the dynamics of SL with optical feedback choosing a bulk Fabry-Pérot SL as paradigmatic example. However, different device structures

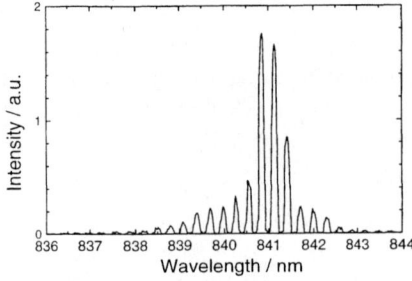

FIGURE 13. Optical spectrum of a Fabry-Pérot laser subject to optical feedback. A large number of diode modes is excited by the optical feedback.

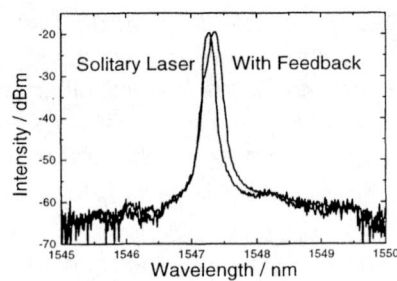

FIGURE 14. Optical spectrum of a DFB laser with and without optical feedback. The DFB laser maintains single mode operation with a side mode suppression ratio of approximately 40dB even when subjected to optical feedback.

may lead to substantial changes in the emission characteristics of external cavity configurations. We now discuss these effects. In particular, we will concentrate on the effect of changes in the optical spectrum: Fabry-Pérot lasers subject to delayed optical feedback operate on a large number of longitudinal diode modes, whereas DFB lasers rigorously restrict the emission of the laser to a single diode mode even under strong feedback conditions.

Figure 13 gives typical examples for the optical spectra of a Fabry-Pérot laser with optical feedback, and Fig. 14 for a DFB laser with and and without optical feedback, respectively. Figure 14 demonstrates that the linewidth of the DFB laser with optical feedback is drastically broadened towards longer wavelengths compared to the solitary laser emission line. This broadening is consistent with theoretical predictions [23–25]. In addition, Fig. 14 represents an instructive illustration of the notion 'coherence collapse' in the spectral domain. We note that the width of the solitary laser line is determined by the resolution of the grating spectrometer.

Interestingly, the typical feedback induced instabilities like LFF and fully developed coherence collapse occur for DFB and Fabry-Pérot lasers in a very similar manner despite of the differences in the optical spectra. This has been confirmed by statistical investigations of LFF behavior in DFB and Fabry-Pérot lasers [26,27]. Nevertheless, the exact consequences of single mode and multi mode operation, respectively, for the dynamics of SL with optical feedback, and the question whether there is a dynamical interplay between the modes under multi mode operation are still under debate [28–30].

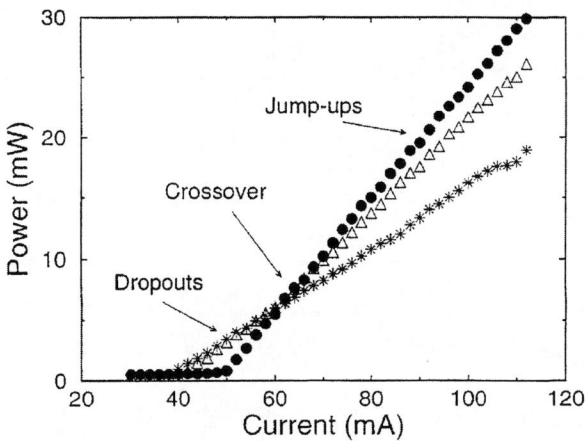

FIGURE 15. P-I-curves of an asymmetrically coated Mitsubishi 64111R laser diode. Closed circles: solitary laser. open triangles and stars: with optical feedback. The stars denote the maximum feedback case. From [31] ©1997 OSA.

Influence of Asymmetric Reflectivities

In many applications of SL, it is advantageous to achieve high power emission from one laser facet, and low power emission from the other facet. Lasers for such applications therefore are produced with an asymmetric coating: one facet with a high reflection (HR) coating, i.e. larger than 90% reflectivity, and a facet with a anti reflection (AR) coating, i.e. less than 10% reflectivity. These asymmetric coatings lead to asymmetries of optical field and carrier density within the laser cavity, and, as requested, to very different output powers from the two facets. Furthermore, they result in an altered appearance of the feedback induced instabilities in asymmetrically coated SL. We discuss these changes in the following.

Placing the the external cavity mirror on the AR coated facet side of the laser strongly influences the P-I-curve: first, a very strong reduction of the threshold current, and second, a substantial flattening of slope of the P-I-curve can be found. Thus, the P-I-curves with and without optical feedback may even cross for a certain value of the injection current which is referred to as cross over point [31]. This situation is illustrated in Fig. 15. We note that the P-I-curves depicted in Fig. 15 represent the power recorded from only the AR-coated laser facet. As the effective power reflectivity on the AR-coated side of the laser markedly increases due to the external cavity, the output power on the HR-coated side of the laser diode will strongly increase. Thus, the total output power of the laser does not decrease in the presence of optical feedback.

Another striking phenomenon associated with an asymmetric coating of the laser

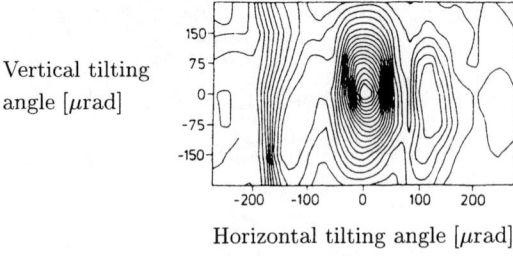

FIGURE 16. Contour plot exhibiting the average output power depending on horizontal and vertical mirror tilting. Maximum power is at (0,0). From [35] © 1991 IEEE.

facets is the appearance of the LFF. For injection currents below the cross over point, the LFF occur in the usual way as power drop-outs. However, above the cross over point, the LFF appear as power jump ups [31]. Thus, the laser always returns to its solitary output power in the course of a LFF event. The asymmetric coating leads to the different appearance of the LFF. Again, we point out that this behavior is observed for measurements taken on the AR-coated side of the laser facet. A more complete picture of the intensity dynamics of asymmetrically coated SL can be expected when the output power of both facets is detected simultaneously.

Influence of Mirror Tilting

Asymmetries in external cavity configurations can easily occur due to a misalignment of the mirror which forms the external cavity. Whereas optimum alignment is defined by the maximum possible threshold reduction [32], tilting the mirror away from this optimum alignment position has the following effects. First, the coupling efficiency of the reflected light back into the laser is reduced. Second, the phase condition determining the interference between emitted and returning optical field changes [33]. Finally, multiple reflections need to be considered. The last two points lead to an undulation of the laser power with increasing mirror tilt. Figure 16 depicts the output power of the laser in dependence on both, vertical and horizontal mirror tilt. Figure 16 shows that even the time averaged laser intensity does not decrease linearly with increasing mirror tilt. Furthermore, tilting the feedback mirror introduces a characteristic dynamical phenomenon to the system: for slight misalignment the system enters a state which is almost periodic in its low frequency behavior, i.e. the LFF events occur more regular in time [34]. Figure 17 gives a typical example for this asymmetry induced dynamical behavior.

We conclude that the method of tilting the feedback mirror in order to reduce the optical feedback strength is critical since the resulting asymmetry may introduce additional features to the dynamics of the system.

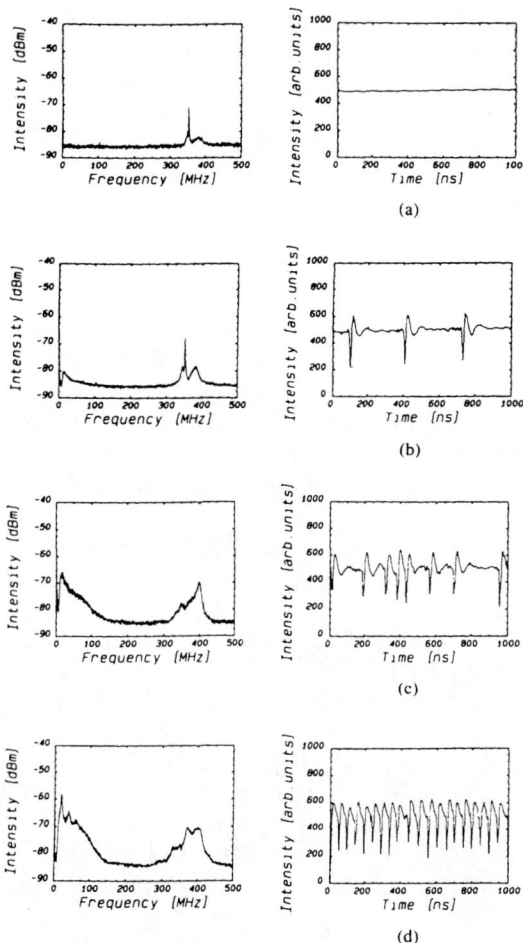

FIGURE 17. Power spectra (left column) and temporal evolution (right column) of the optical output power for different vertical tilting angles of the mirror, namely for 0 μrad (a), 4 μrad (b), 8 μrad (c), and 17 μrad (d). From [35] © 1991 IEEE.

CONCLUSION

In this chapter, we presented a purely experimental approach in order to investigate and analyze the dynamics of semiconductor lasers subject to optical feedback in dependence on various system parameters as injection current, feedback strength, and linewidth enhancement factor. In the different sections, we described the influence of detection apparatus, device structure, and asymmetries on the actually observed experimental data. In each of these sections, we presented characteristic phenomena which must be taken into account for an accurate interpretation of the experimental data. Our intention was to give a broad overview on what happens to the observed dynamical behavior, if one of the numerous parameters including detection, device, geometry, and system aspects is changed. For the physical mechanisms underlying this great variety of dynamical phenomena and the corresponding theoretical considerations we refer to the chapters by *Erneux, Gavrielides, Lenstra & Yousefi,* and *Verduyn Lunel & Krauskopf.*

REFERENCES

1. Risch, Ch., Voumard, C., Reinhardt, F.K., and Salathe, R., *IEEE J. Quant. Electron.* **QE-13**, 692 (1977).
2. Lenstra, D., Verbeek, B.H., and den Boef, A.J., *IEEE J. Quant. Electron.* **QE-21**, 674 (1985).
3. Tkach, R.W., and Chraplyvy, A.R., *IEEE J. Lightwave Tech.* **LT-4**, 1655 (1986).
4. Mørk, J., Mark, J., and Tromborg, B., *Phys. Rev. Lett.* **65**, 1999 (1990).
5. Ye, J., Li, H., and McInerney, J.G., *Phys. Rev.* **A 47**, 2249 (1993).
6. Jiang, S., Pan, Z., Dagenais, M., Morgan, R.A., and Kojima, K., *IEEE Photon. Technol. Lett.* **3**, 597 (1994).
7. Jiang, S., Dagenais, M., and Morgan, R.A., *IEEE Photon. Technol. Lett.* **7**, 739 (1995).
8. Masoller, C., and Abraham, N.B., *Phys. Rev.* **A 59**, 3021 (1999).
9. Hohl, A., van der Linden, H.J.C., and Roy, R., *Opt. Lett.* **20**, 2396 (1995).
10. Haug, H., and Haken, H., *Z. Physik* **204**, 262 (1967).
11. Henry, C.H., *IEEE J. Quant. Electron.* **QE-18**, 259 (1982).
12. Agrawal, G.P., and Dutta, N. K., *Long-wavelength semiconductor lasers*, Van Nostrand Reinhold, 1986.
13. Hofmann, M., Koch, M., Heinrich, H.J., Weiser, G., Feldmann, J., Elsäßer, W., Göbel, E.O., Chow, W.W.,and Koch, S.W., *IEE Proc. Optoelectron.* Vol. **141**, no. 2, (1994).
14. Vahala, K., Chiu, L.C., Margalit, S., and Yariv, A., *Appl. Phys. Lett.* **42** , 631 (1983).
15. Henning, I.D., and Collins, J.V., *Electron. Lett.* **19**, 927 (1983).
16. Hakki, B.W., and Paoli, T.L., *J. Appl. Phys.* **46**, 1299 (1975).
17. Osinski, M., and Buus, J., *IEEE J. Quant. Electron.* **QE-23**, 9 (1987).
18. Heil, T., Fischer, I., and Elsäßer, W., *Phys. Rev.* **A 60**, 634 (1999).
19. Ryan, A.T., Agrawal, G.P., Gray, G.R., and Gage, E.C., *IEEE J. Quant. Electron.* **QE-30**, 668 (1994).
20. Masoller, C., and Abraham, N.B., *Phys. Rev.* **A 57**, 1313 (1998).
21. Heil, T., Fischer, I., and Elsäßer, W., *J. Opt. B Quant. Semiclass. Opt.* **2**, 413 (2000).
22. Yamanaka, T., Yoshikuni, Y., Lui, W., Yokoyama, K., and Seki, S., *IEEE Photon. Technol. Lett.* **4**, 1318 (1992).
23. Sano, T., *Phys. Rev.* **A 50**, 2719 (1994).

24. van Tartwijk, G.H.M., Levine, A.M., and Lenstra, D., *IEEE Sel. Top. Quant. Electron.* **1**, 466 (1995).
25. Levine, A.M., van Tartwijk, G.H.M., Lenstra, D., and Erneux, T., *Phys. Rev.* **A52**, R3436 (1995).
26. Sukow, D.W., Gardner, J.R., and Gauthier, D.J., *Phys. Rev.* **A 52**, R3370 (1997).
27. Heil, T., Fischer, I., Elsäßer, W., Mulet, J., and Mirasso, C.R., *Opt. Lett.* **24**, 1275 (1999).
28. Huyet, G., Balle, S., Guidici, M., Green, C., Giacomelli, G., and Tredicce, J.R., *Opt. Comm.* **149**, 341 (1998).
29. Vashenko, G., Giudici, M., Rocca, J.J., Menoni, C.S., Tredicce, J.R., and Balle, S., *Phys. Rev. Lett.* **81**, 5536 (1998).
30. Huyet, G., White, J.K., Kent, A.J., Hegarty, S.P., Moloney, J.V., and McInerney, J.G., *Phys. Rev.* **A 60**, 1534 (1999).
31. Pan, M., Shi, B., and Gray, G.R., *Opt. Lett.* **22**, 166 (1997).
32. Besnard, P., Meziane, B., and Stephan, G.M., *IEEE J. Quant. Electron.* **QE-29**, 1271 (1993).
33. Seo, D., Park, J., McInerney, J.G., and Osinski, M., *IEEE J. Quant. Electron.* **25**, 2229 (1989).
34. Sacher, J., Elsäßer, W., and Göbel, E., *Phys. Rev. Lett.* **633**, 2224 (1989).
35. Sacher, J., Elsäßer, W., and Göbel, E., *IEEE J. Quantum Electron.* **QE-27**, 373 (1991)

Excitability in Laser Systems: the Experimental Side

Jorge R. Tredicce

Institut Non-Linéaire de Nice, UMR 6618, 1361 Route des Lucioles, Sophia Antipolis, 06560 Valbonne, France

Abstract. The main goal of this contribution is to give an insight into one of the most beautiful experiences: to be an experimentalist in Nonlinear Dynamics. In particular, I will refer to my field of expertise of Experimental Nonlinear Dynamics in Optical Systems. However, the great property of Nonlinear Dynamics is that the results are usually valid for different subjects ranging from biology to economics, and from physics to administration (to mention just those that we may consider opposite in spirit). With this in mind, you will be able to transfer many things I will write about to other fields, even to theoretical physics!

The second goal is to describe how to interpret data and how to link it to general concepts of Nonlinear Dynamics. Third and last goal (but not less important) is to show that Experimental Nonlinear Dynamics is one of the most *exciting* fields in Physics and Mathematics. For this reason I will limit the discussions to a phenomenon called *excitability*.

GENERAL CONCEPTS AND MODELS OF EXCITABLE SYSTEMS

Excitability is a very old concept, so old that our children study it in elementary school. In fact, excitability is the process describing the wave propagation in muscles and nerves as it was described by N. Wiener and A. Rosenblueth [1] in 1946, and Hodgkin & Huxley [2] in 1951. There we learn how excitable pulses, which are almost identical in shape and duration, travel through the cardiac and nervous system. We learn also that there is a two channel communication: one to send and another to receive the pulses, because if two excitable waves traveling in opposite direction meet, they annihilate each other. These authors made a very good qualitative description of the processes taking place in the cells of striated muscles and in neurons. As a result, excitability belonged to the domain of biology for a long time.

In 1970 excitation waves were observed in chemistry by Zaikin & Zhabotinsky [3]. From that on chemical reactions were used to study the properties of excitable waves. Later they were observed in nematic liquid crystals [4,5]. However, just recently there was a revival in the interest in excitable media. This is due to an advance in the construction of models and theories able to explain not only phenomenologically but quantitatively the appearance of excitable pulses and excitable waves [5–7]. Finally, their introduction in optics came in relation to lasers with optical feedback [8], lasers with injected signal [9], and lasers with saturable absorber [10].

The description of the temporal behavior of an excitable medium has been given a long time ago. If we consider a dynamical system in a quiescent state (at a stable steady-state) and we apply a perturbation, the system is activated only if the perturbation exceeds a critical value. It generates a pulse whose amplitude is independent of the strength of the perturbation. The response is followed by a non-excitable period of definite duration, called the refractory time, until it reaches the original quiescent state. If the system is spatially extended, as some striated muscles, the pulse generated by the initial perturbation in a small area will propagate through the entire system almost unperturbed. We then say that we have an excitable wave. This qualitative description was formulated mathematically by FitzHugh and Nagumo et al. [6]. Their paradigmatic model considered a set of two variable reaction-diffusion equations of the type:

$$\frac{\partial u}{\partial t} = u - u^3 - v + \nabla^2 u, \qquad (1)$$
$$\frac{\partial v}{\partial t} = \varepsilon(u - u_0),$$

where u is commonly called activator and v inhibitor, ε and u_0 are positive control parameters where ε is very small compared to unity. The diffusion term gives the connection between different points in space and it allows the study of the propagation of pulses. Equations (1) have a single steady-state solution represented by $u = u_0$ and $v = u_0 - u_0^3$. This solution is stable if $|u_0| > \sqrt{1/3}$. At $|u_0| = \sqrt{1/3}$ there is a Hopf bifurcation because the eigenvalues of the characteristic equation are complex and the solution is oscillating in time. For $1/3 < u_0^2 < 4/3$, the system is excitable. This model, which just comes from heuristic hypothesis on the behavior of muscle cells or neurons, was widely used as a paradigmatic model to describe all kinds of excitable systems. In fact, it reproduces quite well the qualitative description we gave above. If the parameters are such that the steady-state solution is stable, then the system is in a quiescent state. If we apply a fast perturbation to the control and/or to the order parameters, the system may come back very fast to the strongly attracting manifold of v to which the stationary solution belongs. However, if the perturbation is big enough, then the trajectory in phase space will follow three processes as shown in Fig. 1. First, it evolves very fast in phase space toward the branch of the S-shaped attracting manifold to which the steady-state does not belong. Second, it follows the manifold until the fold and then switches quickly to the branch of the S-shaped attracting manifold containing the steady-state. Third, it slowly returns to the quiescent state, i.e.

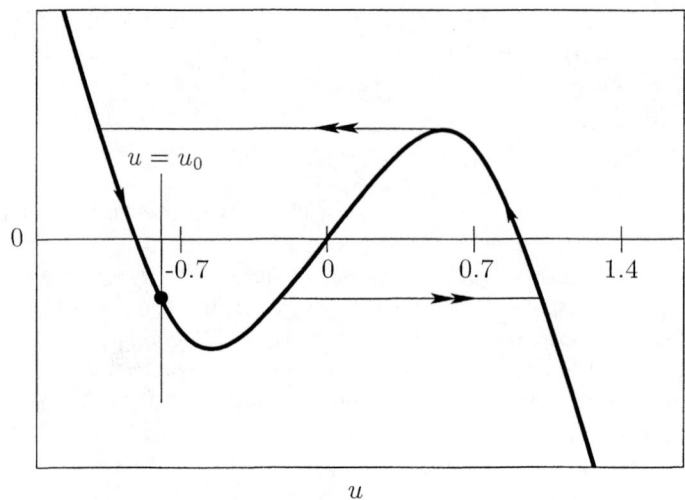

FIGURE 1. The FitzHugh-Nagumo model.

the attracting steady-state. The time required during this third leg to reach the stationary solution is called the refractory time

For many years this phenomenological model was considered the most simple model explaining the behavior of excitable systems. However, recently P. Coullet [5] developed a theoretical model based on a modified Ginzburg-Landau equation:

$$\frac{\partial B}{\partial t} = \mu B - |B|^2 B + \epsilon + \nabla^2 B, \qquad (2)$$

where B is a complex order parameter and μ and ϵ are control parameters. Under the approximation that $|B| = R$ is constant, it is possible to write an equation for the phase ϕ of the type:

$$\frac{\partial \phi}{\partial t} = \mu_i - \frac{\epsilon}{R} \sin \phi + \nabla^2 \phi, \qquad (3)$$

where μ_i is the imaginary part of the control parameter μ. The Laplacian term is again the spatial coupling. Equation (3) may describe excitable behavior. In fact, it possess two stationary solutions for $|\mu_i| < \epsilon/R$, which are given by the solutions of the equation:

$$\sin \phi = R\mu_i/\epsilon, \qquad (4)$$

but only one solution is stable.

This leads to the three cases sketched in Fig. 2. For $\mu_i = \epsilon/R$ a saddle-node bifurcation occurs leading to the disappearance of the two solutions. If $\mu_i > \epsilon/R$

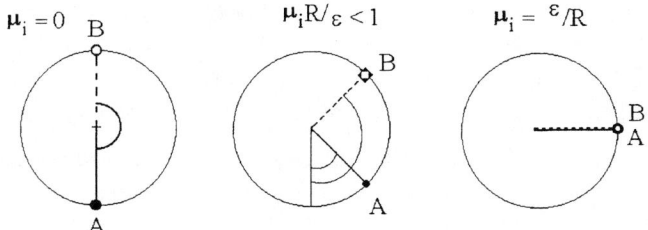

FIGURE 2. Coullet's model of excitability, where A is a stable steady state and B is a saddle point.

the phase oscillates periodically in time with a period that goes to infinity very close to the bifurcation point.

For $\mu_i < \epsilon/R$ the system is excitable: a small perturbation of the phase yields a fast return back to the stable steady-state. However, if the perturbation is big enough to place the system initially on the other side of the unstable steady-state, the phase will change by 2π before coming back to the stable state. Thus the trajectory is independent of the perturbation. We can in principle define the refractory time as the time required to return to the quiescent state.

Equation (2) is rather similar to the equation used to describe a Class A laser with injected signal in the cubic approximation except for the real coefficient in front of the Laplacian. However, this coefficient will control the propagation of the pulse in the transverse direction but it does not affect the excitable character of the system. Equation (3) is the well-known Adlers equation [12] describing the phase locking among coupled oscillators. Coullets model of excitability is rather simple and, in principle, it could be used as predictive of excitability in a laser with injected signal [9].

It is important to notice that in the case of the FitzHugh-Nagumo equation, the associated bifurcation that leads to excitability is a Hopf bifurcation, while in Coullet's model it is a homoclinic saddle-node bifurcation (also called saddle-node bifurcation on a limit cycle; see the chapter by *Krauskopf*).

The third model of excitability is based on the so-called avalanche-collapse process. The equations of motion are of the form:

$$\frac{\partial X}{\partial t} = X(X - Y + k),\qquad(5)$$
$$\frac{\partial Y}{\partial t} = \alpha(-Y + X^2),$$

where X and Y are the order parameters, and α and k are control parameters. This system possess three coexisting steady-states for a given range of the control parameters. It can give bistability, oscillations and excitability depending on the value of α in an interval of k values [7].

We will not analyze this model in detail here because it is not the aim of this paper, but it is worthwhile to note that this model was presented as a possible simple model describing a laser with saturable absorber [7].

The three models share some well defined characteristics of excitable media:

(a) the system reacts to a perturbation generating an excitable pulse whenever the perturbation overcomes a well defined threshold value; the threshold depends on the values of the control parameters,

(b) there is a time required for the system to relax to the quiescent state; this time defines the refractory phase,

(c) if the parameter values are well chosen, the excitable pulse has the same amplitude and width independently of the applied perturbation. The minimum perturbation able to generate an excitable pulse is called the ignition solution.

If we consider the propagation of excitable pulses in space, we must note that two pulses propagating in opposite direction would, in principle, annihilate each other at the position where they meet. For a long time this behavior was considered an intrinsic property of excitability.

LASER MODELS AND THEIR PREDICTIONS

The models introduced above are general enough to be able to describe different kinds of physical systems. We may ask ourselves if some more specific models in optics, like the Maxwell-Bloch equations can be reduced to one of these 'excitable models'.

Single-Cavity Laser

The Maxwell-Bloch equations describing two-level homogeneously broadening atoms interacting with an electromagnetic field in the slowly varying and plane wave approximations are [11]:

$$\frac{\partial E}{\partial z} + \frac{1}{c}\frac{\partial E}{\partial t} = gP + i\frac{(\delta\omega)}{c}E,$$
$$\frac{\partial P}{\partial t} = -\gamma_\perp \left[(1+i\delta)P - EN\right], \qquad (6)$$
$$\frac{\partial N}{\partial t} = -\gamma_\parallel \left[N - N_0 + \frac{1}{2}(E^*P + EP^*)\right],$$

where E, P, and N are the electric field, the atomic polarization and the population inversion, respectively, $\gamma_\perp$ and $\gamma_\parallel$ are the decay rates for the polarization and population inversion, respectively, g is the gain of the medium per unit of length,

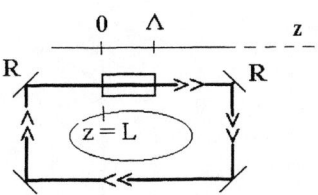

FIGURE 3. Sketch of a unidirectional ring laser, where R is the reflection coefficient, L is the total length of the cavity, and Λ is the length of the active medium.

N_0 is the pumping rate, and δ is the detuning between the atomic resonance (ω_a) and the operating frequency ($\delta\omega$) of the laser normalized to $\gamma_\perp$.

For the sake of simplicity we consider a unidirectional ring cavity imposing the boundary conditions [11]:

$$\rho(0) = R\rho(\Lambda), \tag{7}$$

$$\phi(0) = \phi(\Lambda) - (\delta\omega)\left(\frac{\Lambda+L}{c}\right),$$

where $\rho(0)$, $\rho(\Lambda)$, $\phi(0)$, and $\phi(L)$ are the moduli and phases of the field at the input $z=0$ and output $z=\Lambda$ positions of the active medium, respectively, R is the reflection coefficient of each of two semi-transparent mirrors of the cavity, and L is the length of the cavity; see Fig. 3.

The stationary solutions of the equations of motion with the restriction imposed by the boundary conditions, are:

$$\rho_j^2(\Lambda) = \frac{2}{1-R^2}\left[gN_o\Lambda - \left(1+\delta_j^2\right)|\ln R|\right],$$

$$(\delta\omega_j)\left(1+\frac{\kappa}{\gamma_\perp}\right) = \delta_{ac}\frac{\kappa}{\gamma_\perp} + 2\pi\frac{c}{L}, \tag{8}$$

$$\kappa = \frac{c}{L}|\ln R|,$$

where $\delta_{ac} = (\omega_a - \omega_c)/\gamma_\perp$ is the detuning between the atomic frequency and its nearest cavity resonance ω_c.

Therefore, an infinite set of monochromatic solutions exist. Each of them is characterized by the index j, which in term defines the corresponding frequency and intensity. Solutions having negative intensity ($\rho^2 < 0$) are clearly unphysical. The solutions for which ρ^2 is greater than zero have different intensity values because of their different operating frequencies, and their separation in the phase ϕ is always an integer multiple of 2π; see Fig. 4. They correspond to the well-known longitudinal modes of the laser.

It was shown in Ref. [11] that only one of these solutions is stable for a large range of parameter values. The unstable solutions have a complex eigenvalue whose real

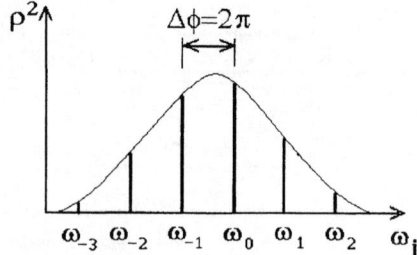

FIGURE 4. Intensity as a function of frequency for the modes of a single cavity laser. The phase difference between the steady-state solutions is constant and equal to 2π. Usually the mode with the highest intensity is stable while the other modes are unstable.

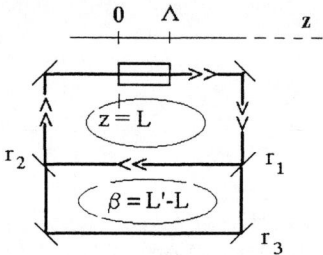

FIGURE 5. Sketch of a double-cavity unidirectional ring laser.

part is positive. The imaginary part of the eigenvalue is the frequency separation with respect to the stable steady-state. Between the stable steady-state and the unstable ones there are no saddles, and the only saddle point corresponds to the trivial solution ($E = 0$). Thus, if we accept that the model of a single-cavity laser cannot be reduced to any of the three existing models of excitability, we must draw the conclusion that a laser by itself is not excitable.

Double-Cavity Laser

Let us now consider a double cavity as shown in Fig. 5. The equations of motion are unaltered, but the boundary conditions for the steady-state solutions become:

$$\rho(0) = \rho(\Lambda)\sqrt{a^2 + b^2 + 2ab\cos\left[(\omega_c + \delta\omega)\frac{\beta}{c}\right]} = \Re(\delta\omega)\,\rho(\Lambda), \quad (9)$$

$$\phi(0) = \phi(\Lambda) + (\delta\omega)\frac{L}{c} - (\delta\omega)\frac{\Lambda}{c} + \arctan\left\{\frac{b.\sin\left[(\omega_c + \delta\omega)\beta/c\right]}{a + b\cos\left[(\omega_c + \delta\omega)\beta/c\right]}\right\} + 2\pi j,$$

where $a = \sqrt{r_2(1-r_1)}$ and $b = \sqrt{r_3 r_1(1-r_2)}$ with r_1 and r_2 being reflection coefficients of the mirrors of the principal cavity, and r_3 that of the secondary or

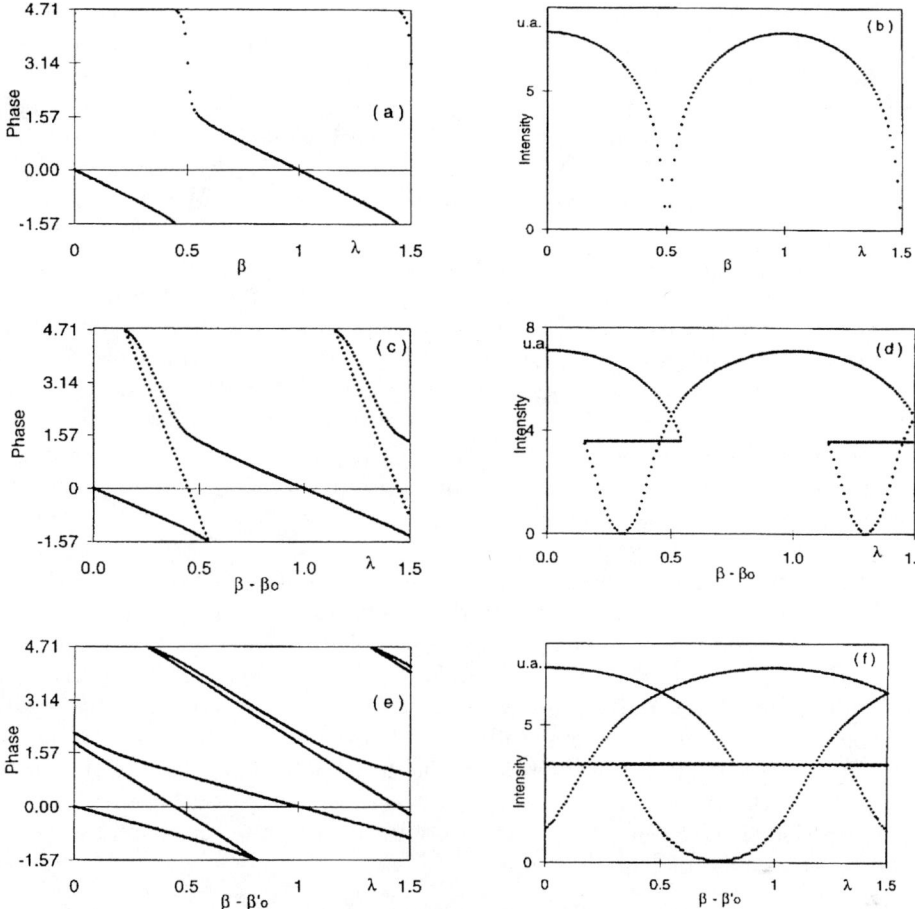

FIGURE 6. Steady-state solutions for the double-cavity laser model, plotted as phase and intensity as a function of the length difference (in units of wavelength) of the two cavities. The parameters are $\gamma = 10^{13}$, $r_1 = 0.6$, $r_2 = 0.4$, $r_3 = 0.5$, $L = 0.1$, $(\omega_a - \omega_c) = 0$, $n = 10^5$, $N_0 = 3.75$, $\beta_0 = 0.4$, and $\beta'_0 = 0.15$.

external cavity. Furthermore, $\beta = L' - L$ is the length difference between the two cavities. In principle we consider the case that the gain of the medium is strong enough for lasing even with $r_1 = 0$. Then the boundary conditions describe a double-cavity laser.

A simple integration of the equations of motion leads to steady-state solutions having the same form as Eqs. (7), except that the constant reflectivity R is replaced by the functional $\Re(\delta\omega)$. For the longitudinal mode whose operating frequency lies closer to the maximum gain and an appropriate value of β, we found three different steady-state solutions. One is a stable steady-state and its neighbor is a saddle point.

If β is an integer multiple of the laser wavelength λ and smaller than L, then there exists only one steady-state solution for each value of j, as in a single-cavity laser. In this case the effect of the second cavity is essentially to change the amount of the total losses for the field, decreasing or even increasing the laser threshold pump.

Three steady-state solutions appear if β is around an integer multiple of a half-wavelength and between $0.1\,L$ and L. Their domain of existence increases as β increases. The relative phase among them changes as the parameter β changes and it may become much smaller than 2π; see Fig. 6. In this case the system may behave as an excitable medium mainly for lengths differences of the order of $(m+1/2)\lambda$ where m is an integer. It is worthwhile to note that its excitable character would disappear if the lengths of the two cavities differ by an integer number of wavelengths. However, if β becomes greater than L, multiple solutions appear even for lengths differences equal to $m\lambda$, and the possibility of excitable behavior always exists.

On the other hand, the model is general enough to include the laser with optical feedback, because this corresponds to decreasing r_3 in order to lower the Q-factor of the second cavity, such that the cavity losses are greater than the unsaturated gain when $r_1 = 0$ and $r_2 = 0$. Therefore, we can imagine that both systems, a double-cavity laser and a laser with optical feedback, are equivalent because a smooth change in the parameter values will allow to switch from one to the other. Of course, the change in parameter may induce the disappearance of the excitable character of the system.

The physical model of a laser with a double cavity may correspond in principle to Coullet's model of excitability.

Laser with Injected Signal

A detailed model of a ring laser with an injected coherent field in the slowly varying and plane wave approximations includes the same equations of motion as before but with the boundary conditions:

$$E(0,t) = RE(\Lambda,t)\exp i\omega\left(L-\Delta\right)/c + A\,. \qquad (10)$$

This model is difficult to analyze, even qualitatively, if we do not perform some simplifying assumptions. If we assume a single-mode laser, a uniform field inside the medium, and that the frequency of the external field is very close to the laser operating frequency, then the model can be reduced to [16,17]:

$$\frac{\partial E}{\partial t} = -k\left(1 + i\theta\right) E + \xi P + A,$$
$$\frac{\partial P}{\partial t} = -\gamma_\perp \left[\left(1 + i\delta\right) P - EN\right], \quad (11)$$
$$\frac{\partial N}{\partial t} = -\gamma_\parallel \left[N - N_0 + \frac{1}{2}\left(E^* P + E P^*\right)\right],$$

where $\theta = (\omega - \omega_c)/k$ and $\delta = (\omega_a - \omega)/\gamma_\perp$. Furthermore, ω is the frequency of the external field of amplitude A, and $k = c\left|\ln R\right|/L$ is the cavity loss rate. It is important to note that this system has a very important property different from the previous laser models: it has no phase symmetry. The phase or time symmetry has been broken by the external field which imposes an *external clock*. Steady state solutions can only be found operating at the frequency of the external field. From the dynamical point of view, the system given by Eqs. (11) is five dimensional, relatively simple, and at the same time it presents a large variety of behaviors and phenomena, like Hopf bifurcation, homoclinic saddle-node bifurcation, conservative and dissipative behavior, generalized multistabilty, phase locking, chaos, etc. [23,16,9]. However, it is important to always keep in mind what approximations were made, because the description we could get for parameter values outside the limits of validity of these equations may be highly misleading. Therefore, almost every analytical or numerical result for this simplified model must be validated by a numerical simulation of the complete set of equations.

In Eqs. (11) it is easy to find a parameter region where a single stable steady-state exists. It corresponds to a solution that is phase-locked to the external field and usually appears when the amplitude of the external field is greater than a critical value. For a given interval of the external field the equations show excitable behavior as was shown in Ref. [9]. In fact, under more restrictive approximations (very small and constant amplitude field), Eqs. (11) can be reduced to:

$$\frac{d\phi}{dt} = \eta \sin \phi + D, \quad (12)$$

where ϕ is the phase of the electromagnetic field E, and η and D are constants. This is essentially Adler's equation and it corresponds to the model of excitability proposed by Coullet.

Therefore, we may expect that a laser with optical feedback, a laser with a double cavity, or a laser with injected signal may display excitable behavior in some region of the parameter space.

Finally, we have to remark that excitability in optics was theoretically predicted in a laser with injected signal [9], in the interaction among different polarization of the field [24], in a passive system with a temperature dependent parameter [19], in the well-known Lang-Kobayashi model [18] and in a laser with saturable absorber [10]. Experimentally we showed evidence of excitability in a laser with optical feedback [8,21], in a laser with a double cavity [20], and in a laser with injected signal.

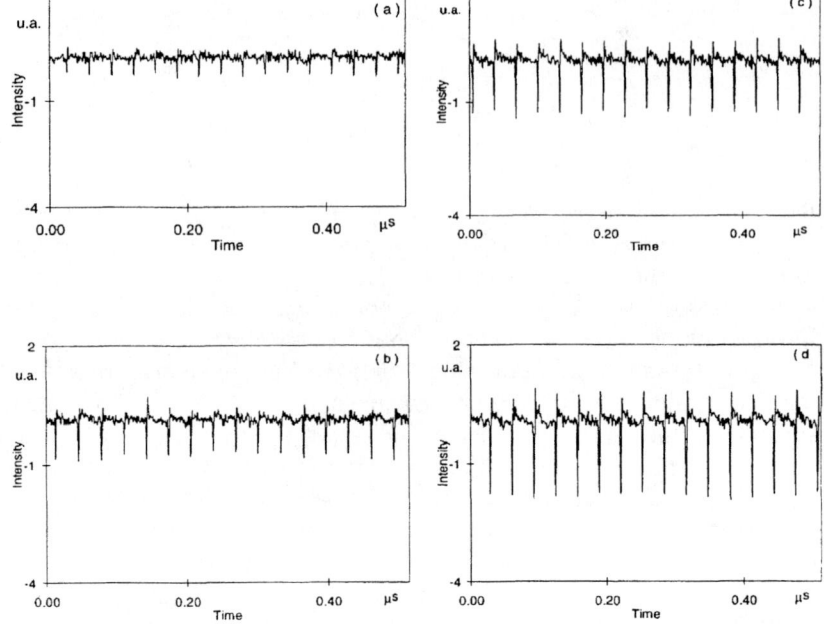

FIGURE 7. Intensity of a single-cavity laser as a function of time when we apply perturbations at a reate of 33 MHz and 600 ps width for the different perturbation amplitudes of 6 mA (a), 10 mA (b), 14 mA (c), and 18 mA (d). The amplitude of the laser intensity fluctuations increases as we increase the amplitude of the perturbation.

EXCITABILITY FROM THE EXPERIMENTAL POINT OF VIEW

From the experimental point of view, we can use the following properties to identify an excitable system, independently of the particular underlying model.

1. The response of the system becomes 'macroscopic' if the excitation amplitude overcomes a critical value, usually called *threshold for excitation*.

2. Width and amplitude of the 'macroscopic' response are independent of the characteristics, amplitude and duration, of the excitation.

3. The system requires a well-defined time to recover its steady-state and, therefore, will not respond to two consecutive excitations separated in time by less than the refractory time.

4. The threshold for excitation decreases as the bifurcation point is approached.

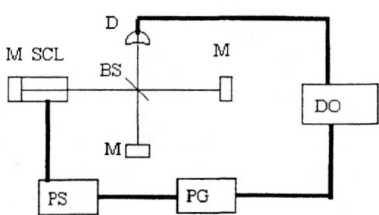

FIGURE 8. Experimental setup, where M is a mirror, BS a beam splitter, SLC the semiconductor laser, D a detector, PS a dc current supply, PG a pulse generator, and DO a digital oscilloscope.

Measurements in Single-Cavity and Double-Cavity Lasers

As we explained above, single-cavity lasers by themselves cannot be excitable. (Of course, assuming that our model is correct!) We experimentally checked this theoretical prediction using an AR coated semiconductor medium within an external cavity. For a pumping current above the laser threshold the output intensity is stable. We added periodic perturbations to the pumping current consisting of square pulses of 100 ps duration and variable amplitude at a rate of 33 MHz. This repetition rate is low enough to allow the system the time to relax to the steady-state intensity value before the next perturbation is applied. The pulse is shorter than the characteristic times of the medium, so that it can be considered as a method to change the initial condition or equivalently to generate a hard excitation. The amplitude of the response increases monotonically as we increase the perturbation amplitude, as is shown in Fig. 7. There is no evidence of the existence of a threshold value of the excitation pulse for which there is a response independently of the excitation itself. Thus, a usual laser by itself does indeed not behave as an excitable medium.

An important test is given also by a phase space plot. In Fig. 11(a) (which is compared with the excitable case below) we show the derivative of the intensity (dI/dt) as a function of the intensity itself for different values of the amplitude of the excitation. The trajectories in phase space cover uniformly the two-dimensional space because the medium response is almost proportional to the perturbation. The laser steady-state value can be found in the middle of the black region.

To study the behavior of a double-cavity laser, we construct the experimental set-up shown in Fig. 8. A Michelson interferometer is used as external mirror. The length difference between the two arms of the interferometer is of the order of 20%. The laser output intensity is constant in time for all values of available pumping currents. We apply perturbations to the system by adding a 60 ps pulse to the pump current at a rate of 33.1 MHz. The amplitude of the excitation is varied from 2 mA up to 20 mA, which represents 3% and 30% of the DC current respectively. An avalanche photo-diode with a bandwidth of 6 GHz detects the

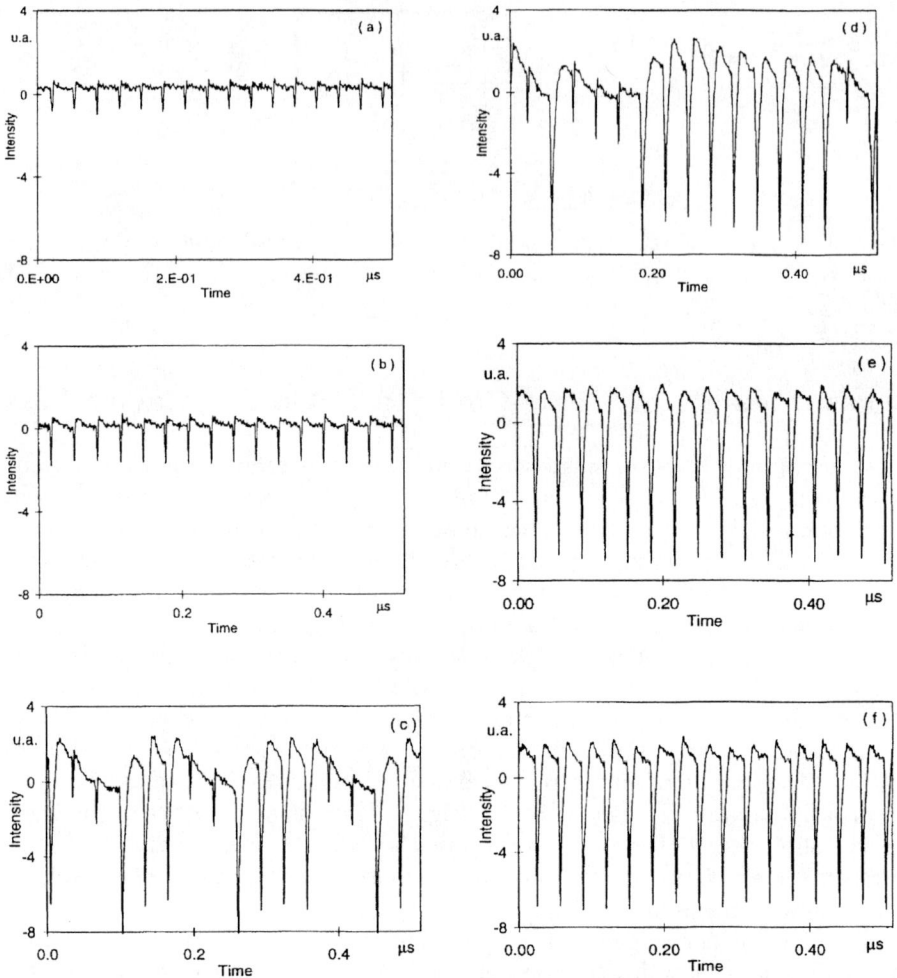

FIGURE 9. Response of the laser with double cavity as a function of time for different values of the perturbation amplitude of 2 mA (a), 6 mA (b), 7 mA (c), 8 mA (d), 9 mA (e), and 12 mA (f). Frequency and width of the current pulses are kept constant at 33 MHz and 600 ps, respectively.

laser intensity. A LeCroy7200 digital oscilloscope allows the analysis of the output signal.

We show the response of the system for increasing values of the perturbation amplitudes in Fig. 9. For an excitation amplitude smaller than 7 mA the laser intensity fluctuations are equivalent to the single-cavity case. However, the amplitude of the response of the system increases by one order of magnitude when the

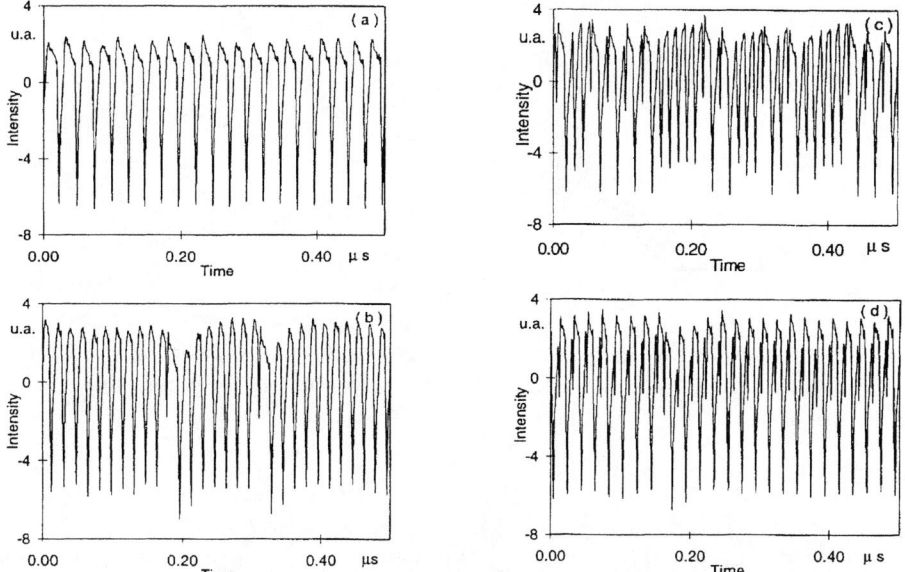

FIGURE 10. Intensity as a function of time for different frequencies of the perturbation of 40 MHz (a), 60 MHz (b), 80 MHz (c), and 100 MHz (d). Amplitude and width of the current pulses are kept constant at 12 mA and 600 ps, respectively.

perturbation amplitude exceeds 7 mA. This is proof that the perturbation amplitude has to exceed a well defined critical value before a macroscopic response of the system is observed.

A plot of the amplitude of the output intensity fluctuations as a function of the amplitude of the perturbation shows that the excitation amplitude has almost no influence in the response of the system when it exceeds 7 mA. The evolution in phase space follows a trajectory which became independent of the perturbation itself.

To check the third experimental property of excitable systems, we increase the frequency of the excitation pulses. At 70 MHz the response of the double cavity laser becomes erratic and at 90 MHz the laser intensity shows giant pulses only after two consecutive perturbations, as is shown in Fig. 10. This is proof that the laser does not react if two consecutive perturbations are separated in time by less than the refractory time. We estimate that the refractory time is of the order of 10 ns for the active medium we used. This time is of the order of magnitude of the carrier lifetime in a semiconductor.

The fourth and last experimental test is performed by changing periodically the length difference β between the two arms of the Michelson. This parameter controls

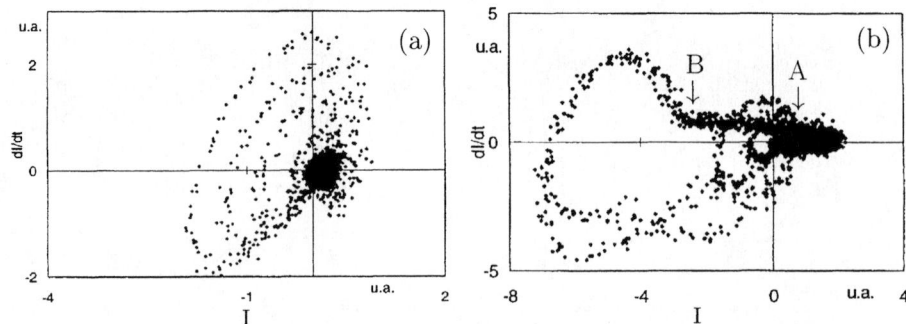

FIGURE 11. Time derivative of the intensity dI/dt as a function of the intensity I for the laser in a single cavity (a) and in a double cavity (b). The data from Fig. 7 is superimposed to obtain (a), and the data from Fig. 9(a),(b),(e),(f) is superimposed to obtain (b).

the phase difference between the stable steady-state and the saddle point in the theoretical model. The critical value of the excitation amplitude required to obtain giant pulses of the intensity changes from the minimum possible value with our set-up (1 mA) up to 10 mA for different values of β.

Finally it is interesting to observe the dynamical evolution of the system in a phase space plot. In Fig. 11 we show the derivative of the output intensity as a function of the intensity for different values of the perturbation amplitude. As mentioned earlier, in the laser of a single cavity the size of the orbit in phase space increases as we increase the amplitude of the perturbation; see Fig. 11(a).

The same behavior is observed in the double-cavity case when the perturbation amplitude is below the critical value; see Fig. 11(b). On the other hand, above this threshold the system performs a much larger excursion in phase space. The excitation current pulse drives the system out of the stable steady-state, changing the initial condition. This can be seen by the clouds of points close to the point A in Fig. 11(b). The dynamical evolution converges towards a well-defined orbit which, after a fast evolution, almost stops at point B (where the time derivative of the intensity approaches zero). From there on the dynamical evolution slows down (as indicated by the high density of points in the phase space plot) until it returns to the steady-state solution. This indicates the position of at least one saddle point in phase space. In some of the trajectories for an amplitude of the perturbation close to threshold, we have a similar situation for another point closer in time to the beginning of the pulse. It seems to indicate that the position of the saddle point is closer to the stable steady-state. This last saddle point is not observed for all trajectories. Comparing the number of points in the phase space plot of a single pulse with the digitizing time of the oscilloscope we can give an estimate of the time needed for convergence towards the final orbit in phase space. In our double-cavity experiment this time was evaluated to be 250 ps.

In conclusion we showed experimentally and theoretically that an optical system consisting of a laser with a double cavity behaves as a typical excitable medium. By a simple theoretical analysis of the Maxwell-Bloch equations we showed that the modifications introduced by a double cavity in a laser are enough to create saddle points lying very close in phase space to the stable steady-state, which is a requirement of excitability. Our experimental measurements performed on a single- and double-cavity semiconductor laser confirm the theoretical predictions. Thus, a spatially extended system of this kind may be able to generate excitable waves whose propagation properties still need to be explored.

Laser with an External Signal: Bistability, Instabilities and Excitability

An active medium, as well as a passive one, with injected signal was extensively studied experimentally due to its scientific and practical interest. In fact, a laser with injected signal is one of the best ways to obtain high-power stabilized coherent sources, and it is used in many scientific and industrial applications. Passive systems with injected field have been conceived as a practical optical bistable systems and since several years they are promoted as potential optical computing elements.

This optical system can display a large variety of behavior depending only on the parameter values. For the same experimental system the parameter space can be divided into (roughly) three different interesting regions:

- region **B** showing bistable behavior,

- region **I** showing instabilities; periodic or chaotic oscillations of the variables of the system,

- region **E** showing excitability.

The same experimental set-up can show, depending on the operating parameter value, all the above mentioned physical effects. For example, a semiconductor material with injected coherent field will show bistable behavior if pumped below the laser threshold. On the other hand, it will be unstable if pumped above threshold when the injected field is not strong enough to lock the slave, or if it is pumped below transparency and the detuning between the cavity resonance and the external field frequency is strong enough. Finally, it will be excitable if the master and the slave are locked but the external field is not very strong.

It is important to note that **E** in general excludes **B** and **I** once the parameters are fixed, but **I** does not exclude **B** if we use the concept of generalized multistability introduced in reference [14]. In fact different stable orbits may coexist in parameter space. This concept can be explained as the result of the finite dissipation of any real system [15].

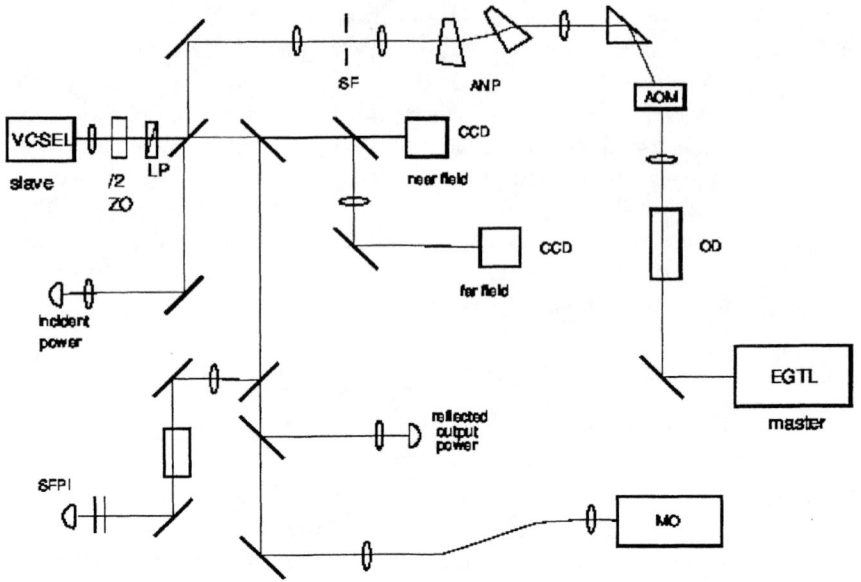

FIGURE 12. Setup for the experiment on the laser with injected signal.

The bistable and unstable behavior of a medium with injected coherent field has already been explored extensively [13]. The experimental set-up we used is represented in Fig. 12 and it is similar to previous ones.

Our master and slave oscillators are top emitting vertical cavity surface emitting lasers (VCSEL) operating in the 830 nm region of the optical spectrum. This type of laser presents the advantage of being single longitudinal mode and, when injected, also single linear polarization. Furthermore, the high reflectivity of the mirrors and the small length of the active medium ensure the validity of the uniform field limit, which is one of the main assumptions of the model. Both lasers are temperature stabilized better than 0.1 degrees, and they are independently electrically pumped. Both temperature and current amplitude are used in order to change the frequency detuning between the two lasers.

Two optical isolators are used to avoid re-injection of the electromagnetic field into the master. A collimator, a spatial filter and a couple of lenses are used to phase match the injected field in the slave cavity. Polarizer beam splitters are used to analyze the polarization of the master and the slave. Fast optical detectors (up to 6 GHz bandwidth) and a digitizer oscilloscope (2 Gsamples/sec, 500 MHz bandwidth) let us know the temporal evolution of the master and slave output intensities. Two Fabry-Perot resonators of 100 Ghz and 4 THz of FSR allow to identify the locking region and to measure the frequency detuning between the two lasers. A spectrum analyzer (22 GHz bandwidth) gives us the frequency components of the intensity fluctuations. The main parameters of the system are the amplitude and frequency of the electromagnetic field generated by the master, and the pumping current of the slave.

We distinguish two interesting operating regimes when changing the pumping current of the slave. If the the medium is above transparency but below the laser threshold, the output intensity is bistable with respect to the amplitude of the injected field. The heuristic interpretation is simple: for low amplitudes of the external field there is almost no influence of the nonlinearities of the medium, and the external field is almost linearly amplified by the active medium. The slave cannot generate its own electromagnetic field because the losses are higher than the gain at all frequencies, but it amplifies the external field because there is a positive population inversion. As the amplitude of the external field increases, the nonlinear term becomes more relevant. Finally, the external field is used as pump in order for the system to overcome losses and to give a high intensity output which requires a lower inversion than the low intensity output. As the inversion remains lower than the pump level, the output intensity remains high, until the injected field decreases enough to a level that cannot sustain such a level of inversion and the field goes back down to the lower branch. This behavior was previously observed in several experiments including edge emitting semiconductor lasers.

As we stated above, the existence of bistability excludes the possibility of excitability. In fact, the application of an excitation pulse induces only two types of response: the system evolves in phase space depending on the strength of the perturbation or, if the perturbation overcomes a given critical value, the system jumps towards the other branch and remains there. In the parameter region where only one branch exists the response of the system is always commensurate with the perturbation. In the bistable region, it may jump to the other branch and it remains there without coming back to the original steady-state. This may introduce in some cases a confusion between excitability and bistability or multistability. Some systems possess different stable steady-states which are distinguishable only by the value of their phase but not by their intensity. If we observe only the intensity the initial and final state are undistinguishable and we may think that the behavior of the system is excitable instead of bistable. An example in the literature is the case of the well known real equations for the laser (the Lorenz model): initial and final state differ in the phase by π, and the system is not excitable. In the full laser model, the phase invariance induces the existence of an infinite set of coexisting solutions with different phase values. The Lorenz model with real variables becomes unrealistic when we interpret the solution with $|E| < 0$ as a solution corresponding to a change of phase of π for the field. This solution is connected to the $|E| > 0$ solution by an infinity of solutions with different phases. The case of the inverted medium with injected signal breaks the phase symmetry, the two states acquire a real meaning and are easily observable as two different intensities like in our experiment. The low and high intensity solutions becomes two stable steady-states that also differ in phase.

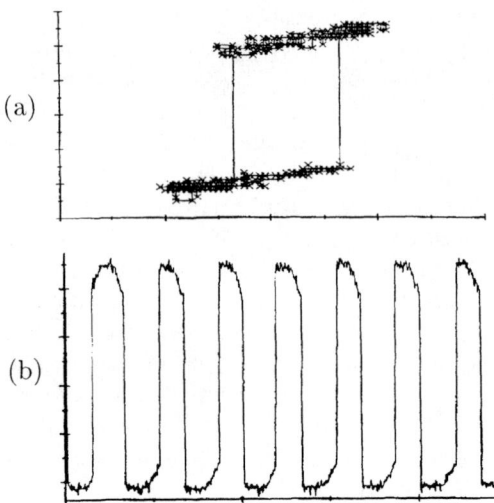

FIGURE 13. Bistability in a VCSEL with injected signal pumped below threshold, shown as output intensity as a function of the injected field mplitude (a), and as output intensity as a function of time as the injection amplitude is swept (b).

However, if we pump the slave above its laser threshold, the low intensity solution becomes a saddle point and, therefore, unstable. In this situation we do not observe bistability, and every stable solution acquires the frequency and phase of the external field. We say that the slave is phase-locked to the external field. Any other situation will correspond to an oscillatory output intensity. If the amplitude of the external field is small enough, the interaction between the fields generated by the slave and the master is very weak and we observe just the beating frequency. As the amplitude of the external field increases, the nonlinear terms play a relevant role in the dynamical behavior and different situations have been reported both theoretically and experimentally. It is necessary to remark here that theoretically a laser with injected signal shows an incomparable richness from the dynamical point of view [22]; see also the chapter by *Krauskopf*. Unfortunately, the experimental situation is technically complicated because most of the interesting oscillatory behavior happens at very low frequency and, therefore, its observation requires a very stabilized master, at the limits of the actual state-of-the-art in frequency stabilization. The only available measurements on semiconductor lasers are based on the observation of the power spectrum. However, while it is a very useful tool, the power spectrum cannot show the exact temporal behavior, and the interpretation of the results is based on arguments of ergodicity which are no longer valid for irregular signals.

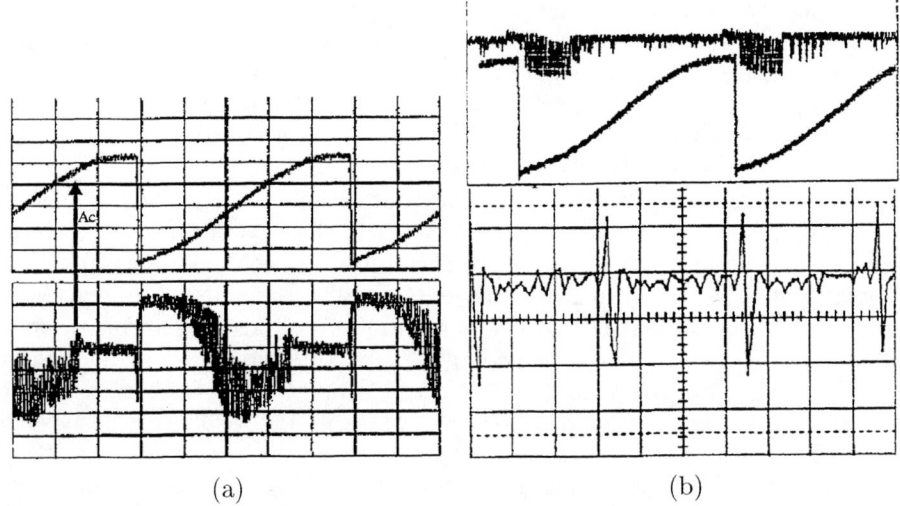

FIGURE 14. Panel (a) shows in the upper trace the injected intensity as a function of time, and in the lower trace the output intensity as a function of time. The time scale is 20 μs/div and the value Ac is indicated by the arrow. Panel (b) shows the intensity as a function of time for a VCSEL with injected signal. In the upper trace the injected intensity is swept with values grater than Ac, where the time scale is 20 ms/div, and periodic pulses at 33 MHz repetition rate are added to the pump current. The lower trace shows the output intensity as a function of time in the excitable region for a fixed value of the injection intensity greater than Ac.

Increasing the amplitude of the external field, the slave laser is locked and the output intensity becomes high and constant in time. It is in this region of parameter space where most of the injected lasers used in other branches of research and industrial processing operate. A high intensity laser acquires the amplitude and frequency stability of a low intensity master oscillator. It is also in this region where we operate our system. The amount of injected power required to operate in the locking state depends of course on the frequency detuning between the master and the slave. For vanishing detuning the locking is achieved with vanishing amplitude of the external field because both frequencies, the slave and master operating frequencies, coincide. As the detuning is increased, the amplitude of the external field necessary to get a stable output intensity, increases. For small detunings the critical field is directly proportional to the detuning, as in Adler's equation.

If the slave is above the laser threshold, the total output intensity may have oscillatory behavior not depending on the amplitude and frequency of the master. We select the situation in which the frequency of the master is slightly detuned with respect to the operating frequency of the slave, and we change the amplitude A of the external field. At very low amplitude the intensity is unstable showing oscillations which become strongly nonlinear as we increase A. At a critical value A_c the slave locks in frequency to the master and the output intensity becomes constant in time. A further increase in A will not alter this situation; see Fig. 13.

We choose A such that the slave locks to the master, but not far away from A_c. Then we apply very short pulses to the pumping current of the slave. The reaction of the system is shown in Fig. 14. As in the case of a laser with a double cavity, if the excitation overcomes a critical amplitude and duration then the output intensity shows strong pulsing. Furthermore, smaller excitations produce almost no reaction of the system. All other characteristics of excitability can be measured in this system. We do not give here a complete report because it is not the aim of this paper. We only want to remark that if the field amplitude of the master becomes smaller than A_c, then the intensity becomes unstable and we observe positive pulses in the total intensity. The frequency decreases as we approach the bifurcation point. The value of A_c depends strongly on the detuning between the frequency of the master and the frequency of the slave without injection. For vanishing detuning we did observe neither instabilities nor excitable behavior.

ACKNOWLEDGEMENTS

The author wants to acknowledge the collaborations with M. Giudici, S. Balle, G. Huyet, U. Nespolo, and P. Coullet on the subject and on the analysis of the results presented in this article. Furthermore, the author thanks B. Krauskopf for helpful discussions and C. Taggiasco for helping with the preparation of this manuscript.

REFERENCES

1. Wiener, N., and Rosenblueth, A., *Archivos del Instituto de Cardiologia de Mexico* **16**, 205 (1945).
2. Hodgkin, A.L., and Huxley, A.F., *J. Physiol.* **117**, 500 (1952).
3. Zaikin, A.N., and Zhabotinsky, A.M., *Nature* **225**, 535 (1970).
4. Frisch, T., Rica, S., Coullet, P., and Gilli, J., *Phys. Rev. Lett.* **72**, 1471 (1994).
5. Coullet, P., Frisch, T., Gilli, J.M., and Rica, S., *Chaos* **4**, 485 (1994).
6. FitzHugh, R., *Biophys. Journal* **1**, 445 (1961); Nagumo, J., Arimoto, S., and Yoshizawa, S., *Proc. I.R.E.* **50**, 2061 (1962).
7. Plaza, F., and Velarde, M., *Int. J. Bif. Chaos* **6**, 1873 (1996).
8. Giudici, M., Green, C., Giacomelli, G., Nespolo, U., and Tredicce, J.R., *Phys. Rev.* **E 55**, 6414 (1997).
9. Coullet, P., Daboussy, D., and Tredicce, J.R., *Phys. Rev.* **E 58**, 5347 (1998).
10. Dubbeldam, J.L.A., Krauskopf, B., and Lenstra, D., *Phys. Rev.* **E 60**, 6580 (1999).
11. Narducci, L.M., Tredicce, J.R., Lugiato, L.A., Abraham, N.B., and Bandy, D.K., *Phys. Rev.* **A 33**, 1842 (1986).

12. Adler, R., *Proc. IRE* **34**, 351(1946).
13. Abraham, N.B., Mandel, P., and Narducci, L.M., *Progress in Optics* **XXV**, 1 (1988).
14. Arecchi, F.T., Meucci, R., Puccioni, G.P., and Tredicce, J.R., *Phys. Rev. Lett.* **49**, 1217 (1982).
15. Meucci, R., Poggi, A., Arecchi, F.T., and Tredicce, J.R., *Opt. Commun.* **65**, 151 (1986).
16. Tredicce, J.R., Arecchi, F.T., Lippi, G.L., and Puccioni, G.P., *JOSA B* **2**, 173 (1985).
17. Lugiato, L.A., Narducci, L.M., Bandy, D.K., and Pennise, C.A., *Opt. Commun.* **46**, 64 (1983).
18. Mullet, J., and Mirasso, C., *Phys. Rev.* **E 59**, 5400 (1999).
19. Lu, W., Yu, D., and Harrison, R.G., *Phys. Rev.* **A 58**, R809 (1998).
20. Tredicce, J.R., Balle, S., Coullet, P., Giacomelli, G., and Giudici, M., *Proc. SPIE* **3944**, 444 (2000).
21. Giacomelli, G., Giudici, M., Balle, S., and Tredicce, J.R., *Phys. Rev. Lett.* **84**, 3298 (2000).
22. Wieczorek, S., Krauskopf, B., and Lenstra, D., *Opt. Commun.* **172**, 279 (1999).
23. Solari, H.G., and Oppo, G.L., *Opt. Commun.* **111**, 173 (1994).
24. San Miguel, M., Feng, Q., and Moloney, J.V., *Phys. Rev.* **A 52**, 1728 (1995).

Coherence, Chaos and Communication: Exploring and Applying Nonlinear Laser Dynamics

Rajarshi Roy

Department of Physics and
Institute for Physical Science and Technology
University of Maryland
College Park MD 20742

Abstract. Surprising conceptual connections between quite different physical systems are traced. Nonlinear dynamics is the common formalism that provides the basis for understanding many puzzling observations, including laser instabilities. Experiments in our laboratory to investigate coupled laser systems and communication with chaotic waveforms are described.

LASERS, COHERENCE AND CHAOS

A theme that many speakers at this Spring School have focused on is the nonlinear dynamics of laser devices relevant to communications. Somewhat unexpectedly, in our discussions on the final day of the workshop, it became very clear that most of us are also inspired in our scientific studies by the dynamics of living matter in its immense, awe-inspiring variety. We hope to contribute in some way to an understanding of phenomena that occur in biological systems, even if only by analogy. The goal of this article is to outline the amazing confluence of ideas and concepts that often occurs in the course of scientific research, and illustrate this through examples that relate to nonlinear laser dynamics.

In our studies of laser instabilities, we hope that the principles of nonlinear dynamics will guide us in revealing connections and relationships between what may appear to be unrelated systems. It is often necessary to use the mathematical tools of nonlinear dynamics to identify and explore the astonishing range of dynamical spatio-temporal behaviors exhibited in different parameter regimes and different conditions of operation of the same system. This is true even for systems that are described by (deceptively) simple sets of difference or differential equations.

In 1950, von Neumann and Ulam studied difference equations and may have encountered some aspects of "chaotic" dynamics [1]. In ref. [1], Rudolf Kalman reminisces about von Neumann and Ulam, and describes his own mathematical explorations, in the mid-fifties, of deterministic piecewise linear map equations (the tent map in particular) that touched upon chaos, in the context of the control theory of discretely sampled systems. He was interested in showing that there was no analog of the Poincare-Bendixson theorem for difference equations even in one dimension [2].

The events leading to the invention of the maser and the laser in roughly the same years are beautifully described by Charles Townes in his recent book, *How the Laser Happened*. His major motivation was to find an intense source of monochromatic (coherent) light for spectroscopy [3]. Townes emphasizes that we often do not have any idea how an invention or discovery may be used in the future, and the path leading to the invention itself may be quite random and full of apparent coincidences. His inspiration for writing down the equations for maser action came from a biologist studying population dynamics. He also relates that he told von Neumann about his idea for the maser at a cocktail party. Von Neumann's initial skepticism turned into acceptance and enthusiasm by the end of the party. It is interesting to note that von Neumann considered the possibility of coherent laser action in semiconductors in 1954, in a letter to Edward Teller [4].

In the earliest days of lasers, scientists and engineers noted that continuously pumped ruby lasers sometimes behaved in a highly unstable manner – the light emitted consisted of a series of intense spikes, of irregular amplitude and occurrence - even though the pump excitation was constant [5]. These observations took a long time to explain, since neither the computational tools nor the mathematical concepts necessary for their understanding were available at the time. Computers were instrumental in the development of a new discipline - nonlinear dynamics - initiated by Poincaré in his studies of the stability of the solar system, which today gives us the means for understanding such phenomena. Initially regarded as the very embodiment of stability and regularity, the solar system has provided some of the most striking examples of chaotic dynamics [6]. It seems that this situation is found to occur frequently in science. The maser and laser were devices that made possible the transition from light sources dominated by *incoherent, spontaneous emission* of radiation by atoms and molecules to those governed by *coherent, stimulated emission.* However, it was found before too long that even these most stable of oscillators could display an astonishing variety of dynamical instabilities and chaos. Today, nonlinear dynamics provides a unified framework for exploring and understanding irregularities in the dynamics of planetary systems, the chaotic dynamics of lasers, as well as the dynamics of heart fibrillation and epileptic seizures.

A major motivation for the creation and development of computers was the hope of long term weather prediction and control. In fact, John von Neumann dreamed that computers would change the world we live in and played a major role in many aspects of their birth and applications. He is quoted as saying - "All processes that are stable we shall predict. All processes that are unstable we shall control" [7]. Today, it is still a major goal of scientists and engineers interested in nonlinear dynamics to predict and control spatio-temporal phenomena in a great variety of systems ranging from the weather to the human heart and brain, as well as lasers, molecules and atoms. The processing of large amounts of information is clearly necessary to achieve these aims.

Edward Lorenz, a meteorologist interested in weather prediction, used an early computer and studied a simple model system of nonlinear ordinary differential equations to describe a cell of fluid heated from below. These equations were derived from the Navier -Stokes equations of hydrodynamics by making appropriate approximations [8]. He discovered through numerical computations that the solutions were irregular in time and highly sensitive to tiny differences in initial conditions; we now recognize these features as important signatures of chaotic dynamics [9]; see also the chapter by *Broer and Krauskopf.*

A quite surprising discovery and unexpected connection between the Lorenz system and lasers was made by Hermann Haken in 1975 [10]. He showed that the three equations often used to describe the time evolution of the laser field, inversion and polarization (dipole moment) of the active medium were similar in form to those derived by Lorenz if one made the right transformations of variables and parameter combinations. The nonlinear ordinary differential equations typically used to describe laser operation comprise a simplified model derived from the partial differential equations (Maxwell-Bloch equations) of quantum optics; a process analogous to the derivation of the Lorenz equations from the Navier-Stokes equations.

Haken's discovery initiated a major effort to identify similarities between the dynamics of fluids and of optical systems. A comprehensive review of these similarities (and of the inherent differences) is given by Newell and Moloney [11]. The discovery of chaotic dynamics in laser emission demonstrates that even reasonably coherent, stimulated emission may occur in an irregular, aperiodic fashion. Often, there are several orders of magnitude separating the time scales for the optical oscillations (10^{14} Hz) and the chaotic dynamics of the slowly varying field envelope. Under some conditions, e.g., feedback of light into the laser cavity from a distant reflector, this separation may be drastically reduced. *Coherence collapse* may occur, and the laser field may exhibit a reversion to much larger linewidths [12].

Lorenz chaos was not easy to find experimentally in lasers. Many experimental groups began a search, and it was eventually determined that only very exotic systems such as the ammonia laser possess the right combination of decay rates for the field, inversion and polarization to display Lorenz-like chaos [13]. In the search process, however, it was found that even the most widely used laser systems could display chaotic dynamics under certain conditions. For example,

- Reflective feedback of light into the laser cavity from an external reflector could cause instability and the coherence collapse phenomenon referred to above.
- Multimode operation of the laser could result in chaos and instabilities, both temporal and spatio-temporal.
- Modulation of the laser gain or loss could initiate chaotic pulsations and spikes of laser intensity.
- Intracavity nonlinearities could result in chaotic dynamics.
- Injection (one-way) or mutual coupling of lasers can destabilize them.

The more one looked, the more situations one found where lasers could display instabilities and chaos [14]. The situation was rather similar to the study of planetary systems - what appeared initially to be the most stable periodic systems revealed a plethora of possible irregular behaviors when studied carefully.

Much of our current interest in laser nonlinear dynamics is motivated by their use as light sources for communication and information processing. The concurrent and interlinked development of electronic and optical materials, devices and systems are primarily responsible for these applications, which have had an enormous global impact. Semiconductor and fiber lasers are important light sources for these applications. Dynamical instabilities observed in these systems have therefore been studied extensively and much effort has been made to eliminate them and develop stable sources of light for these applications.

Though electronic and optical systems for communication have been extensively developed over the past century, the processes by which information is communicated in biological systems remains largely a mystery. Neurons emit irregular trains of pulses (spikes) that encode information from visual, auditory, olfactory and other sensory stimuli; other neurons are coupled to these signals and act as receivers. The way in which information is encoded in neuronal spike trains and how it is decoded and interpreted in biological systems is not well understood [15] The dynamics of coupled neurons has begun to be studied, and nonlinear dynamics is clearly relevant in understanding this fascinating problem [16].

Now that we have seen how nonlinear dynamics, computers, the weather, lasers and neurons are all intimately connected (!?), we will trace another set of conceptual connections that is directly relevant to the study of dynamical systems. It is obvious that to study the dynamics (or time evolution) of any kind of system, we need to measure time accurately. While Galileo measured the time taken by objects to fall from the leaning tower by counting the beats of his pulse, Christian Huygens developed the most accurate pendulum clocks of his day and recommended their use for navigation to King Louis XIV of France [17]. The study of dynamical systems may thus be said to have begun with Huygens, for it was after him that really accurate and consistent measurements of dynamical behavior could be made.

COUPLED LASER DYNAMICS: MUTUAL COHERENCE AND CHAOS

Huygens also made the acute observation that two slightly out of step pendulum clocks synchronized when placed close to each other, but not when kept far apart [18]. This discovery of the entrainment of coupled clocks has been seminal in the development of synchronized oscillators of all types. Today, some of the most stable oscillators in existence are masers and lasers. It would thus seem natural to perform a modern version of Huygens' experiment with coupled lasers. The original suggestion

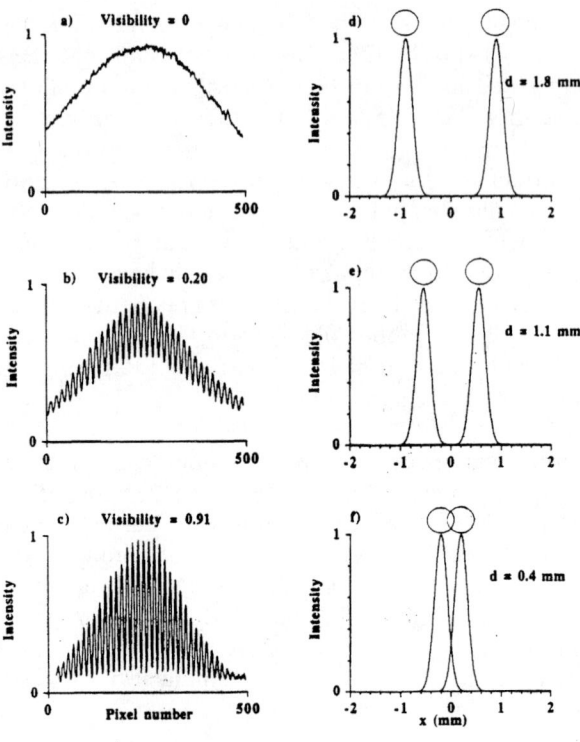

FIGURE 1. Interference patterns for different coupling strengths of the two lasers. Incoherence, partial coherence and full mutual coherence are seen in the three plots. The Gaussian beam profiles are shown to the right, indicating the amount of beam overlap. The circles are the ($1/e^2$) diameters of the beams [19].

for this experiment was made to me in 1979 by Peter Ortoleva, a physical chemist, who was interested in oscillatory Belousov-Zhabotinsky reactions.

A dozen years later, Larry Fabiny and I set out to do such an experiment, in collaboration with theorists Pere Colet and Daan Lenstra [19]. These experiments were refined and made much more precise by Michael Moeller and Scott Thornburg [20]. We decided that if we could generate two lasers parallel to each other in the same crystal, then we could study their interaction as a function of coupling strength by varying the distance between them, as described below. Nd: YAG lasers are stable and well studied lasers with well understood dynamics and established models; we chose this system for our experiments. The oscillation frequency of the Nd: YAG lasers was of the order 10^{14} Hz, and their detuning of the order of 1 MHz. To have almost identical resonators we decided to generate two Nd:YAG lasers in the same crystal with two pump beams from an Argon laser at 514.5 nm. The beam profiles for the pump and laser beams and the distance between them were carefully mapped by measurements with a beamscan instrument – this is a device with a photodetector behind a cylinder with a narrow rotating slit. The overlap of the Gaussian profiles of the laser beams is estimated from these measurements and gives us a measure of the coupling strength of the two lasers. The detuning of the laser oscillation frequencies was varied by slightly tilting the common mirror of the laser resonator.

Huygens was able to detect whether his pendula were synchronized just by looking at them. Our oscillator frequencies were optical, so we observed the phase relationship of the lasers by recording the interference fringes formed by the overlap of the laser beams on a CCD camera. Our experiment was, in fact, also a modern version of Young's famous interference experiment from the early eighteen hundreds, where he illuminated two pinholes in an opaque screen with a point source of light. In modern terminology, Young's experiment measures the *mutual coherence* of the light from the two pinholes. The visibility or contrast of the fringes formed by the superposition of the beams from the two light sources (pinholes or two lasers) gives a direct measure of the correlation function of the light fields. We explored what happened when we tried to form interference fringes, first with two independent lasers, and then gradually increased their coupling. Our experimental results are shown in Fig. 1. There were no fringes detected when the coupling was negligible (Fig. 1(a)). It should be noted that there are always fringes formed for times of the order of the coherence time of the laser light (microseconds). These fringes were washed out since the phase relationships of the lasers changed on the longer time (milliseconds) over which the CCD camera recorded the interference pattern. As the coupling approached a part per million overlap for the laser beams, partial fringes were recorded by the camera (Fig. 1(b)), providing evidence that the laser phases were locking intermittently. For stronger coupling, fully modulated fringes were recorded (Fig. 1(c)); the lasers were now consistently locked in frequency and phase, and stable, high contrast fringes were formed.

Pere Colet and Daan Lenstra developed the theoretical and computational models for this system and showed that if we assumed that the intensities and gains of the coupled laser system remained approximately constant, then the phase dynamics of the lasers could be described by a single Langevin equation. They showed that the equation for the phase difference of the coupled lasers was identical in form to that for the phase of a Josephson junction [21]; in fact, it was an Adler equation [22] with noise sources that represented the spontaneous emission present in both lasers. The detuning of the lasers and the coupling strength were the two relevant parameters that determined the nature of the phase dynamics. When the detuning (measured in dimensionless units of the cavity round trip time) was large compared with the coupling strength (measured by the Gaussian beam overlap), the phase evolved rapidly. The analogy in this case is to that of an overdamped particle rolling down an incline. When the detuning and coupling were comparable, the potential in which the particle moves becomes a modulated washboard. The particle is temporarily trapped when it is in one of the shallow wells, but noise often kicks it out of the well and it slips to the next well. This lock-slip dynamics of the phase results in partial visibility of the fringes. Finally, when the detuning is very small compared with the coupling strength, the particle gets trapped for very long times in one of the potential wells, and we have phase-locked operation of the two oscillators.

Thomas Erneux soon pointed out to us that we should also observe the dynamics of the laser intensities, particularly in the regime of partial fringe visibility, since he and his collaborators Tom Carr and R-D. Li predicted there would be an amplitude instability, when the coupling strength and detuning were comparable to each other. They showed that this instability existed by relaxing the too-stringent constraints that we had placed on the intensities and gains of the lasers. If these quantities were not assumed to be constant in the model equations, but only to be equal to each other for

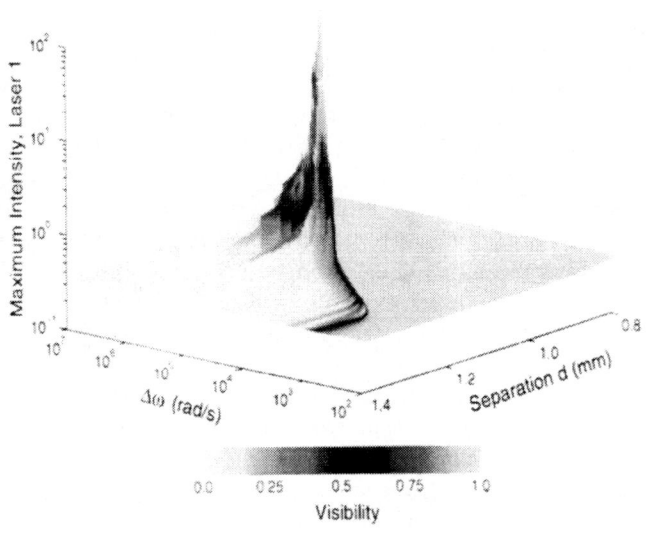

FIGURE 2. Amplitude of two coupled lasers shown as a function of the detuning and separation of the lasers. The visibility of the fringes is shown on a grayscale. The extent of amplitude fluctuations is shown by the height of the graph. The pump parameters of the lasers are taken to be slightly different in the computations [20].

the two lasers, then one obtained a set of three equations – one each for the amplitude, gain and phase difference of the lasers. The coupling of the lasers now appeared in an effective loss modulation term in the equation for the amplitude. They concluded that the amplitude instability resulted from a subharmonic resonance of this modulation frequency with the *relaxation oscillation* frequency of the lasers.

Relaxation oscillations in a laser arise from the exchange of energy between the field and the atoms or molecules of the active medium. If the system is perturbed away from the steady state by a fluctuation, it relaxes back to the steady state, performing damped oscillations at a frequency determined by a combination of the decay time of the optical cavity of the upper lasing level of the atoms. A modulation of the laser loss or pump excitation at approximately the relaxation oscillation frequency or its harmonics and subharmonics may destabilize the laser, as described above and in the next section. Many systems in nature display relaxation oscillations, including, on a very macroscopic scale, the geysers of Yellowstone National Park in the U.S.A. Old Faithful is a particularly famous geyser; its somewhat regular eruptions have been studied over many years [23].

A detailed numerical computation based on the laser equations without the approximations above confirmed their predictions fully. The results of this computation by Scott Thornburg showed the amplitude instability that could occur in this coupled laser system in a striking way, reproduced here as Fig. 2 [24]. In a careful experiment, Thornburg and Moeller showed that the intensities of the lasers displayed intermittent bursting exactly in the range of coupling strengths predicted by Erneux. Figure 2 reveals something else that is very interesting - the coupled lasers can be mutually coherent in phase (have a visibility close to 1) even while their amplitudes are unstable or chaotic (show large intensity spikes). It shows the complexity of dynamics possible with just two coupled nonlinear oscillators.

TAKING ADVANTAGE OF CHAOS: SYNCHRONIZATION OF CHAOTIC DYNAMICS

Chaotic dynamics has been viewed so far as something that is very fundamental and basic to the nature and behavior of dynamical systems. The geometric structures and patterns that one may construct from studies of chaotic dynamics are indeed very beautiful. But one may look at chaos from a very different perspective and ask - what good is it? Is it only something that complicates our lives, or is there some reason for its existence? Does it possibly serve a useful function in some way? It is only recently that scientists have begun to ask this question and think of possible applications of chaos. Of course, such an approach also has the potential for giving us insight into the role of chaotic behavior in biological systems, where it may well serve a useful role.

The discovery of *synchronization of chaotic systems* over a decade ago gave us hope for new applications of chaotic behavior. The possibility that chaotic systems could synchronize seems quite counterintuitive at first, since the signature of chaos is an exponential divergence of initially close trajectories in the phase space of the dynamical system. It was therefore quite a surprise when Pecora and Carroll [25] showed that chaotic systems could synchronize with one another under appropriate conditions of coupling. This may be viewed as a generalization of the entrainment or synchronization of Huygens' coupled clocks. Now, the synchronized motion considered was not periodic, but chaotic.

One of the first examples of synchronization of chaos was that of coupled Lorenz systems. In this example, there exists a one-way coupling of a "transmitter" to a "receiver" system. The transmitter x-variable is sent to the receiver and replaces the value of the receiver x-variable in its y and z equations. Now, the surprising result is that even when the transmitter and receiver have very different initial conditions, the receiver ends up synchronized to the transmitter; i.e., the values of its y and z variables evolve to become identical to those of the transmitter y and z variables. Sending only the x variable from the transmitter to the receiver synchronizes the dynamics of all the variables of the two systems.

Pecora and Carroll suggested that this property of synchronization could be used to transmit information from one chaotic system to another. The information could be analog or digital. The scheme they suggested consisted of adding a message signal to the z variable of the transmitter and sending the sum to the receiver, along with the

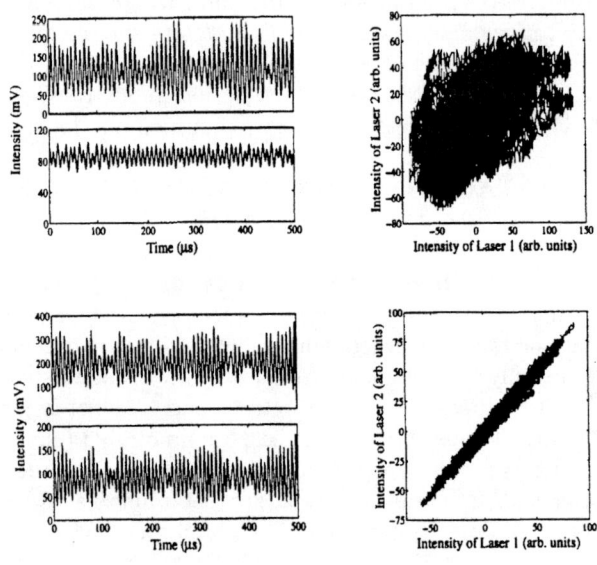

FIGURE 3. Intensity time traces of unsynchronized (top) and synchronized (bottom) operation of mutually coupled lasers [24].

x variable. The transmitted x variable synchronized the receiver, and the z variable thus generated by the receiver was subtracted from the sum of message and z variable transmitted to recover the message [26]. A further development by Cuomo and Oppenheim [27] was to show that there was no necessity to transmit the x variable for synchronization. Just the z variable with the message added could be used both for synchronization and message recovery if the message amplitude was not too large compared to the chaotic variable. They demonstrated this elegant scheme for chaotic communication through experiments on electronic circuits that emulated the dynamics of the Lorenz system.

Very soon after the publication of Pecora and Carroll's paper, Winful and Rahman [28] showed numerically that it should be possible to observe synchronization of chaotic semiconductor lasers. Unlike the coupling required for the Pecora-Carroll system, which is very difficult to implement in an optical system, these lasers were coupled to each other by sharing light between their cavities. This prediction remained untested since the time scale for evolution of the semiconductor light intensities was in the picosecond range and made direct observation of the dynamics quite difficult.

We realized in 1993 that our system of mutually coupled Nd lasers described above was ideal to test chaotic synchronization, since their dynamics evolved on a microsecond time scale, and could therefore be recorded accurately with simple photodetectors and a digital oscilloscope. The lasers could be made chaotic in their behavior by modulating their losses or pumps at a frequency close to the relaxation oscillation frequency. Our experiments showed clearly that chaotic laser synchronization was indeed possible with very small coupling between the lasers, and this synchronization was robust and persistent over long times [29]. An example of such synchronized chaos is shown in Fig. 3. Unknown to us, and at the same time as we were doing our experiments on synchronization, Sugawara et al. in Japan performed experiments on CO_2 lasers that showed synchronization of that system in a one-way coupling scheme [30]. The next step was to demonstrate optical communication with chaotic lasers.

Pere Colet and I worked on a simple numerical model of Nd:YAG lasers to try to test a scheme for digital communication with chaotic lasers [31]. Loss modulation at the appropriate frequency produced an irregular sequence of pulses of varying amplitude. These pulses were then attenuated by 5% above or below a set level depending on whether we wished to send a zero or one, and then injected into the receiver, which was an identical laser with the same loss modulation. The amazing result of our simulations was that the receiver laser emitted a sequence of pulses that reproduced quite accurately the transmitter pulse sequence *before* the modifications for digital information encoding were performed. This meant that subtraction of the output of the receiver from its input reproduced the encoded sequence of bits. We simulated mismatches between the receiver and the transmitter parameters, found that the communication scheme was quite robust, and succeeded in sending large numbers of bits of information and recovering them accurately. This numerical experiment revealed two important facts. First, synchronization of chaos could be used as a means of recovering information encoded on a chaotic waveform (consisting of an irregular pulse train, in this case). Second, we needed to find a laser system that was much faster in its chaotic dynamics; the Nd: YAG laser dynamics was quite slow (microsecond time scale) for digital communications in practical systems.

ERBIUM LASER DYNAMICS: COMMUNICATION WITH CHAOTIC WAVEFORMS

Quinton Williams had been studying erbium doped fiber ring lasers in our lab, and found that it was relatively simple to observe irregular intensity fluctuations in this laser system. These fluctuations were clearly faster than the instruments we could find to measure them with; 1 nsec single-shot sampling with a digital oscilloscope was the best we could do at the time. We tried to understand the dynamics of these lasers, and developed a model based on delay-differential equations that appeared to reproduce dynamics similar to what we observed in the laboratory [32].

Delay-differential equations were a good compromise - they included basic aspects of the fiber ring laser that could not be described by a set of ODEs, which are normally derived from the Maxwell-Bloch PDEs with the assumption that the round trip time of the light in the laser cavity is very short compared to the time scale for fluctuations of the field. Our fiber cavity round trip time was of the order of a hundred nanoseconds, and the field fluctuations were clearly on a much faster time scale, so we used delay equations to describe the field dynamics. Two coupled equations accounted for the coupled polarization components of the field, which was observed not to be linearly polarized. The population inversion was described by a coupled differential equation. The decay time for the upper lasing level in erbium is about 10 ms - very long indeed compared to the field fluctuations. A visitor to our lab in Atlanta, Jordi Garcia-Ojalvo, extended the computations and added sources of noise to the model to simulate many of the observed features of the laser dynamics [33]. One such feature was the slow drift of the irregular pattern that repeated over many round trips but gradually drifted and changed into different patterns over the course of milliseconds. In this process, he also developed a nice extension of the Ikeda model to include noise sources [34]. Henry Abarbanel and his group have made significant advances to the modeling of these erbium doped fiber ring lasers, and produced increasingly precise descriptions and comparisons of model predictions with experimental observations [35].

We decided that we would try to build an experiment to communicate information using chaotic erbium lasers, since these lasers emit light at 1.55 microns and are widely used in communication applications. This meant developing a technique to encode information on the chaotic waveforms from the transmitter, and then designing a receiver to synchronize and recover the information from the waveforms.

We began a collaboration with Henry Abarbanel and Steve Strogatz to accomplish this work; our collaboration has lasted over a period of several years and has been most fruitful in many ways. As Greg VanWiggeren began to set up the experimental system, our groups decided to meet three times a year at our respective locations. At Cornell, in the summer of 1997, we became aware through a news report in Science- "Chaos Keeps Data Under Wraps" - that a group in France led by Jean-Pierre Goedgebuer was working on optical communication with chaos as well [36]. Their effort was based on tunable semiconductor lasers, and they used an electro-optic feedback loop where the optical signal was converted into an electronic signal with a photodetector. A major difference in approach was that they encoded the information on frequency fluctuations of the tunable laser. We could conjecture what they were trying to do, but none of the details of their system were available to us since there were no published results at this point – so we were tantalized by the report and went back to work with a great deal of motivation.

One of our first important steps, as Lou Pecora, a guest at our Cornell meeting,

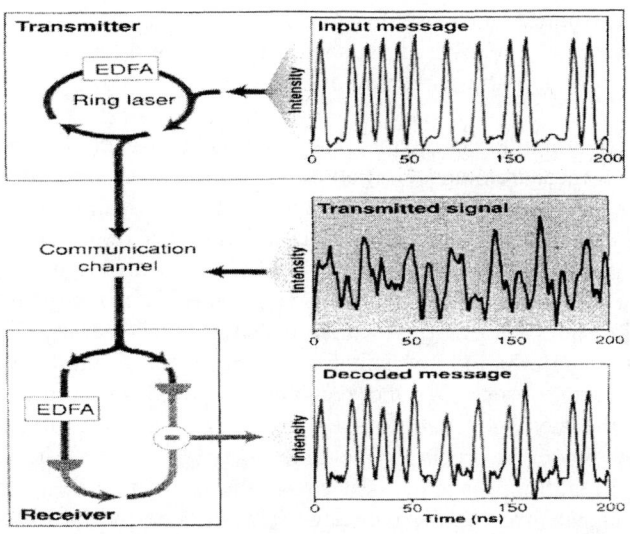

FIGURE 4. Injecting a message into the transmitter laser "folds" the data into the chaotic amplitude fluctuations. The receiver reverses this process, thereby recovering a copy of the message. EDFA: erbium doped fiber amplifier.

pointed out, was to realize that it was not necessary to synchronize two *lasers* for the recovery of the information. Only the transmitter needed to be a laser. An identical amplifier to that in the transmitter could be used as an open-loop receiver, as pointed out by Volkovskii and Rulkov some years ago [37]. This made matters less difficult and more robust as far as communication was concerned. The transmitter and receiver time delays still needed to be quite precisely matched, to a small fraction of a nanosecond. To introduce the message information, we used an intensity modulated tunable single frequency semiconductor laser to inject light into the erbium ring laser. The message consisted of a simple square wave modulation at 10 MHz. The receiver consisted of an EDFA (erbium doped fiber amplifier) with pump excitation to match that of the transmitter and a length of fiber to create a delay equal to the round trip time in the ring of the transmitter. The transmitter output consisted of irregular intensity fluctuations which contained the information, but this was not evident through a simple examination of the signal. The receiver contained two detectors - one directly detected the signal transmitted, and the other detected the waveform after the signal was sent through the EDFA and fiber delay. The output of the second detector was subtracted from that of the first to obtain the information transmitted, and the result reproduced the square wave message reasonably well [38]. Our experiments were soon extended to much higher bit rates, limited only by our detection electronics and oscilloscopes. The chaotic waveforms used for transmission as well a recovered message consisting of a random bit sequence at 125 Mbits/sec are shown in the schematic of Fig. 4 [39].

It is interesting to compare this method of detection with conventional autocorrelation measurements. In that case, two detectors measure the signal with a variable delay between them. The product of these values is averaged over an ensemble of measurements and the autocorrelation function so constructed provides us with insight into the dynamics. The purpose here is different. For communication purposes, we are interested not so much in the specific waveforms generated by the system as in the perturbations which constitute the message. Our goal is to reconstruct the perturbations by suitably observing the irregular waveforms emitted by the transmitter laser. At the receiver, we need to unfold the perturbations (injection of the message) made to the system. This is done with two detectors that have a delay which is precisely the time taken for a signal to travel around the transmitter. In addition, it is necessary to replicate the nonlinear transformation that the transmitter performs on the message. This is done by the EDFA in the receiver which is carefully designed to be as identical as possible to the one in the transmitter in geometry and operation. If the perturbations are added to the transmitter signal in the ring, then a subtraction of the detector outputs is necessary to recover the message. If the message perturbations are multiplied into the transmitter light field with a modulator in the ring, as was done in later schemes, we needed to perform a division operation at the receiver.

Many different techniques for encoding and decoding messages on irregular waveforms were developed and examined by Greg VanWiggeren over the next

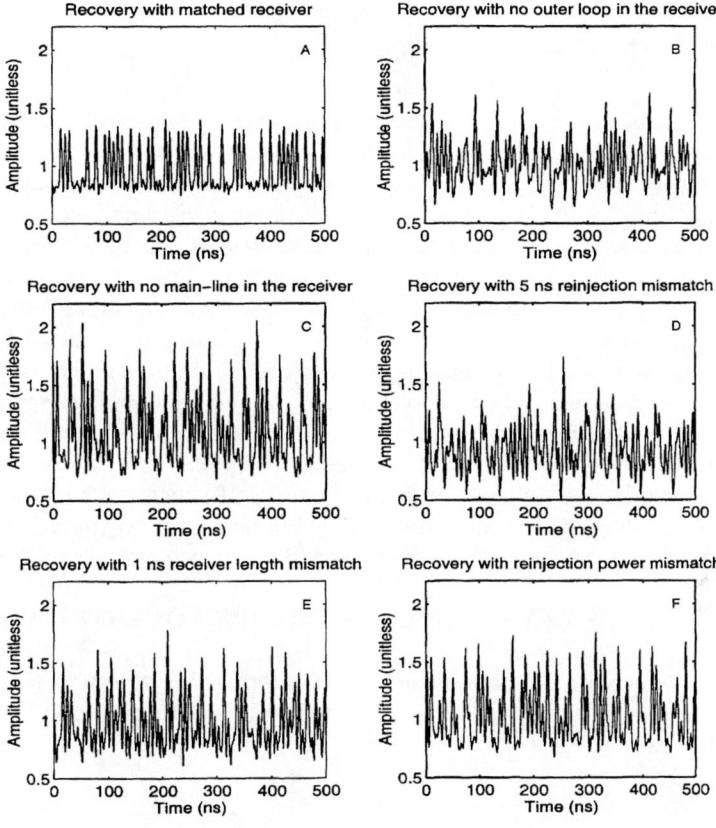

FIGURE 5. (A) Recovery of the random bit sequence when the transmitter and receiver are matched. (B) – (F) illustrate failure to recover the information with a variety of different mismatched parameters and operating configurations (from Ref. [41], with permission of World Scientific Publishing Co Pte Ltd.)

couple of years. He showed that the message could be encoded directly in the transmitter laser without a semiconductor laser to inject light into the cavity. A scheme for creating interference between two EDFAs in the transmitter was tested; for this case a receiver geometry in which the interference was recreated was necessary for message recovery [40]. Greg succeeded in transmitting and recovering messages at hundreds of Mbits/sec quite reliably, and showed that the information could be sent over a 35 km fiber at 250 Mbits/sec in the laboratory. He also showed that mismatched parameters in the transmitter and receiver, or differences in configuration, resulted in non-recovery of the message (Fig. 5) [41]. Our discussions with Henry Abarbanel and Steve Strogatz and their groups led to many clarifications and developments in our understanding of the laser system and chaotic communication schemes.

At the same time as we were working on these experiments, Jean-Pierre Goedgebuer and his colleagues made many advances in their research with semiconductor lasers [42]. Their approach was to use a specially constructed wavelength filter as the nonlinear element in their transmitter feedback loop. The message was coded as a sequence of excursions in frequency by their tunable semiconductor laser, while the intensity of the output remained reasonably constant. It is amusing to note that the related idea of sending messages while switching frequencies at random times with a tunable transmitter was invented by the actress Hedy Lamarr in the early forties along with George Antheil, a maker of player pianos [43]. She is regarded as the inventor of frequency-hopping spread spectrum techniques. Goedgebuer's and our methods for chaotic communication were closely related to frequency hopping and direct sequence spread spectrum methods [44]. A useful characteristic of chaotic systems is the self-synchronizing property of the receivers to the transmitted waveforms. It is not necessary to share a pseudorandom code and externally synchronize transmitter and receiver for message recovery.

NEW DEVELOPMENTS AND OPEN QUESTIONS

Today there are at least a dozen groups around the world working on optical communications with chaotic lasers, experimentally and theoretically. At this time it is important to ask - what are the advantages of sending messages using chaotic waveforms?

A major motivation for several groups is the possibility of enhancing security or privacy in communications by hiding messages in the chaotic waveforms. Of course, as new methods for encoding information are developed, so are methods for breaking the codes. Immediately after the early work of Cuomo and Oppenheim and others, Perez and Cerdeira showed that messages masked by low dimensional chaotic dynamics could be recovered [45]. After Pere Colet and I published our paper on digital communication with chaotic lasers, Paul Alsing, Tom Gavrielides and Vassilios Kovanis showed that a good computational model of the Nd:YAG laser system and time-delay embedding techniques could be used to recover the message form the chaos [46]. Kevin Short and his colleagues showed recently that they could construct a reasonably accurate model of the laser system used in ref. [38] and recover the information encoded from one of the waveforms that Greg VanWiggeren provided them [47]. This success in code-breaking is an excellent sign that good progress has been made in modeling the dynamical systems used for encoding messages, at least for certain parameter ranges.

The question of cryptographic security for chaotic communications has, however, not really been addressed (to my knowledge) except in an anecdotal sense and remains an open topic for discussion and research. It is well worth remembering what

Shannon had to say about security as early as 1949 [48]. He identified three distinct possible elements – concealment of a message, the use of customized or special hardware, and cryptographic security – in any scheme for secret communication. Chaotic communications possess aspects of all three elements. It is clear that any digital cryptographic scheme can be used concurrently with the techniques reviewed here for chaotic communication.

Perturbations (temperature fluctuations and mechanical stresses, for example) of the optical fiber communication channel are usually very disruptive to communications. Greg VanWiggeren has recently shown how to encode information on the polarization fluctuations of an erbium ring laser and recover it with a suitably designed receiver. He recently developed a technique for measuring the Stokes parameters of the light from the erbium fiber lasers [49]. He found that the polarization state of the light fluctuates from point to point on the Poincaré sphere that used to visualize the polarization dynamics. A major difference of this scheme with previous ones for polarization shift coding is that it uses dynamical changes of polarization states to encode bits - previous methods assigned a given polarization to a bit. The use of polarization to encode information increases the number of variables that may be used for communication with light - amplitude, frequency and polarization may all be employed. It was recently pointed out to me that Carl Sagan had suggested that polarization be used as a variable for information transmission in his science fiction novel *Contact*, though not with a chaotic system [50]. A potential advantage of the scheme that Greg has demonstrated is that a message can be recovered even when the communication channel fiber is perturbed (stressed or twisted) during transmission.

Another possible advantage of chaotic optical communication may be in the realm of multiplexing. How does one send several messages over the same channel and recover them separately? Many schemes are used in communications today - wavelength division and time division multiplexing are well developed [51]. A very interesting possibility recently raised by Tsimring [52] and by Liu and Davis [53] is illustrated by dual synchronization - two chaotic waveforms generated by different transmitters are added to give a single scalar signal. This signal is communicated to two matched receivers. A feedback scheme is used, and the two receivers synchronize separately to the two transmitted waveforms based on their dynamical characteristics. Can information be recovered from two carriers that are transmitted by a single scalar variable using the dynamical properties of the transmitters and receivers?

An obvious extension of the work on chaotic communications reviewed so far is to spatio-temporal carrier waveforms and messages. Several groups have investigated the synchronization of chaotic spatio-temporal systems, but little work has been done to explore the nonlinear dynamics of communication with such systems. This is how biological systems seem to work - the inherent parallelism of optics is exploited exquisitely by our visual system, for example.

Generalized synchronization, where two systems that are not necessarily identical to each other but have a functional relationship, has yet to be investigated in detail. A couple of experiments have examined generalized synchronization in optical systems [54] but no applications to optical communications have been implemented. Such an approach may provide enhanced security [55]. I should also mention the issue of sporadic or impulsive coupling for synchronization - two systems need not be continuously coupled together to be synchronized [56]. The literature on synchronization has grown rapidly over the past years and there are many mathematical studies of elegant dynamical schemes that have yet to be implemented in optical systems.

Issues of data compression, signal to noise ratio, bit-error rates and practical aspects of communication systems such as error control have hardly been mentioned here. This is not because they are unimportant or uninteresting - they are absolutely crucial to the success of new schemes for chaotic communication and involve beautiful mathematical concepts - but because this would take us too far afield at this time.

Finally, I can safely say that the most interesting new developments in the area of laser nonlinear dynamics will undoubtedly be the most unexpected ones. How these developments will relate to the general goal of controlling spatio-temporal phenomena on both microscopic and macroscopic levels is a fascinating story waiting to unfold. Connections to other fields of science (biology and information theory in particular) and applications in new contexts will surely be made that will reshape even the qualitative ways in which we approach this field.

ACKNOWLEDGMENTS

It is a pleasure to acknowledge support from the Office of Naval Research and the National Science Foundation, and to thank the many talented collaborators over the years without whom none of this work would have been possible.

REFERENCES

1. Kalman, R.E., *IEICE Trans. Fundamentals* **E82-A**, 1686 (1999).
2. Strogatz, S.H., *Nonlinear Dynamics and Chaos*, Addison-Wesley, Reading, Mass. (1994).
3. Townes, C.H., *How the Laser Happened*, Oxford University Press, N.Y. (1999).
4. Von Neumann, J., *Collected Works*, Volume V, p. 420, Macmillan, N.Y. (1963).
5. Nelson, D.F., and Boyle, W.S., *App. Optics* **1**, 181 (1962).
6. Diacu, F., and Holmes, P., *Celestial Encounters*, Princeton University Press, Princeton, (1996).

7. J. von Neumann as quoted in Dyson, F.J., *Infinite in All Directions*, p. 182, Harper and Row, New York (1988).
8. Hilborn, R.C., *Chaos and Nonlinear Dynamics: An Introduction for Scientists and Engineers*, Oxford University Press, New York (1994); Schuster, H.J., *Deterministic Chaos*, p. 233, 3rd edition, VCH, Germany (1995); Lorenz, E.N., "Deterministic nonperiodic flow," *J. Atmos. Sci.* **20**, 130 (1963).
9. Lorenz, E.N., *The Essence of Chaos*, University of Washington Press, Seattle (1993).
10. Haken, H., *Phys. Lett.* **53A**, 77 (1975).
11. Newell, A.C., and Moloney, J.V., *Nonlinear Optics*, Addison-Wesley, Redwood City (1992).
12. Lenstra, D., Verbeek, B.H., and Den Boef, A.J., *IEEE J. Quantum Electron.* **QE-21**, 674 (1985).
13. Weiss, C.O., and Vilaseca, R., *Dynamics of Lasers*, VCH Publishers, New York (1991).
14. See ref. 13 and Van Tartwijk, G.H.M., and Agrawal, G.P., *Prog. In Quantum Electron.* **22**, 43 (1998).
15. Rieke, F., Warland, D., De Ruyter van Steveninck, R., and Bialek, W., *Spikes: Exploring the Neural Code*, MIT Press, Cambridge, Mass. (1997).
16. Rabinovich, M.I., Varona, P., and Abarbanel, H.D.I., *Int. J. Bif. Chaos* **10**, 913 (2000).
17. Huygens, C., *Horologium Oscillatorium*, trans. Richard Blackwell, University of Iowa Press, Ames, Iowa (1986).
18. Letter from C. Huygens to his father, quoted in Sargent III, M., Scully, M.O., and Lamb Jr. W.E., *Laser Physics*, p.52, Addison-Wesley, Reading, Mass. (1974).
19. Fabiny, L., Colet, P., Roy, R., and Lenstra, D., *Phys. Rev.* **A47**, 4287 (1993).
20. Thornburg Jr., K.S., Moeller, M., Roy, R., Carr, T.W., Li R.-D., and Erneux, T. *Phys.Rev.* **E55**, 3865 (1997).
21. Ambegaokar, V., and Halperin, B.I., *Phys. Rev. Lett.* **22**, 1364 (1967).
22. Adler, R., *Proc. IRE* **34**, 351 (1946), reprinted in *Proc. IEEE* **6**, 1380 (1973).
23. Pippard, A.B., *The Physics of Vibration*, p.359, Cambridge University Press, Cambridge (1989).
24. Thornburg Jr., K.S., *Ph.D. thesis*, Georgia Institute of Technology (1998).
25. Pecora, L.M., and Carroll, T.L., *Phys. Rev. Lett.* **64**, 821 (1990); *Phys. Rev.* **A44**, 2374 (1991). See also Fujisaka, H., and Yamada, T., *Prog. Theor. Phys.* **69**, 32 (1983); Afraimovich, V.S., Verichev, N.N., and Rabinovich, M.I., *Radiophys. and Quantum Elect.* **29**, 795 (1986), and Blekhman, I.I., *Synchronization in Science and Technology*, ASME Press, New York (1988).
26. Ditto, W.L., and Pecora, L.M., *Sci. Am.* **269**, 62 (1993).
27. Cuomo, K.M., and Oppenheim, A.V., *Phys. Rev. Lett.* **71**, 65 (1993).
28. Winful, H.G., and Rahman, L., *Phys. Rev. Lett.* **65**, 1575 (1990).
29. Roy, R., and Thornburg Jr., K.S., *Phys.Rev. Lett.* **72**, 2009 (1994).

30. Sugawara, T., Tachikawa, M., Tsukamoto, T., and Shimizu, T., *Phys. Rev. Lett.* **72**, 3502 (1994).
31. Colet, P., and Roy, R., *Opt. Lett.* **19**, 2056 (1994).
32. Williams, Q.L., and Roy, R., *Opt. Lett.* **21**, 1478 (1996).
33. Williams, Q.L., Garcia-Ojalvo, J., and Roy, R., *Phys. Rev.* **A55**, 2376 (1997).
34. Garcia-Ojalvo, J., and Roy, R., *Phys. Lett.* **A 224**, 51 (1996).
35. Abarbanel, H.D.I., and Kennel, M.B., *Phys. Rev. Lett.* **80**, 3153 (1998); Abarbanel, H.D.I., Kennel, M.B., Buhl, M., and Lewis, C.T., *Phys. Rev.* **A60**, 2360 (1999).
36. Hellemans, A., *Science* **276**, 679 (1997).
37. Volkovski, A.R., and Rulkov, N., *Tech. Phys. Lett.* **19**, 97 (1993).
38. VanWiggeren, G.D., and Roy, R., *Science* **279**, 1198 (1998).
39. The author thanks G.D. VanWiggeren and D. Gauthier for the preparation of this figure.
40. VanWiggeren, G.D., and Roy, R., *Phys. Rev. Lett.* **81**, 3547 (1998).
41. VanWiggeren, G.D., and Roy, R., *Int. Jour. Bif. Chaos* **9**, 2129 (1999).
42. Goedgebuer, J.-P., Larger, L., and Porte, H., *Phys. Rev. Lett.* **80**, 2249 (1998); Larger, L., Goedgebuer, J.-P., and Delorme, F., *Phys. Rev.* **E57**, 6618 (1998).
43. Braun, H.-J., *Invention and Technology* **12**, 10 (1997).
44. Peterson, R.L., Ziemer, R.E., and Borth, D.E., *Introduction to Spread Spectrum Communications*, Prentice –Hall, Englewood Cliffs, NJ (1995).
45. Perez, G., and Cerdeira, H.A., *Phys. Rev. Lett.* **74**, 1970 (1995).
46. Alsing, P.M., Gavrielides, A., Kovanis, V., Roy, R., and Thornburg Jr., K.S., *Phys. Rev.* **E56**, 6302 (1997).
47. Geddes, J.B., Short, K.M., and Black, K., *Phys. Rev. Lett.* **83**, 5389 (1999).
48. Shannon, C., *Bell Sys. Tech. J.* **28**, 657 (1949).
49. VanWiggeren, G.D., and Roy, R., *Opt. Comm.* **164**, 107 (1999).
50. Sagan, C., *Contact*, p. 83, Pocket Books, Simon and Schuster, New York (1985). The author thanks M. Swisdak for pointing out this interesting reference.
51. Agrawal, G.P., *Fiber-Optic Communication Systems*, Chapter 7, John Wiley, New York (1997).
52. Tsimring, L.S., and Susschik, M.M., *Phys. Lett.* **A213**, 155 (1996).
53. Liu, Y., and Davis, P., *Phys. Rev.* **E61**, R2176 (2000).
54. Tang, D.Y., Dykstra, R., Hamilton, M.W., and Heckenberg, N.T., *Phys. Rev.* **E57**, 5247 (1998); Terry, J.R., Thornburg Jr., K.S., DeShazer, D.J., VanWiggeren, G.D., Zhu, S., Ashwin, P., and Roy, R., *Phys. Rev.* **E59**, 4036 (1999).
55. Murali, K., and Lakshmanan, M., *Phys. Lett.* **A241**, 303 (1998).
56. Stojanovski, T., Kocarev, L., and Parlitz, U., *Phys. Rev.* **E54**, 2128 (1996).

Vertical Cavity Surface Emitting Lasers for Communications

J. M. Rorison

*Electrical and Electronic Engineering Department,
Merchant Ventures Building, Woodland Road., Bristol BS8 1UB*

Abstract. Vertical Cavity Surface Emitting Lasers (VCSELs) are a novel type of laser in which the lasing light is emitted from the surface of the device, perpendicular to the gain layer, rather than from the edge of the laser, parallel to the gain layer. These lasers show interesting behaviour, particularly involving mode dynamics and polarisation. They show promising characteristics for use in communications applications but their behaviour needs to be further understood before they can reach their potential. This chapter attempts to review VCSELs generally and discuss their use in communications systems.

INTRODUCTION

Vertical Cavity Surface Emitting Lasers (VCSELs) are a fairly novel type of laser whose applications are just being assessed. They offer the potential of a very cheap, low power consumption laser, with a single longitudinal mode output and a circular output beam. In addition, they offer fast modulation speed and may be fabricated into two dimensional arrays. These latter three advantages are of particular application in the communications field. The circular beams are easy to couple into fibres or optical elements and the array nature allows parallel optical interconnections. Their applications in communications are as: fibre optic data links, short range communications links via plastic fibres, and all-optical terabit data networks. In addition, they are also being investigated as components for quantum communications systems, either through optical fibres or through free-space.

To understand why VCSELs are so useful we need to understand the principle of operation and fabrication of these devices compared with the standard edge-emitters. This is undertaken in the first section of this chapter. This is followed by a discussion on the dynamic behaviour of VCSELs, important for communication signal encoding. The issues of optical fibres as the transmission medium are then discussed. The development of VCSELs for 1.3 and 1.55 micron wavelengths, appropriate for transmission through optical fibres, is discussed. The advantages of using resonant cavity structures in detectors is subsequently addressed, as both sources and detectors are required in a communications system. Finally, general communications issues are discussed, such as how to maximise the information rate through the channel by interleaving the information through time division multiplexing or wavelength division multiplexing.

VERTICAL CAVITY SURFACE EMITTING LASERS VS EDGE EMITTING LASERS

Vertical Cavity Surface Emitting Lasers were proposed by the Iga group in 1979 [1]. The principle behind them was to develop a low threshold laser based on the quantisation of the longitudinal mode of the VCSEL cavity. This quantisation of the optical cavity modes arises from creating a mirror with high reflectivity on either side of a short optical cavity containing a gain medium. Because the cavity is short the frequency spacing between adjacent optical modes supported by the cavity is increased. If properly designed the desired optical longitudinal mode of the cavity chosen to be the lasing mode is the only longitudinal mode that overlaps with the material gain of the cavity. Therefore there is no modal competition between different longitudinal modes, meaning that all the current injected into the device goes into this longitudinal mode. This makes for a very low threshold laser. It also somewhat confuses the usual signature of lasing in edge-emitting lasers, the line-narrowing, since this does not occur in VCSELs. In VCSELs all the photons generated go into the lasing mode creating a *thresholdless* laser [2,3]. A schematic figure of a VCSEL is shown in Fig. 1a and compared to a standard schematic edge-emitting Fabry-Perot laser shown in Fig. 1b.

The dimension of an edge emitting laser is typically 200-500 microns in length and 0.25-3 microns in width while the length of the VCSEL is typically 0.01-0.02 microns and the width typically 5-15 microns, defined by it output aperature. The lateral extent of the VCSEL can be viewed as the lateral extent of current flow. The confinement of the current may be done through several methods and results in the 4 main types of VCSELs: *pillar VCSELs* in which the VCSEL is defined by etching away the material surrounding the top mirror, *proton implanted VCSELs* in which the neighbouring material is damaged through implanation by protons thereby reducing its ability to conduct carriers, *oxide-confined VCSELs* in which aluminium containing layers are inserted in the structure and controllably oxidized to reduce conductivity and channel carriers, and *heterostructure VCSELs* in which layers are etched away and other layers subsequently grown to create a structure with lateral and longitudinal control. A schematic of some of the VCSEL structures are shown in Fig. 1a.

In addition, altering the surrounding material not only affects the carrier flow but alters the refractive index which alters the optical mode of the structure. Generally in VCSELs the fully three-dimensional optical mode is modelled by factoring it into a longitudinal and a transverse part. The longitudinal modes are well separated in frequency but the transverse modes are close together in frequency due to the large lateral extent of the standard VCSEL. By varying the lateral profile of the refractive index (ie. air pillar VCSEL, oxide VCSEL or heterostructure VCSEL), the transverse mode behaviour can be altered.

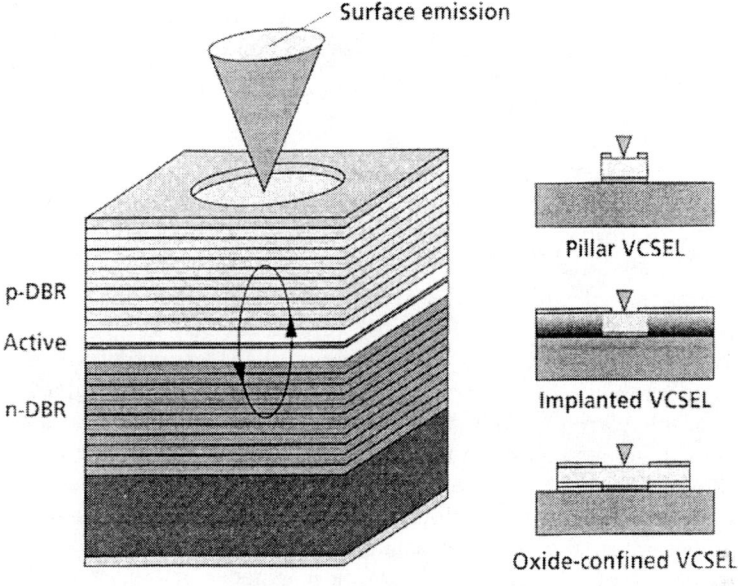

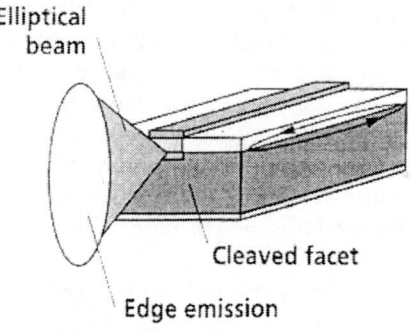

FIGURE 1. Comparison of a schematic VCSEL (a) with a schematic Fabry-Perot edge-emitting laser (b) [4] (© 1998 IEEE).

To fabricate a VCSEL two highly reflective mirrors, a cavity and an active gain region are required. It is possible to make very accurate thickness semiconductor structures using molecular beam epitaxy (MBE) or metal-organic chemical vapour deposition (MOCVD) semiconductor growth systems, which grow the materials using a process which lays down one atomic layer at a time. The thickness of the layers can therefore be carefully controlled. Active gain mediums have been well studied for the edge-emitting lasers. The problem with many VCSEL devices is the fabrication of high quality mirrors. In edge-emitting lasers the semiconductor/air interface has a reflectivity of about 30% which is usually sufficient since the optical wave is in contact with the gain material throughout its passage in the cavity (as shown in Fig 1b). Therefore the photon wave only needs to spend a few passes within the cavity for gain to exceed loss and for lasing to occur. In the VCSEL, because of the small width of the gain region, the photon wave must pass many times through the gain region for the gain to exceed the losses and for lasing to occur (as shown Fig 1a). This necessitates high reflectivity mirrors of greater than 99.5% reflectivity. Dielectric stacks of two materials with largely differing refractive indices $n_{1,2}$, such as SiO_2 and TiO_2, are grown with layer thicknesses of $\lambda/(4n_{1,2})$, where λ is the wavelength of light, to form a mirror. These mirrors are known as distributed Bragg reflectors (DBRs).

In semiconductors it is convenient for fabrication if at least one of the mirrors can be made from compounds of the semiconductor materials themselves which are conductive for the carriers. To minimise the number of mirror pairs required the refractive index difference between the constituent material should be maximised. The materials chosen must be lattice matched to the substrate material or strain compensated in some way. In addition, the generated light must pass through one of the mirrors to get out so one mirror is usually made less reflective. The VCSELs can be made either top or bottom emitting. Therefore it is necessary that the DBR materials must be non-absorbing at the emission wavelength. These constraints are not easy to match in all possible VCSELs but in GaAs or InGaAs VCSELs operating at 850 nm and 980 nm wavelengths respectively, the possibility to use the $Al_xGa_{1-x}As$ alloy scheme for the DBRs is possible. This is a very well characterised material system in which the lattice mismatch between the GaAs and AlAs is very small and provides sufficient Δn so that 20-30 pairs of DBRs is sufficient to get reflectivities of 99.5%. DBRs made from these materials have been shown to be highly successful and can be produced uniformly and reproducibly. To improve conductivity in the mirrors they are doped either n or p type. To improve the transport of the carriers between the different bandgap materials used to construct the DBRs the semiconductor layers are graded at the interface and the doping is increased here. This reduces the operating voltage to below 2V for 850 nm VCSELs.

GaAs and InGaAs VCSELs have been shown to operate successfully in the single longitudinal mode as predicted. with a threshold current density J_{th} at 300K (room temperature) of 240 A cm^{-2} (corresponding to 0.1 mA threshold current for a small VCSEL) having been reported for a single InGaAs quantum well [2]. This is advantageous for modulation properties and the small size of the device minimises capacitive effects making VCSELs inherently high speed lasers.

VCSELs do not necessarily operate in the lowest order transverse optical mode regime. In some cases the VCSELs turn-on in the lowest order transverse optical mode while in other cases they may turn on in higher-order transverse optical modes or in a multitude of modes. Lasers lasing in many higher order transverse optical modes are termed multimode VCSELs. The nature of the VCSEL is very apparent from the light output of the VCSEL. A circular spot of light emanating from the centre of the VCSEL signifies a VCSEL lasing in the lowest order transverse optical mode while a ring mode or off-centre number of spots of light signifies a VCSEL lasing in its multimode state. Various VCSEL outputs are shown in Fig. 2.

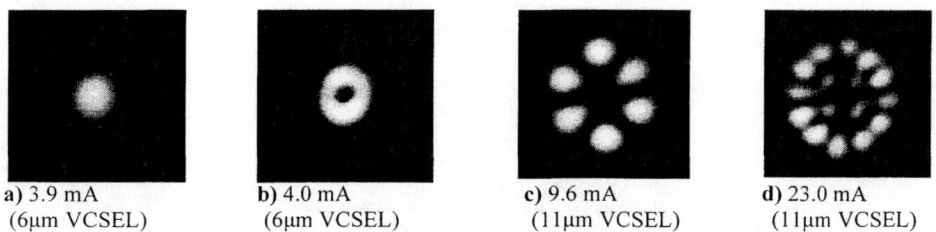

a) 3.9 mA   b) 4.0 mA   c) 9.6 mA   d) 23.0 mA
(6μm VCSEL) (6μm VCSEL) (11μm VCSEL) (11μm VCSEL)

FIGURE 2. Nearfield light output from VCSELs at the current inputs indicated showing the transverse modes involved, from[5]. (© 2000 IoP Publishing).

The light output against current (L-I) curves often shows kinks when the higher order transverse optical modes switch in. For communications applications it was originally thought that lowest-order transverse mode lasing operation was desired due to the resulting good coupling into the central core of the optical fibre. Recently it has been shown that multimode operation may be advantageous since it excites higher order modes within the fibre that have their maximum intensity away from the centre of the fibre. These modes suffer less spreading as the signal travels down the fibre because their propagation constants are fairly similar. It is advantageous to have predictable linear L-I curves without kinks for signal encoding.

The light output against current plots also show deviation from the linear L-I output curve expected due to polarisation effects. Because the aperature of the VCSELs are circularly symmetric the orientation of the TE emission within the <011> plane is arbitrary and an orientation is usually chosen because of random strains in the device or underlying crystallographic orientations. Since the selection of emission angle within the plane is so weakly selected the VCSEL easily switches from one mode orientation to another polarisation (at 90 degrees to the first polarisation). This polarisation switching has generally been thought to be a nuisance and there are several schemes in operation to control the polarisation, chiefly through introducing anisotropic strain, either during growth [6] or after fabrication, via post processing techniques [7]. A typical VCSEL output showing polarisation switching is shown in Fig. 3a along with a post processed result, shown in Fig 3b, in which the ratio of the maximum polarisation to the minimum polarisation is shown as a function of current, after post-processing, to stabilise one of the polarisations. Recently there have been schemes proposed to encode the signal within the polarisation itself [8]. This can be done by biasing the laser near the polarisation switching position.

VCSELs are very different from their edge emitting counterparts in their L-I temperature dependence as can be seen in Fig. 4. In an edge emitting laser above threshold the light output increases approximately linearly with increasing current for a wide range of current. In VCSELs, although the light output increases linearly initially, it quickly rolls over. The sublinear dependence on output with current is due in part to

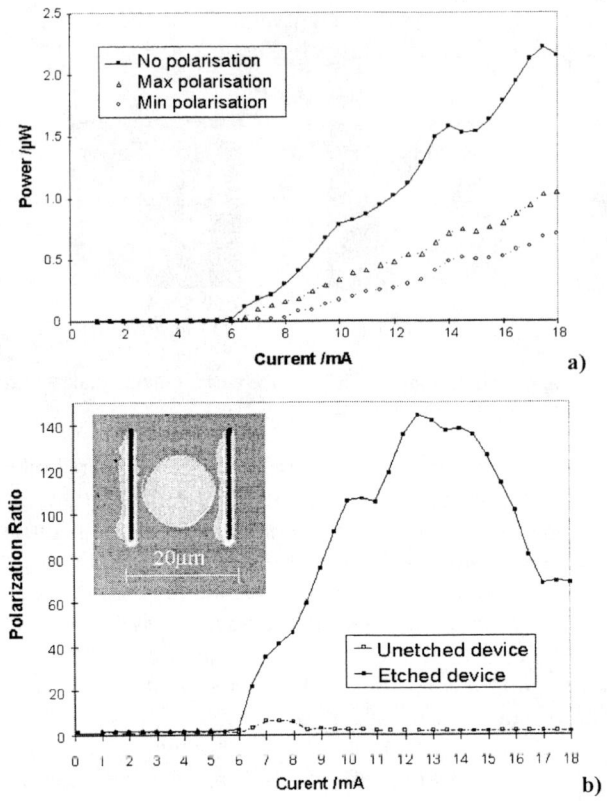

Figure 3. A polarised light output against current input is shown before polarisation –fixing post processing (a) and after post-processing polarisation-fixing (b). The polarisation ratio is the power in the maximum polarisation over the power in the minimum polarisation. The insert shows a figure of the post-processed VCSEL [9].

heating of the laser with increasing current. In an edge emitting laser as the temperature is increased the lasing output power decreases, essentially mirroring the temperature dependence (decrease) of the gain and the temperature dependence (increase) of carrier loss processes. At a sufficiently high temperature, the gain may overlap with a different longitudinal mode and start lasing from that mode causing a shift in the wavelength of the light which essentially follows the temperature dependence of the gain shift. In VCSELs the output power decreases rapidly with increasing temperature but the wavelength is shifted only slightly with temperature. In VCSELs the output wavelength and its power is determined by the Fabry Perot optical cavity mode and its overlap with the gain at that value. The peak of both the FP mode and the gain can be engineered to be optimal at one operating temperature, but since both move at different rates with temperature they detune as the temperature varies

a)

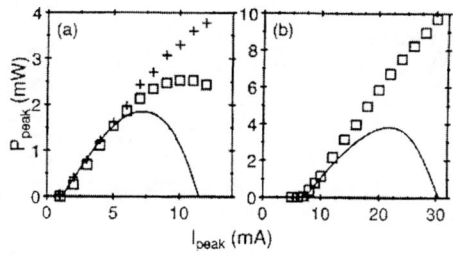

b)

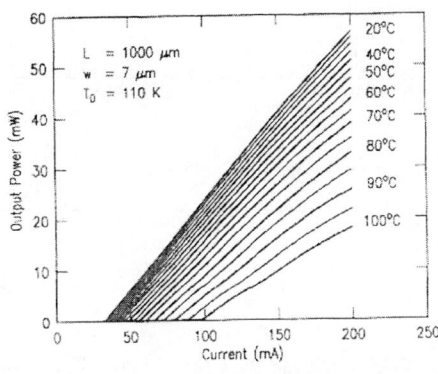

FIGURE 4. Typical light output against current input for a VCSEL under CW (continuous wave), shown as a solid line, and pulsed operation is shown in **a)** [10]. (CW operation allows the VCSEL to heat up showing the effect of heating the VCSEL.) The typical light output against current input curves for an edge emitting laser in CW operation as a function of temperature is shown in b). From [11] (© 1993 IEEE).

from the tuned value. This causes a substantial roll over as the temperature is increased. Because the FP modes move more slowly than the gain with temperature, the VCSELs show less wavelength change with temperature which is an advantage in communications.

In this section the general principles of VCSELs have been introduced and the GaAs and InGaAs 850 nm and 980 nm VCSELs have been discussed. These VCSELs are currently being used in short multimode communications systems. In addition, there are VCSELs being investigated for communications use in plastic fibres at 650 nm (using GaAsP/AlGaAsP material systems) and for use in silica fibres at 1300nm and 1550 nm using various material systems. Several material systems are being considered as basis for VCSELs operating at various wavelengths as shown in Table 1.

TABLE 1. Material systems used for VCSELs

Material system	DBR materials	avelength	Application
GaAs/ AlGaAs	AlGaAs/AlAs(lattice matched)	680-890 nm	Solid state laser pumps Optical data storage (CDs) General purpose infra-red source
InGaAs/ GaAs	AlGaAs/AlAs(lattice matched)	950-1100 nm	Low power sources Erbium fibre laser pump
InGaAsP/ InP	AlGaAs/AlAs(fused) InGaAs/InP(lattice matched) SiO_2/TiO_2	1.0-1.7 µm	Optical fibre applications in transmission windows at 1.3 µm and 1.55 µm
GaInP/AlGaInP/ GaAs	AlGaAs/AlAs	600-700 nm	Visible applications Polymer short haul fibres
CdZnSe/MgZnSSe	MgZnSSe alloys	450-550 nm	Potential uses in colour displays
GaInN/AlGaN	GaN/AlGaN	200-640 nm	Has the potential to span the visible and UV spectrum

DYNAMICS OF VCSELs

Small and Large Signal Modulation using Ideal Rate Equations

In the last section the principles of VCSELs were considered. To derive both the steady state and dynamical response of the VCSELs a rate equation approach is usually used [3]. These equations link the generation of photons with the recombination of carriers. In its most simple form the rate equation for the carriers is

$$\frac{dN}{dt} = \frac{\eta_i I}{qV} - \frac{N}{\tau_c} - R_{st}$$

where N is the number of carriers, I the injected current, q the charge of the electron, V the volume of the active region, η_i is the injection efficiency, τ_c is the carrier lifetime and R_{st} is the stimulated recombination of carriers to generate photons defined as R_{st} = $(R_{21}-R_{12})$ where R is the transition rate between states 1 and 2. The injected current $\eta_i I/qV$ provides the generation term while radiative and nonradiative recombination plus carrier leakage provide the recombination contained in the second term. In this simple form $N/\tau_c = BN^2+(AN+CN^3)$ where BN^2 is the spontaneous recombination R_{sp}. The simplest form of the photon rate equation is

$$\frac{dP}{dt} = \Gamma R_{st} + \Gamma \beta_{sp} R_{sp} - \frac{P}{\tau_p}$$

where P is the number of photons, R_{st} is the stimulated emission, R_{sp} is the spontaneous emission created by recombination of carriers, and the last term represents photon decay with the photon lifetime τ_p. The term β_{sp} is the spontaneous emission factor which is the reciprocal of the number of optical photons in the bandwidth of the spontaneous emission, while Γ is the confinement factor which relates the volume of the electron to that of the photon. It is common to consider the stimulated emission as a gain process for photons and put $R_{st} = v_g g P$, where v_g is the group velocity and g the gain, into the rate equations where g may be represented by a linear or logarithmic expression. It is possible to solve these coupled rate equations to find the steady state solution of output power against the applied bias current and to derive the L-I curves which can be compared with the experimental values. This gives the steady state L-I curves for lasers discussed previously.

For communications the dynamical response to a perturbation of the input signal is required. This model can be used to analyse the small signal and large signal response of the laser. The response to a small level of modulation superimposed on the bias current I_0 is the small signal response. Consider the application of an above threshold dc current I_0 on which is superimposed an ac modulation I_1 according to $I = I_0 + I_1 e^{i\omega t}$. Then N and P will be of the same form. Substituting I, N and P into the rate equations using $g = a(N - N_{tr})$ where a is the differential gain and N_{tr} is the carrier density at transparency (when the gain=0), assuming that R_{sp} is negligible and grouping the steady state contributions and setting them equal to zero produces

$$i\omega N_1 = \frac{\eta_i I}{qV} - \frac{N_1}{\tau_c} - \frac{P_1}{\Gamma \tau_p} - v_g a N_1 P_0$$

$$i\omega P_1 = \Gamma v_g a N_1 P_0$$

where $a(N_0 - N_{tr})$ is replaced by $[\Gamma v_g \tau_p]^{-1}$ and higher harmonics have been deleted. This can be solved for the transfer function $P_1(\omega)/I_1(\omega)$. Multiplying the equations together and ignoring all but the final term in the first equation yields $\omega_R^2 = a v_g P_0 / \tau_p$ where ω_R is the relaxation resonance frequency. An alternative form of the resonance frequency is $\omega_R^2 = (\Gamma v_g a / qV) \eta_i (I - I_{th})$. Through manipulation the transfer function can be written in the form

$$\frac{P_1(\omega)}{I_1(\omega)} = (\frac{\eta_d h v}{q}) \frac{1}{\{1 - (\frac{\omega}{\omega_R})^2 + i(\frac{\omega}{\omega_R})[\omega_R \tau_p + \frac{1}{\omega_R \tau_c}]\}}$$

where hυ is the energy per photon, η_d is the differential quantum efficiency and for low modulation frequencies $P_1=(\eta_d h\upsilon/q)(I-I_{th})$ where I_{th} is the current threshold. For higher modulation thresholds the term $1-(\omega/\omega_R)^2$ creates a strong resonance. The damping of the resonance depends on both ω_R and $1/\omega_R$. Beyond the strong resonance the transfer characteristics degrade significantly. Thus the effective modulation of the output power can only be achieved over a modulation bandwidth of $\sim \omega_R$. Of particular interest is the –3dB frequency $\omega_{3dB} = \sqrt{1+\sqrt{2}}\omega_R$, the point at which the input signal power is reduced in power by –3dB. The 3dB output power is

$$f_{3dB} \cong \frac{1.55}{2\pi}\sqrt{\frac{\Gamma v_g a \eta_i}{V\eta_d h\nu}}\sqrt{P_0}.$$

This corresponds to the frequency where the received electrical power is one half its dc value. The modulation bandwidth can be increased by increasing the output power, but the increased damping at high powers limits the operating power. This can be seen in Fig. 5. In more elaborate models at high powers the maximum bandwidth becomes independent of ω_R and is related to the damping factor (K) which is affected by gain compression and transport effects. This analysis provides an insight into the frequency response of the laser. The frequency response can be calculated and compared with the experimentally measured values. Experimentally each curve is produced by applying the small signal modulation at one frequency for a finite period of time and measuring the maximum and minimum values of output power produced. These are then used to calculate a root mean square value of output power and this output power is then normalised to the power at DC. Because of the small size of the VCSELs the carrier transport effects are very small and there are low parasitics, meaning that they can be successfully modulated at 10GB/s.

The current density modulation causes a change in the refractive index of the active region of the laser. This causes the optical length of the cavity to change causing the resonant mode to shift back and forth in frequency. This frequency modulation (FM) of the laser can be used to dynamically tune the laser. For intensity modulation (IM) applications this frequency variation, termed frequency chirping, effectively broadens the modulated spectrum of the laser, reducing its effectiveness in optical fibre communications. The frequency shift in response to the change in carrier density is

$$\Delta \nu = \frac{\alpha}{4\pi}\Gamma v_g a \Delta N$$

where α is the linewidth enhancement factor defined as $\alpha \equiv \dfrac{dn/dN}{dn_i/dN}$ where n is the real part and n_i the imaginary part of the refractive index of the material. The α

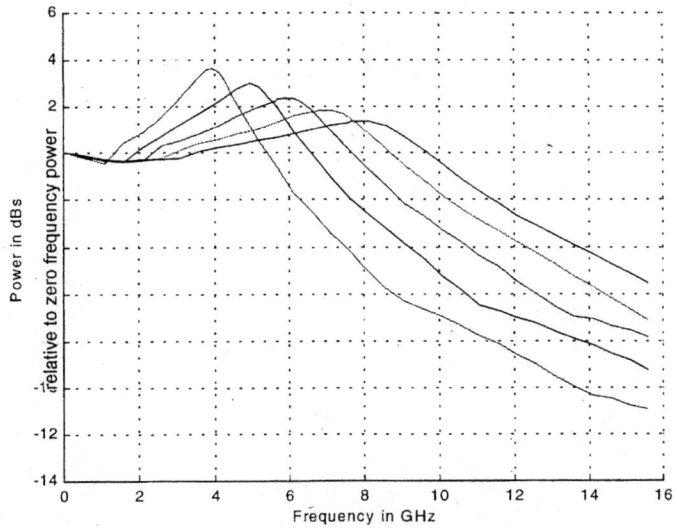

FIGURE 5. Small signal frequency response of an idealised laser diode for several output powers showing the resonance frequency and its subsequent damping at higher power. The resonance peak at 3.91 GHz results from a bias at 26.23 mA, the resonance peak at 4.91 GHz results from a bias at 35.23 mA, the resonance peak at 5.91 GHz arises from a bias at 46.23 mA, the resonance peak at 6.91 GHz results from a bias at 59.23 mA and the resonance peak at 7.91 GHz results from a bias at 74.23mA [12].

parameter is typically 4-6 but may be as low as 2 in some active materials. In addition to its dependence on carrier density N, the refractive index is also dependent upon temperature. For VCSELs, because of the high resistance of the DBR mirrors, the temperature impedance of the laser is about 10 times larger that for the comparable edge emitter making this effect significant in VCSELs.

It is also possible to apply an input current waveform of any form within this model. This allows a realistic large signal modulated signal to be applied to the laser. The laser output response to a large signal response is known as the large signal response. In practice the laser will be modulated by digital data most commonly in communications. This data takes the form of a non-return-to-zero (NRZ) bit stream. A pseudo random bit sequence can be generated and shifted to a suitable bias level. This bit stream will have an infinitely short rise and fall time between successive bits. It is possible to generate another bit stream from the first by increasing the rise and fall time (usually to about 10% of the bit period). This is more realistic and also reduces ringing in the model output. Usually a period at the start of the waveform before modulation is applied to allow the laser output to stabilise at the bias current. A waveform can be applied to the model and the simulation produces an output file containing output power and time step. Before observing the output a 5th-order Bessel filter is applied to the data with a cutoff frequency of 75% of the bit rate. This removes high frequency oscillations from the output. (In practice the detector would filter out these oscillations). At lower modulation rates the laser output follows the input modulation

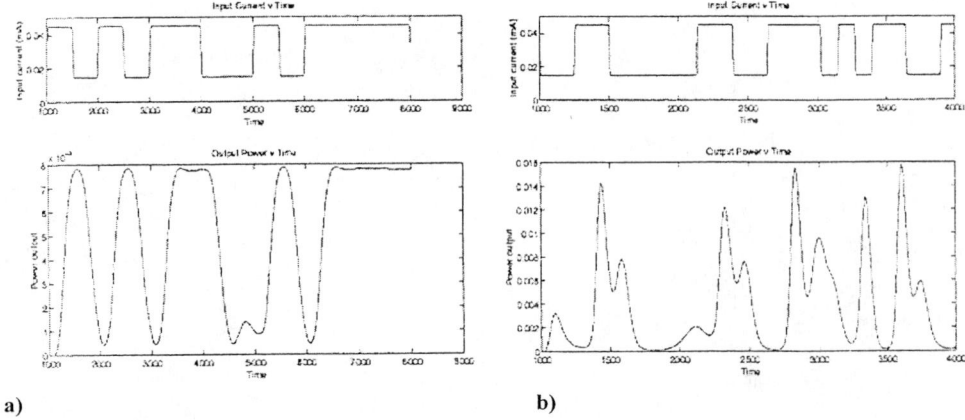

FIGURE 6. The top diagrams show the input modulation current of 15 mA as a function of time (in arbitrary units) while the lower diagrams show the corresponding output power as a function of the time (in the same time units). a) shows the response to a lower modulation rate where the laser is able to follow the modulation while b) corresponds to a higher modulation rate where the laser is unable to follow the modulation [12].

closely as can be seen in Fig. 6a. (Some ripple can be noticed connected with relaxation oscillations in the device.) At the higher modulation rate the laser output is very distorted as can be seen in Fig. 6b. In this case the useful bandwidth of the device has been exceeded and there is a clear delay in the output response to an input change.

Modelling Real VCSELs

In reality these idealised rate equations are too simple to describe the VCSEL. As mentioned earlier there is competition between different transverse modes. The spatial distribution of carriers across the VCSEL and thermal gradients which vary in space and time must also be included. To include these effects the simple equations need to be extended to the form

$$\frac{dN(r,t)}{dt} = \frac{\eta_i I}{qV} - \frac{N}{\tau_c} - v_g \{\sum_m g_m(t) P_m(t) | \Psi_m(r) |^2\} + \frac{D}{r} \frac{\delta}{\delta r}(r \frac{\delta N(r,t)}{\delta r})$$

where the carrier equation has become spatially resolved and includes a sum over the transverse modes m where the coupling to the modes depends on the spatial overlap of the carriers with the optical transverse modes $| \Psi_m(r) |^2$. The final part of the equation is a carrier diffusion term that attempts to equilibrate the carrier density laterally across the structure where D, the diffusion constant, has a temperature dependence. There is now a photon rate equation for *each* of the transverse modes considered where each mode has its own characteristic lifetime $\tau_{p,m}$. The equations are of the form

$$\frac{dP_m(t)}{dt} = v_g \Gamma g_m(t) P_m + \frac{\beta_{sp} N(t)}{\tau_n} - \frac{P_m(t)}{\tau_{p,m}}$$

where the carrier density is averaged spatially for the evaluation of the spontaneous emission rate in the second term and τ_n is the spontaneous emission lifetime. The thermal conduction equation in cylindrical symmetry is

$$\frac{\delta^2 T(r,z,t)}{\delta r^2} + \frac{1}{r}\frac{\delta}{\delta r}T(r,z,t) + \frac{\delta^2}{\delta z^2}T(r,z,t) + \frac{1}{th_c}Q(r) = \frac{1}{th_f}\frac{\delta}{\delta t}T(r,z,t)$$

where T is the temperature in the lateral r and longitudinal z direction, th$_c$ is the hermal onductivity, th$_f$ is the thermal diffusion and Q is the heat density .This system of equations can be solved iteratively. Some interesting new behaviour results from this spatially resolved model. Initially an antiguiding effect is observed due to carriers which lower the refractive index. The laser turns on in the lowest order transverse mode when thermal effects cause a significant increase of the refractive index for guiding. Carriers are depleted from the centre of the VCSEL because of the large intensity of the lowest order transverse optical mode at the centre. This is termed spatial hole burning. Before the carriers can equilibrate the distribution the laser may switch to a higher order transverse mode which has more spatial overlap with the existing carrier distribution.

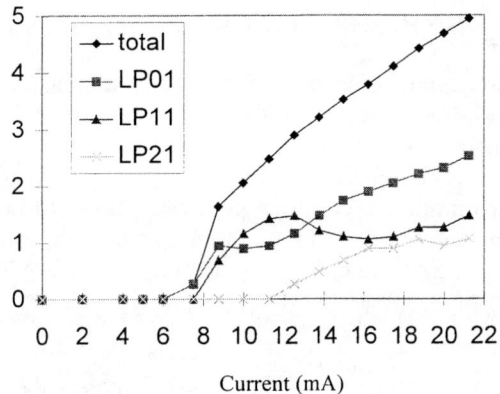

FIGURE 7. The modelled lasing characteristics of a 8 micron VCSEL are shown where outpower in mW is plotted against the input current in mA, and the contributions of the individual transverse modes LP are shown. Strong mode competition is shown particularly between transverse modes LP$_{11}$ and LP$_{21}$ [13].

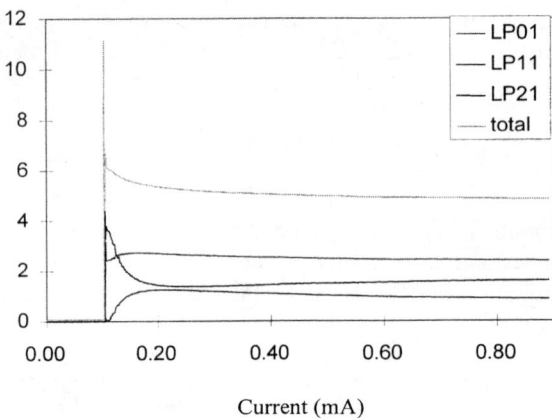

FIGURE 8. The modelled power output in mW is shown as a function of time in microseconds with the contribution of the various transverse modes LP shown. During a single electrical pulse LP_{21} causes a reduction of power in LP_{11}, but as the heating effects become stronger LP_{11} gains power over LP_{21} [13].

As the heating effects become more important the fundamental mode gains power from the higher order modes. There is a significant temperature gradient laterally across the VCSEL with the centre of the VCSEL being about 20 degrees hotter than at the edge for a VCSEL with a radius of 15 microns. The effect of aperature size, differing photon lifetimes (through altering the cavity by Gaussian etching the top most pairs) and the effect of adding oxidized layers to control current injection and spreading are currently areas being investigated using such spatially dependent models [13,14]. Fig. 7 and Fig. 8 show results from modelling VCSELs including these effects.

SNR, RIN and Eye Diagrams

The intensity and frequency response of the laser due to small and large scale modulation have been discussed above. In the steady state it has been assumed that the carrier and photon densities are constant. There are instantaneous fluctuations in the densities even without modulation. The variations in photon density produce variations in the magnitude of the output power which provides a noise floor while the variations in carrier density result in variations in the output wavelength. For analog applications the noise is quantified using the electrical signal-to-noise ratio (SNR). For a sinusoidal signal $P(t)=P_0+P_1\sin(\omega t) +\delta P(t)$ where $\sqrt{<\delta P(t)^2>}$ is the root mean square of the noise

$$SNR = \frac{<i_s^2>}{<i_N^2>} = \frac{<(P_1\sin(\omega t))^2>}{<\delta P(t)^2>}$$

where s refers to signal and N to noise and < > denotes a time average. For digital applications a decision at the midpoint defines whether a '0' or '1' is recorded. If the noise exceeds $P_0/2$ where P_0 is the separation between the '0' and '1' then a false reading may be made. If the noise has a Gaussian distribution about the mean power levels then, in order to reduce the probability of finding $|\delta P(t)| > P_0/2$ to less than 1 in 10^9 (ie a bit error rate (BER) $<10^{-9}$), we require that

$$\frac{P_0^2}{<\delta P(t)^2>} > (11.89)^2.$$

The relative intensity noise (RIN) is defined as

$$RIN = \frac{<\delta P(t)^2>}{P_0^2}$$

and is usually described in decibels or $10 \log_{10}(RIN)$. For analog applications, if a given SNR > 50 dB with $P_1/P_0=1$, then the laser must have a RIN < -53 dB. For a BER < 10^{-9} in digital applications the laser must have a RIN < 21.5 dB. It is often useful to consider an alternative RIN defined as

$$\frac{RIN}{\Delta f} = \frac{2S_{\delta P}(\omega)}{P_0^2}.$$

where Δf is the filter bandwidth of the measurement apparatus. In this definition the RIN is in units of dB/Hz or RIN per unit bandwidth. In designing a communications system the desired SNR or BER sets a maximum limit on the total RIN of the laser. For example, in a digital transmission link with a 2Gbit/s (1 GHz) system bandwidth and a required BER of 10^{-9} (ie. RIN< -21.5dB) the laser must have an average RIN(dB/Hz) < -21.5dB -90=-111.5 dB/Hz. If the system bandwidth is increased the laser RIN per unit bandwidth must be decreased in order to maintain the same total RIN.

It is common practice to make use of *eye diagrams* when analysing the response of a laser to a large signal modulation. An eye diagram is formed by splitting the output waveform into sections, where each section is a multiple of the bit period in length. These sections are then plotted on top of one another. Fig. 9 show eye diagrams for bit rates of 2, 4, 6 and 8 Gb/s. At 2 GB/s the eye diagram is very clear and the eye formed is open. In this case the two bit states are easily distinquishable leading to few errors at the receiver. In contrast at 8 Gb/s the eye diagram is very distorted and the eye appears closed. At the receiver it would be difficult to distinguish a high bit from a low bit leading to a high BER.

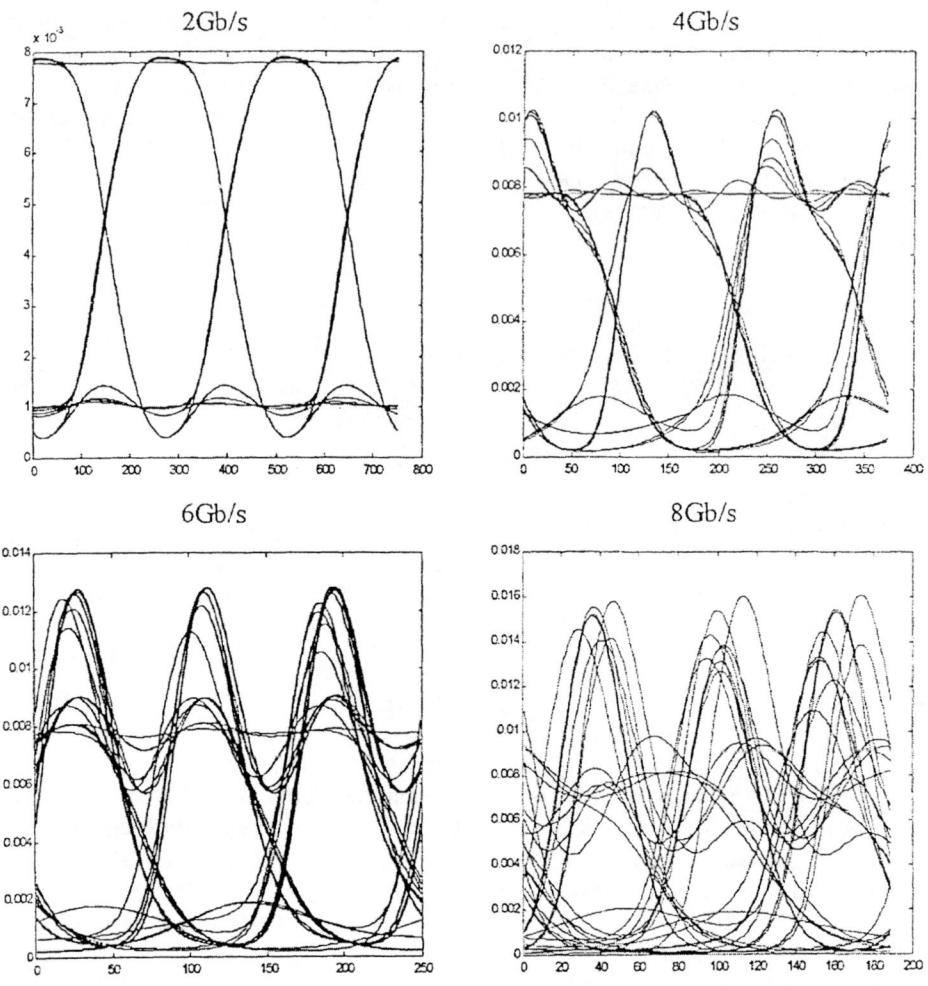

FIGURE 9. Eye diagrams from simulation at bit rates of 2 GB/s, 4 GB/s, 6 Gb/s and 8 GB/s where the power output is plotted relative to the unmodulated signal. It can be seen that as the bit rate increases the eye starts to close (ie the difference between the two output levels decreases and the rise and fall times of the eye merge), signalling a degradation of the signal [15].

The extinction ratio (ER) is defined as the ratio of the output power for a 'one' bit level to that for the 'zero' level in the eye diagrams. Generally a ratio of at least 7dB is required for successful detection of the signal at the receiving end of an optical link. The extinction ratio is calculated as follows ER=10 log P_1/P_0 where P_1 is the power for the 'one' level and P_0 is the power at the 'zero' level. At high power levels the noise level in the receiver is higher making detection more problematic. To maintain a good extinction ratio the absolute power difference between the 'one' and 'zero' levels must increase as the offset from zero power increases. It can be seen that as the bias current is changed a trade-off between small signal bandwidth and extinction ratio occurs. As observed earlier an increase in bias current increases the –3dB frequency of the small signal response. However this increase in bias moves the operating point of the device further up the LI curve leading to an increased optical power offset and a reduced extinction ratio.

OPTICAL FIBRES

In a communications system a transmitter or source, a receiver or detector and a transmission medium is required. In this section the transmission medium is discussed and how its transmission properties impact on the design of the source. Optical fibres are the commonly used channel for guiding light for communications applications. They are usually made of SiO_2 and consist of a small core of a higher refractive index material surrounded by a cladding material of a lower refractive index. The refractive index profile is controlled by the careful control of spatial distribution of dopant such as GeO_2 that serve the purpose of raising the index of refraction of the intermost part of the core, thus confining light there by the same physics as total internal reflection. The dopant does not seem to affect the loss. The guiding of the modes in the fibres may by analysed by a ray picture or by analysis of the cylindrical modes supported by the structure. The lowest order modes of the fibre are peaked towards the centre of the fibre while higher order modes are peaked further from the core. Each mode of the fibre has a different velocity of propagation through the fibre as determined by its propagation constant. Mono-mode fibre supports only one mode in the fibre and has a smaller inner core. It is used in long haul communication. Multi mode fibres, which have a larger central core support several modes simultaneously in the fibre. Multi-mode fibre is only used in short haul communication since mode interference and dispersion will degrade the ultimate distance that the modes can travel down the fibres. A schematic cross-section through a monomode and multimode fibre is shown in Fig. 10.

Losses in the fibre determine the distance over which the fibre can usefully transmit information. The loss is defined as how much a signal decreases in intensity over a given length that the signal travels, so the loss is a number per unit length. Light with a wavelength of 1550nm gets only about 1dB dimmer in 4 km. (10dB is one factor of 10). The loss as a function of wavelength for a glass optical fibre has minimum loss around 1550 nm, although there is another minimum about 1300nm. The three key components of the loss are OH absorption, Rayleigh scattering and the Urbach tail (Si-O vibrations). Hydrogen is an impurity so that the O-H absorption peak at 1400nm can be reduced through improved fabrication of the fibres. In addition to loss there is also *chromatic dispersion* which means that the wavelengths of light travel at different

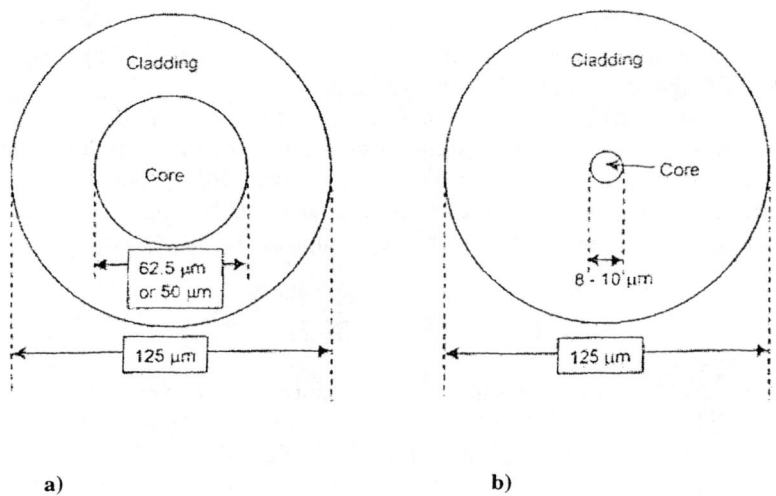

FIGURE 10. A schematic cross-section through a multi-mode fibre a) and a monomode fibre b) showing the differing core size.

velocities through the fibre. This means that a narrow peak spreads out as it travels down the fibre. Chromatic dispersion in a step index or parabolic profile fibre has a minimum at 1300nm, making this wavelength preferable for some applications even though the losses are larger than at 1550nm. Recently triangular profile refractive index fibres, with a surrounding moat of lower refractive index than SiO_2, have been demonstrated to shift the chromatic dispersion to 1550 nm where losses are at a minimum. This type of fibre is called dispersion-shifted fibre and its attenuation can be kept below 0.22 dB/km at 1550nm. Another issue is that of *polarisation mode dispersion* (PMD). There are two polarisation modes in a single mode fibre. The degeneracy between these polarisations can be broken by introducing doping anisotropy (birefringence). Neighbouring frequencies undergo different evolutions that diverge with distance which introduces dispersion. For certain applications polarisation is important and for quantum information it is critical.

For long distance communication it is often necessary to incorporate repeaters or more recently, and preferably all-optical amplifiers. The electronic repeaters create bottlenecks as the optical data is received, converted into an electrical signal, which is then amplified and re-timed before being re-transmitted down the next length of fibre. While the optical fibre offers almost unlimited bandwidth, the electronic repeaters cut the bandwidth potential down significantly. An all-optical replacement for the electronic repeater is the *erbium doped fibre amplifier* (EDFA). It works by using optical transitions between slightly broadened atomic transitions in the f-shell of Erbium atoms, so they only work in the range of energies near 1550nm. It does not need electronic current to operate and it does not reduce the bandwidth of the system.
It can also handle several data streams in the same frequency window without the need for physical upgrading. EDFAs are a significantly important improvement.

There are two other ways of producing optical amplification over a wider range of wavelengths: *semiconductor amplifiers* and *Raman amplifiers* [16]. Optical amplifiers can be used in several ways in a communication system. Firstly in booster amplification, the optical amplifier can boost an optical signal. Secondly, in optical pre-amplification, the optical amplifier can enhance the signal going into a photodetector to improve the operation of conventional or avalanche photodetectors. Thirdly, optical amplifiers can compensate for signal loss along a fibre transmission line. The semiconductor amplifier is structurally similar to the laser-diode region but without mirrors to produce round-trip feedback, the amplifier section can increase the optical flux over one hundred fold in a single pass. Unfortunately, semiconductor amplifiers are relatively noisy.

In the Raman amplification process the entire fibre is turned into the amplification medium, via the transfer of power from a laser pump beam to the signal beam through a process known as stimulated Raman scattering (SRS). Raman scattering is the interaction in a non-linear medium between a light beam of frequency ω and wave vector k and some physical quantity associated with the medium (interatomic displacement or electric density change). These fluctuations modulate the charge polarisation and the subsequent interaction results in an exchange of energy between the incident field and the medium. In an optical fibre internal lattice vibrations (optical phonons) in the medium modulate the charge polarisation seen by the incident electromagnetic wave at the characteristic frequency ω_p. As a result, phonons of frequency ω_p are emitted or absorbed and new light appears at both $\omega+\omega_p$ and $\omega-\omega_p$. The scattering of light from the dynamically modulated charge polarisation is known as spontaneous Raman scattering. Raman amplification relies upon stimulated Raman scattering in which the input light is of sufficiently high intensity to coherently excite the vibrations. This results in a dramatic enhancement in the signal light. The advantage of this process is that it will occur over a broad range of frequencies (over 200 nm for wavelengths around 1400 nm) and that it is also a very fast process (< 1 ps), which is important for high speed optical communications. Lucent technologies have recently demonstated 1.6 Tbs over 400 km of fibre using this approach as opposed to 80 km using an erbium doped amplifier [17].

Plastic Optical fibre is currently being investigated as an alternative to glass optical fibre. Some typical polymer fibres are polystyrene, which has high loss, polycarbonate, which again has high loss but is good at high temperatures, PMMA which has lower attenuation and a transmission window at 650 nm, and fluorpolymer, a graded index fibre with very low loss. These latter two fibres look very promising for low cost data communications. Like glass, polymer fibre has a high bandwidth, but it is cheaper to install, because it is less fragile. In addition its larger size, a diameter of between 750-1000 micron, makes coupling to the light easier (no lenses are required and only inexpensive plastic connectors need be used). A routine glass fibre connection typically takes between 5 and 20 minutes compared with about 1 minute for polymer.

Ideally one would like to match the single mode laser beam into the centre of the monomode fibre to excite the lowest order or fundamental mode in the monomode fibre for maximum transmission along the fibre. This is done in monomode fibres using distributed feedback and Fabry Perot edge emitting lasers. In a multimode fibre a number of modes are excited by the laser depending on the spatial overlap of the intensity of the laser beam with the eigenmodes. These modes travel along different paths within the fibre and either broaden the signal or actually split the signal into different pulses which can interfere with each other. This is termed modal noise. When a transverse multimode VCSEL is used to excite a multimode fibre it excites many modes which propagate down the fibre and spread the pulse. Investigations at Bristol have found that it is advantageous to align the VCSEL off-centre so that it excites a number of higher order modes of the fibre. This is because the difference in the propagation constant between the higher order modes is less, so they spread out less as they travel down the fibre than the lowest order modes excited at the centre. This scheme is known as *offset launch* and results in less spreading of the signal [17]. It was also found that real fibres vary significantly more at the centre of the fibre from fibre to fibre than off-centre, making for more reproducible behaviour if excitation is made off-centre. An extension of this would be to allow the emitting VCSELs to operate in off-centre higher-order transverse modes or in ring modes that would excite the higher order modes in the fibre when aligned with the fibre.

VCSELS FOR 1.3 AND 1.55 MICRON WAVELENGTHS

As discussed in the previous section there are transmission windows for glass optical fibres at 1.3 and 1.5 micron wavelengths and for polymer fibre at about 650nm. While for short distance communication multimode fibre and therefore injection into the fibre by multimode VCSELs operating off these windows is possible for long distance communication single mode fibre and single mode VCSEL sources designed to operate within these wavelength windows are possible. Currently edge emitters are used but it would be very advantageous to have VCSELs for these wavelengths.

The edge-emitter lasers currently used for 1300 nm /1550 nm wavelength have an active gain region of GaInAsP quantum wells grown on InP substrate material. It is **not** straightforward to convert these edge-emitters into VCSELs. The high contrast AlAs/AlGaAs DBRs used for the 850nm and 980 nm GaAs and InGaAs VCSELs cannot be used because their lattice spacing is substantially different from that of InP and GaInAsP. Therefore, we need to investigate materials for the DBRs which can be grown lattice matched to InP. Alloys of GaInAsP can be grown lattice matched to InP but the refractive index differences possible are very small making them very poor DBRs. One approach which has been demonstrated is to fuse AlGaAs/AlAs DBRs onto a GaInAsP active region to form a GaInAsP VCSELs . Unlikely as this sounds, it is possible to do this using pressure, and the resulting junctions between these regions do not inhibit carrier transport much.

Alternative approaches to developing 1.3 and 1.5 micron VCSELs involve developing new gain regions that will lattice match to the GaAs/AlGaAs DBRs. Two such possible gain materials are InGaAsN quantum wells and InAs quantum dots. InGaAsN is a relatively new material where a small (0.06% incorporation of nitride into the material alters the lattice constant and the band gap sufficiently such that a GaInNAs quantum well sytem should emit at 1300nm and be lattice matched to GaAs. The growth has yet to be optimised but is looking promising with a VCSEL having been demonstrated [19]. InAs quantum dots grow in a self organised on GaAs way under the appropriate growth conditions to relax strain.. Quantum dots, because of their peaked density of states should offer the best gain material but practical issues, like getting a high enough density of quantum dots and getting the excellent line-up required between the quantum dot gain and the Fabry-Perot mode present problems. This work is currently being developed [20].

An alternative approach is to use a GaAs 850 nm VCSEL to optically pump a longer wavelength GaInAsP VCSEL for which the DBRs do not need to be conductive for the carriers allowing alternative materials to be used for the DBRs.

VERTICAL CAVITY STRUCTURES FOR DETECTORS

The advantages of using a resonant cavity in laser structures are also applicable to detector structures. It would therefore be very useful to investigate using vertical cavity structures for detectors as well. In addition, the similarity of the two structures should enable arrays to be made with vertical cavity lasers adjacent vertical cavity detectors.

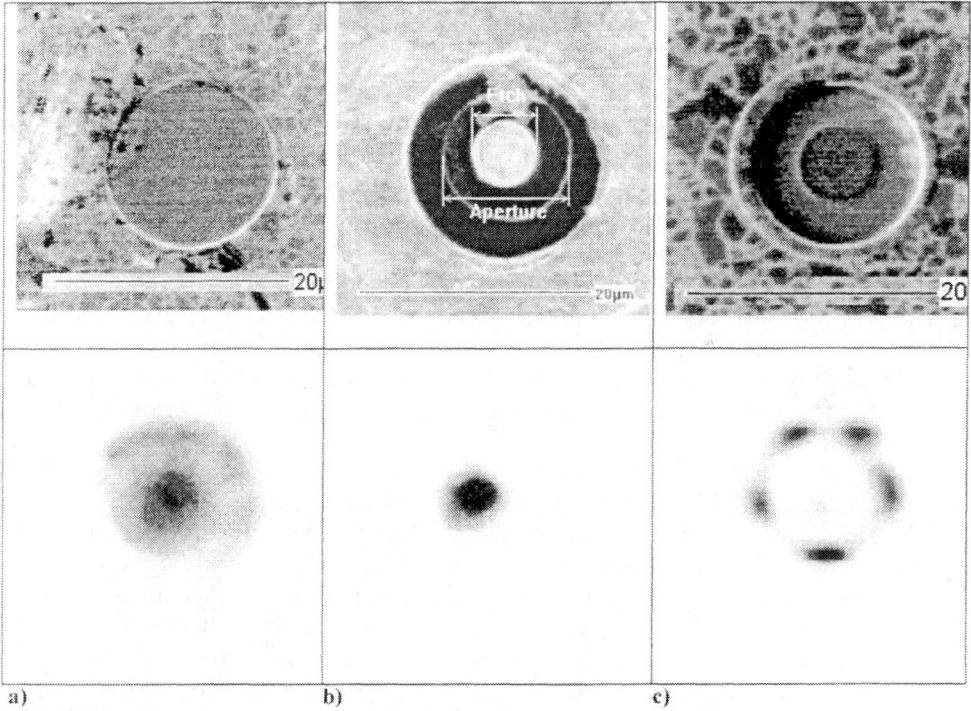

FIGURE 11. The top pictures show the VCSEL structures, unetched **a)**, etched in a centre spot **b)**, and etched in an outside ring **c)**, while beneath each structure is the corresponding near field optical spectra with dark intensity corresponding to highest optical intensity [21].

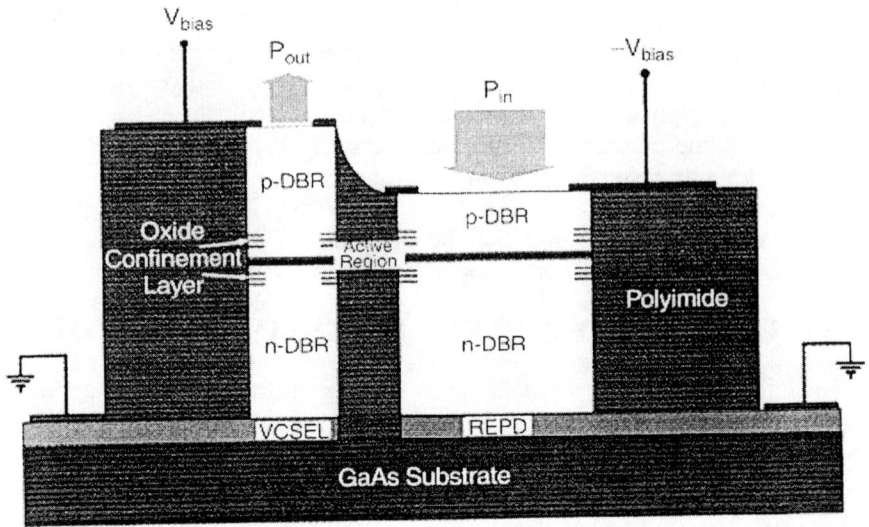

FIGURE 12. Schematic figure of an integrated VCSEL and detector (REPD). Pin are photons coming into the detector and Pout are photons coming out of the laser [22]. (© 1999 IEEE)

Modifications of the VCSEL structure to optimise for detection involve fabricating a structure with fewer mirror pairs to increase the light collected. The vertical cavity structure is still very wavelength specific allowing only photons supported by the cavity structure to exist in the cavity and to excite the quantum well transition involved in the lasing. To collect the carriers a reverse bias is placed across the structure. The transition energy will be effected by the reverse bias but the Fabry Perot mode of the cavity will not and since this mode dominates, as in the laser, the wavelength of the detector is very bias independent. It is also possible to get avalanching of the carriers in reverse bias and this has been observed [21]. The nature of the avalanching will be related to the length of the cavity, doping in the DBR and strength of the field within the structure.

There is the possibility to make a vertical cavity device that can function both as a source (under forward bias) and as a detector (under reverse bias) with different parts of the device optimised for each purpose. This has been investigated at Bristol University where Focussed Ion Beam Etching (FIBE) is used to selectively remove DBR layers in the regions designed to be the detector region while keeping them in the regions designed to be lasing regions. There is a time-delay between the device operating in forward bias (lasing) or in reverse bias (detection) due to charge storage and carrier sweep out times. This has been investigated at Bristol and a switching time. of 3 ns has been found for these devices [21]. A ring etched and a spot etched VCSEL and their respective lasing characteristics are compared with the unetched VCSEL in Fig. 11.

Another approach is to place the detector adjacent the VCSEL where they can share the same resonant cavity and the same lower DBRs but differ in the number of upper

DBRs. They share much of the same fabrication so have closely matched resonance frequencies. A schematic of such a structure is shown in Fig. 12. In this case they can be addressed independently so issues of switching speeds are not important.

CONCLUSIONS

In this chapter the behaviour of VCSELs, both static and dynamic, has been examined and compared to that of edge-emitting lasers. The advantages of VCSELs, particularly for communications where fast modulation speeds are necessary have been discussed.

While the focus of this chapter has been on encoding the information on one modulated laser channel, communication speeds have been increased either: by interleaving signals from different lasers in the time domain, termed *optical time division multiplexing* (OTDM), or by transmitting signals simultaneously down the fibre using different wavelengths, termed *wave division multiplexing* (WDM). For WDM the wavelengths are chosen to be close together to use the transmission windows of the fibre. By growing the VCSELs on treated surfaces the cavity can be tuned across a wafer to produce emitters and detectors designed for a range of wavelengths making VCSELs the ideal candidate for WDM applications.The future of VCSELs in communications is looking very promising.

REFERENCES

1. Sod, H., Iga, K., Kitahara, C., and. Suematsu, Y., *Jpn. J. Appl. Phys.* **18**, 2329-2330 (1979).
2. Sale, T.E., *Vertical Cavity Surface Emitting Lasers*, Research Studies Press Ltd., Wiley, New York, 1995.
3. Coldren, L., and Corzine, S.W., *Diode Lasers and Photonic Integrated Circuits*, Wiley, New York, 1995.
4. Giboney, K.S., Aronson, L.B., and Lemoff, B.E., *The Ideal Light Source for Datanets*, IEEE Spectrum, **43**, February 1998.
5. Degen, C., Krauskopf, B., Jennemann, G., Fischer, I., and Elsasser, W., *J. Opt. B: Quantum and Semiclassical Optics* **2** 517-525 (2000).
6. Mizutani, A., Hatori, N., Nishiyama, N., Koyama, F., and Iga, K., *IEEE Photon. Techn. Lett.* **10**, 633 (1998).
7. Sargent, L.J., Rorison, J.M., Kuball, M., Penty, R.V., White, I.H., Corzine, S.W., Tan, M.R.T., Wang, S.Y., and Heard, P.J., Appl. Phys. Lett.,
8. Panajotov, K., et al., preprint submitted to Appl. Phys. Lett. 2000.
9. Sargent, L.J., Bristol work (unpublished).
10. Degen, C., Fischer, I., and Elsasser, W., *Appl. Phys. Lett.*, **76**, 3352 (2000)
11. Unger, P., Bona, G.L., Germann, R., Roentgen, P., and Webb, D.J., *IEEE J. Quantum Electron.*, **29**, 1880 (1993).
12. Clarke, M., and Penty, R.V., Bristol work (unpublished).
13. Tastavridis, K., Bristol work (unpublished).
14. Vukusic, J.A., Martinsson, H., Gustavsson, J.S., and Larsson, A., *IEEE, J. Quantum Electronics*,
15. Clarke, M. and Penty, R.V., Bristol work (unpublished).
16. Thomas, G.A., Ackerman, D.A., Prucnal, P.R., and Cooper S.L., *Physics in the Communications Whirlwind*, Physics Today.
17. Kalaugher, L., Raman Amplifiers open a wider DWDM window, *FibreSystems*, **4**, 41 (2000).

18. Raddatz, L., White, I.H., Cunningham, D.G., and Nowell, M.C., *IEEE Photonics Technology Lett.*, **10**, 534 (1998).
19. Larson, M.C., Kondow, M., Kitatani, T., Nakahara, K., Tamura, K., Inoue, K., and Uomi, K., *IEEE Photon. Technol. Lett.* **10**,188 (1998).
20. Mirin, R., Gossard, A., and Bowers, J., *Electron. Lett.* **32**, 1732 (1996).
21. Dragas, M., Bristol work (unpublished).
22. Julian Cheng and Yuxin Zhou, *A New Wave in Network Data Links, IEEE Circuits and Devices, November 1999.*

Author Index

B
Broer, H. W., 31

E
Elsäßer, W., 218
Erneux, T., 54

F
Fischer, I., 218

G
Gavrielides, A., 191

H
Heil, T., 218
Hess, O., 128
Houlihan, J., 173

I
Indik, R. A., 173

K
Krauskopf, B., 1, 31, 66

L
Lenstra, D., 87

M
McInerney, J. G., 173
Mirasso, C. R., 112
Moloney, J. V., 149, 173
Mullane, M., 173

O
O'Brien, P., 173
O'Neill, E., 173

R
Rorison, J. M., 279
Roy, R., 260

S
Skovgaard, P., 173

T
Tredicce, J. R., 238

V
Verduyn Lunel, S. M., 66

Y
Yousefi, M., 87